APPLIED
COMPUTATIONAL
FLUID
DYNAMICS

MECHANICAL ENGINEERING

A Series of Textbooks and Reference Books

Editor

L. L. Faulkner

*Columbus Division, Battelle Memorial Institute
and Department of Mechanical Engineering
The Ohio State University
Columbus, Ohio*

114. *Handbook of Thermoplastic Piping System Design*, Thomas Sixsmith and Reinhard Hanselka
115. *Practical Guide to Finite Elements: A Solid Mechanics Approach*, Steven M. Lepi
116. *Applied Computational Fluid Dynamics*, edited by Vijay K. Garg

Additional Volumes in Preparation

Fluid Sealing Technology, Heinz Konrad Muller and Bernard Nau

Friction and Lubrication in Mechanical Design, A. A. Seireg

Machining of Ceramics and Composites, edited by Said Jahanmir and M. Ramulu

Heat Exchange Design Handbook, T. Kuppan

Couplings and Joints, Jon R. Mancuso

Mechanical Engineering Software

Spring Design with an IBM PC, Al Dietrich

Mechanical Design Failure Analysis: With Failure Analysis System Software for the IBM PC, David G. Ullman

APPLIED COMPUTATIONAL FLUID DYNAMICS

EDITED BY
VIJAY K. GARG
AYT Corporation/NASA Lewis Research Center
Cleveland, Ohio

CRC Press
Taylor & Francis Group
Boca Raton London New York

CRC Press is an imprint of the
Taylor & Francis Group, an **informa** business

CRC Press
Taylor & Francis Group
6000 Broken Sound Parkway NW, Suite 300
Boca Raton, FL 33487-2742

First issued in paperback 2019

© 1998 by Taylor & Francis Group, LLC
CRC Press is an imprint of Taylor & Francis Group, an Informa business

No claim to original U.S. Government works

ISBN-13: 978-0-8247-0165-9 (hbk)
ISBN-13: 978-0-367-40045-3 (pbk)

Visit the Taylor & Francis Web site at
http://www.taylorandfrancis.com

and the CRC Press Web site at
http://www.crcpress.com

Library of Congress Cataloging-in-Publication Data

Applied computational fluid dynamics / edited by Vijay K. Garg.
      p. cm. -- (Mechanical engineering ; 116)
    Includes index.
    ISBN 0-8247-0165-8 (alk. paper)
    1. Fluid dynamics. 2. Heat--Transmission. 3. Turbulence.
I. Garg, Vijay K. II. Series: Mechanical engineering
(Marcel Dekker, Inc.) ; 116.
TA357.A657   1998
620.1'064--dc21
                                           98-16640
                                              CIP

Preface

In recent years, growth in computational capability has led to a phenomenal increase in the use of computational methods for engineering design, especially for the design of fluid flow and thermal systems. For such systems, the governing equations are too complex to be solved analytically except in some trivial cases. Thus the use of numerical techniques for problems of practical interest in thermal systems has mushroomed in the last two decades. With ever-increasing costs of experimentation, the need for numerical simulation of the relevant processes is even greater. Of course, experimental data are indispensable for checking the accuracy and validating the numerical model. Computational fluid dynamics (CFD) has become a major tool for designers of fluid flow and thermal systems. Progress in this area has been rapid, and the use of three-dimensional methods is increasingly applied to the design process. The availability of faster and larger computers has enabled us to solve complex fluid flow and thermal problems that arise in a wide variety of applications.

One often finds a lack of valuable applications when teaching CFD to students from a variety of disciplines. Most examples in textbooks on CFD either reach and appeal to few students outside the specific discipline for which the textbook was written or contain rather simple applications. Part of the philosophy behind the preparation of this book was to expose the reader to the latest computational techniques used for the solution of real-life problems in the field of fluid flow and heat transfer. It is our intention that this book, although self-contained, will not be used in isolation. For the prospective student of CFD, there is no substitute for extensive reading and knowledge of the latest developments in order to obtain a true understanding of the subject.

The principal objective in this book has been to bring together a collection of applications of CFD that will be beneficial to workers in diverse fields, such as environmental sciences, energy systems, mechanical engineering, chemical engineering, and aerospace. For each application, there is a contribution from experts who have worked in the particular field for 15–20 years and are at the forefront of the techniques used.

This book explores real-life applications of computational fluid dynamics and heat transfer in various fields such as aeronautics, materials processing and manufacturing, electronic cooling, and environmental control. For example, it enhances the reader's knowledge of how to utilize the modern tools of CFD to increase the efficiency, reduce the fuel consumption, and increase the affordability of aircraft engines; manufacture and process better materials; improve the quality of the air we breathe; and cool computers better so that they can run faster and cheaper, and thus help us solve more complex problems in a shorter time. The reader is exposed to the state of the art in these applications of CFD techniques.

In order to provide the mathematical background for this subject, Chapter 1 begins with a discussion of the nature of governing equations along with the relevant boundary conditions that arise in fluid flow and heat transfer problems. This chapter also describes the governing equations in general curvilinear coordinates, and discusses the mathematical properties of these equations. Chapter 2 describes the three basic techniques—the finite difference method, the finite volume method, and the finite element method—along with considerations of accuracy, stability, and convergence. This is followed by a state-of-the-art review of turbulence modeling, both dynamic and thermal, in Chapter 3. Following the development of basic concepts, this chapter describes the various eddy viscosity and stress transport models for dynamic turbulence and then discusses the various thermal models. This background is supplemented with grid generation techniques in Chapter 4. In fact, CFD has provided much of the impetus for the development of these techniques, which are equally applicable to all physical problems involving field solutions. This chapter describes grid generation via the solution of partial differential equations as well as algebraic, adaptive, and unstructured grid generation techniques. It also provides some details for the various codes currently available for grid generation. The various applications are then discussed in individual chapters as described below.

Chapter 5 discusses inlet, duct, and nozzle flows. After describing the CFD solution process for such flows, it provides details on the application of the process to two examples. The chapter closes with the current status and future directions in terms of modeling, numerical, and procedural issues. Chapter 6 looks at the impact of unsteadiness in turbine flows with a view to distinguish flow parameters that can be modeled with existing steady CFD codes from those that require unsteady codes. This is an important issue since unsteady three-dimensional computation of heat transfer in complex geometries such as a turbomachine is prohibitively expensive, even with present-day computers. Chapter 7 discusses the numerical modeling of heat transfer and fluid flow processes in the thermal processing of materials. It brings out the importance of material properties in an accurate modeling of the process. This chapter also discusses the numerical results and computational problems that arise in various processes, such as plastic extrusion, optical fiber drawing, casting, and heat treatment. Application of CFD techniques in electronic cooling is discussed in Chapter 8. Cost, size, and weight are often the primary constraints on the thermal and physical design of electronic products used in computers; military/aerospace, industrial, and consumer products; business/retail and automotive markets; instrumentation; and telecommunications. Passive thermal control techniques are discussed together with examples. Chapter 8 closes with some thoughts on future directions on the use of CFD for electronic cooling. Chapter 9 discusses the application of CFD to control air quality. Problems such as acid deposition, smog, global climate warming, and stratospheric ozone depletion pose significant threats to both human health and welfare and related ecological damage. The role of air quality models is to identify effective solutions to the complex environmental problems. This chapter describes several advection schemes used by air quality models, and evaluates them for various test cases against the different performance measures.

The chapter authors and I will consider our efforts successful if readers are able to apply CFD techniques to the resolution of problems in their areas of concern.

I wish to thank Dr. Raymond E. Gaugler of NASA Lewis Research Center for his support. Thanks are also due to the staff at Marcel Dekker, Inc., for their help and competent handling of this project. My family has been very understanding and supportive during the writing activity, and I am indebted to them.

Vijay K. Garg

Contents

Contributors

Daniel J. Dorney, Ph.D.* Project Engineer, Pratt & Whitney Aircraft, East Hartford, Connecticut

Vijay K. Garg, Ph.D. Senior Research Engineer, AYT Corporation/NASA Lewis Research Center, Cleveland, Ohio

Yogesh Jaluria, Ph.D. Professor, Department of Mechanical Engineering, Rutgers University, New Brunswick, New Jersey

Philip C. E. Jorgenson, Ph.D. Aerospace Engineer, Engine Systems Technology Branch, NASA Lewis Research Center, Cleveland, Ohio

Yogendra Joshi, Ph.D. Associate Professor, Department of Mechanical Engineering, University of Maryland, College Park, Maryland

* *Current affiliation*: Assistant Professor, Department of Mechanical Engineering, GMI Engineering & Management Institute, Flint, Michigan.

Ron-Ho Ni, Ph.D.   Senior Fellow, Pratt & Whitney Aircraft, East Hartford, Connecticut

Mehmet T. Odman, Ph.D.*   Research Scientist, MCNC—Environmental Programs, Research Triangle Park, North Carolina

Armistead G. Russell, Ph.D.   Georgia Power Professor, Department of Civil and Environmental Engineering, Georgia Institute of Technology, Atlanta, Georgia

John R. Schwab   Fluid and Thermal Analysis Consultant, North Olmsted, Ohio

Om P. Sharma, Ph.D.   Senior Fellow, Pratt & Whitney Aircraft, East Hartford, Connecticut

Seyf Tanrikut, Ph.D.   Senior Design Engineer, Pratt & Whitney Aircraft, East Hartford, Connecticut

Charles E. Towne, Ph.D.   Aerospace Engineer, NASA Lewis Research Center, Cleveland, Ohio

---

* *Current affiliation*: Senior Research Engineer, Department of Civil and Environmental Engineering, Georgia Institute of Technology, Atlanta, Georgia.

1

Governing Equations

Vijay K. Garg

AYT Corporation/NASA Lewis Research Center, Cleveland, Ohio

The equations governing the flow of a fluid and the associated heat transfer are based on the conservation principles for mass, momentum, and energy. These equations are first presented for a compressible, viscous, Newtonian fluid, and then particularized for simpler cases. It is assumed that the reader has some background in this field. Thus, a complete derivation of the governing equations is not included. The reader is referred to Schlichting (1979) for the derivation. For the general case of three-dimensional motion, the flow field is specified by the velocity vector

$$V = ui + vj + wk$$

where u, v, w are the three orthogonal components, by the pressure p, density ρ, and temperature T, all conceived as functions of the space coordinates and time t. For the determination of these six quantities, there exist six equations: the continuity equation (conservation of mass), the three equations of motion (conservation of momentum), the energy equation (conservation of energy), and the thermodynamic equation of state $p = p(\rho, T)$.

1.1 CONTINUITY EQUATION

The continuity equation implies a balance between the masses entering and leaving a control volume per unit time and the change in density within it. For the unsteady flow of a compressible fluid, the conservation of mass applied to a fluid passing through an infinitesimal, fixed control volume yields the following equation of continuity:

$$\frac{\partial \rho}{\partial t} + \nabla \cdot (\rho V) = \frac{D\rho}{Dt} + \rho(\nabla \cdot V) = 0 \tag{1.1}$$

The first term in this equation represents the rate of increase of density in the control volume and the second term represents the rate of mass flux passing out of the control surface (which surrounds the control volume) per unit volume. The symbol $D\rho/Dt$ denotes the substantive derivative, which consists of the local contribution (in unsteady flow) $\partial\rho/\partial t$ and the convective contribution (due to translation) $V \cdot (\nabla\rho)$.

Equation (1.1) is based on the *Eulerian* approach, in which changes to the fluid are recorded as the fluid passes through a fixed control volume. In the alternative *Lagrangian* approach, changes to the properties of a fluid element are recorded by an observer moving with the fluid element. In general, the Eulerian approach is used for fluid flows.

A flow in which the density of each fluid element remains constant is called *incompressible*. This implies that $D\rho/Dt = 0$, which reduces Eq. (1.1) to

$$\nabla \cdot V = 0 \tag{1.2}$$

for a steady or unsteady incompressible flow. The assumption of incompressibility is a good approximation for air flows when the Mach number is less than 0.3.

In indicial notation, Eq. (1.1) can be written as

$$\frac{\partial\rho}{\partial t} + \frac{\partial(\rho V_i)}{\partial x_i} = \frac{D\rho}{Dt} + \rho\,\frac{\partial V_i}{\partial x_i} = 0 \tag{1.3}$$

1.2 MOMENTUM EQUATION

The momentum equation is derived from Newton's second law of motion. Further, in fluid motion it is necessary to consider two types of forces separately: (1) forces acting throughout the mass of the fluid element (e.g., gravitational forces), known as *body* forces, and (2) forces acting on the boundary (pressure and friction), known as *surface* forces. If F_i denotes the body force per unit volume, and P_i denotes the surface force per unit volume, both in direction, i, the momentum equation in indicial notation can be written as

$$\rho\,\frac{DV_i}{Dt} = F_i + P_i, \qquad i = 1, 2, 3 \tag{1.4}$$

where DV_i/Dt denotes the substantive acceleration of the fluid element. The most common body force is the gravitational force. In this case $F_i = \rho g_i$, where g_i is the acceleration due to gravity. While the body forces are regarded as given external forces, the surface forces depend on the *rate* at which the fluid is *strained* by the velocity field in it. In fact, the surface force P_i is related to the stress (force per unit area) by

$$P_i = \frac{\partial \sigma_{ij}}{\partial x_j} \tag{1.5}$$

where σ_{ij} is the stress *on a plane normal* to the i axis *in the direction* of the j axis. The stress tensor is symmetric, i.e., $\sigma_{ij} = \sigma_{ji}$, *except* if body moments are present as in the case of a magnetic body in a magnetic field.

The momentum equation given is quite general and is applicable to both continuum and noncontinuum flows. It loses its generality, however, when approximate expressions are inserted for the stress tensor. Here we restrict ourselves to *isotropic, Newtonian fluids* for which the stress at a point is linearly related to the rate of strain (deformation) of the fluid. All gases which can be treated as a continuum, and most liquids of interest, in particular water, belong to this class. For such fluids, the so-called constitutive equations (relation between *stress* and *rate of strain*) are (Round and Garg, 1986)

$$\sigma_{ij} = -p\delta_{ij} + \mu\left(\frac{\partial V_i}{\partial x_j} + \frac{\partial V_j}{\partial x_i}\right) + \delta_{ij}\mu'\frac{\partial V_k}{\partial x_k} \qquad (i, j, k = 1, 2, 3) \tag{1.6}$$

where the Kronecker delta $\delta_{ij} = 0$ for $i \neq j$ and $\delta_{ij} = 1$ for $i = j$, μ is the coefficient of viscosity (dynamic viscosity), and μ' is the second coefficient of viscosity. The two coefficients of viscosity are related to the coefficient of bulk viscosity (κ) by the expression

$$\kappa = \frac{2}{3}\mu + \mu' \tag{1.7}$$

Except in the study of the structure of shock waves, and in the absorption and attenuation of acoustic waves, it is generally believed that the coefficient of bulk viscosity is negligible. With $\kappa = 0$, the second coefficient of viscosity becomes

$$\mu' = -\frac{2}{3}\mu \tag{1.8}$$

and the stress tensor may be written as

$$\sigma_{ij} = -p\delta_{ij} + \mu\left(\frac{\partial V_i}{\partial x_j} + \frac{\partial V_j}{\partial x_i} - \frac{2}{3}\delta_{ij}\frac{\partial V_k}{\partial x_k}\right) \qquad (i, j, k = 1, 2, 3) \tag{1.9}$$

The stress tensor is also written as

$$\sigma_{ij} = -p\delta_{ij} + \tau_{ij} \tag{1.10}$$

where τ_{ij} represents the viscous (shear) stress tensor given by the bracketed term on the right side of Eqs. (1.9). It is easily seen from Eqs. (1.6) that, for an incompressible fluid, the mean normal stress is equal to the negative pressure; i.e.,

$$-p = \frac{1}{3}\sigma_{ii}$$

due to the continuity equation (1.2).

Substituting (1.5) and (1.9) into (1.4) yields

$$\rho\left(\frac{\partial V_i}{\partial t} + V_j\frac{\partial V_i}{\partial x_j}\right) = F_i - \frac{\partial p}{\partial x_i} + \frac{\partial}{\partial x_j}\left[\mu\left(\frac{\partial V_i}{\partial x_j} + \frac{\partial V_j}{\partial x_i} - \frac{2}{3}\delta_{ij}\frac{\partial V_k}{\partial x_k}\right)\right]$$

$$(i, j, k = 1, 2, 3) \qquad (1.11)$$

These well-known differential equations form the basis of the whole science of fluid mechanics, and are known as the *Navier-Stokes equations*. In this form, they are applicable to a viscous, compressible, isotropic, Newtonian fluid with variable properties. In general, the viscosity μ may be regarded as dependent on the space coordinates, since μ varies considerably with temperature (though little with pressure). In such a case, the temperature dependence of viscosity $\mu(T)$ must be obtained from experiments.

For an *incompressible* fluid, the last term in Eq. (1.11) vanishes identically due to the continuity equation (1.2). Further, since temperature variations are, generally speaking, small in this case, the viscosity may be assumed to be constant.* With this assumption, Eq. (1.11) simplifies to

$$\rho\left(\frac{\partial V_i}{\partial t} + V_j\frac{\partial V_i}{\partial x_j}\right) = F_i - \frac{\partial p}{\partial x_i} + \mu\frac{\partial^2 V_i}{\partial x_j \partial x_j} \qquad (i, j, k = 1, 2, 3) \qquad (1.12)$$

In vector notation, Eq. (1.12) can be written as

$$\rho\frac{DV}{Dt} = F - \nabla p + \mu\nabla^2 V \qquad (1.13)$$

For an *inviscid* fluid, this reduces to

$$\rho\frac{DV}{Dt} = F - \nabla p \qquad (1.14)$$

---

* This condition is more nearly satisfied in gases than in liquids.

which is the well-known *Euler equation*. It is valid for compressible and incompressible inviscid flows.

1.3 ENERGY EQUATION

The energy equation is derived from the first law of thermodynamics, and in the absence of chemical reaction and radiation it can be written in indicial notation as

$$\rho\left(\frac{\partial e}{\partial t} + V_i \frac{\partial e}{\partial x_i}\right) + p \frac{\partial V_i}{\partial x_i} = \frac{\partial}{\partial x_i}\left(k \frac{\partial T}{\partial x_i}\right) + \mu\Phi \tag{1.15}$$

where e is the specific internal energy, k is the thermal conductivity of the fluid, and Fourier's law for heat transfer by conduction through the control surface has been assumed. Also, the total energy per unit volume, E, has been taken to be

$$E = \rho\left(e + \frac{V^2}{2}\right)$$

so that only internal energy and kinetic energy are considered significant. The dissipation function Φ represents the rate at which energy is dissipated per unit volume of fluid through the action of viscosity, and is given by

$$\Phi = \frac{1}{\mu}\,\tau_{ij}\frac{\partial V_i}{\partial x_j} = \left(\frac{\partial V_i}{\partial x_j}\right)^2 + \frac{\partial V_j}{\partial x_i}\frac{\partial V_i}{\partial x_j} - \frac{2}{3}\left(\frac{\partial V_i}{\partial x_i}\right)^2 \tag{1.16}$$

Equation (1.15) enjoys general validity, and can be simplified for special cases.

For a *perfect gas*, using the continuity equation (1.3) together with

$$c_p DT = c_v DT + D\left(\frac{p}{\rho}\right)$$

we can simplify Eq. (1.15) to the form

$$\rho c_p \frac{DT}{Dt} = \frac{Dp}{Dt} + \frac{\partial}{\partial x_i}\left(k \frac{\partial T}{\partial x_i}\right) + \mu\Phi \tag{1.17}$$

where c_p and c_v represent the specific heats per unit mass at constant pressure and at constant volume, respectively.

For an *incompressible* fluid, using the continuity equation (1.2) together with $De = CDT$ simplifies Eq. (1.15) to

$$pC \frac{DT}{Dt} = \frac{\partial}{\partial x_i}\left(k \frac{\partial T}{\partial x_i}\right) + \mu\Phi \tag{1.18}$$

where C is the specific heat per unit mass, and the expression for the dissipation function Φ also simplifies to

$$\Phi = \left(\frac{\partial V_i}{\partial x_j}\right)^2 + \frac{\partial V_j}{\partial x_i}\frac{\partial V_i}{\partial x_j} \tag{1.19}$$

For an *inviscid* fluid, $\mu = 0$, so Eq. (1.15) reduces to

$$\rho\frac{De}{Dt} + p\frac{\partial V_i}{\partial x_i} = \frac{\partial}{\partial x_i}\left(k\frac{\partial T}{\partial x_i}\right) \tag{1.20}$$

For an *adiabatic* flow of an *inviscid* fluid,

$$\rho\frac{De}{Dt} + p\frac{\partial V_i}{\partial x_i} = 0 \tag{1.21}$$

The left side of Eq. (1.15) can be written as

$$\rho\left(\frac{\partial e}{\partial t} + V_i\frac{\partial e}{\partial x_i}\right) + p\frac{\partial V_i}{\partial x_i} = \rho\frac{De}{Dt} + p\frac{\partial V_i}{\partial x_i} = \rho\frac{Dh}{Dt} - \frac{Dp}{Dt}$$

where $h = e + p/\rho$ is the specific enthalpy of the fluid.

Strictly speaking, the term "Navier-Stokes equations" refers to the components of the momentum equation (1.11). However, it is common practice to include the continuity and energy equations in the set of equations referred to as the Navier-Stokes equations. See Appendix A for the Navier-Stokes equations expressed in different coordinate systems. The unsteady Navier-Stokes equations govern the laminar and turbulent flows as long as effective values of the transport coefficients μ and k are used. However, they are not easy to solve for turbulent flows in comparison to that for laminar flows, since the space and time scales of turbulent motion are very small. Thus, a very large number of grid points and a very small time step are required for solution. This puts the computation of many practical turbulent flows via DNS (direct numerical simulation) outside the realm of possibility for present-day computers. The main thrust in the computation of turbulent flows has been through the solution of time-averaged Navier-Stokes equations. These equations are also referred to as the *Reynolds averaged equations*, and are discussed in detail in Chapter 3.

1.4 BOUNDARY CONDITIONS

The solution of the foregoing equations can be determined only when the boundary and initial conditions are specified. For a viscous fluid, the condition of no slip on solid boundaries must be satisfied; i.e., on a solid, impermeable wall, the normal and tangential components of velocity must vanish. If the energy equation is also used, temperature and/or its gradient at the boundaries should also be specified. For an *inviscid* fluid, the tangential component of fluid velocity at a solid wall is *not* required to vanish.

1.5 EQUATION OF STATE

In order to close this system of equations, it is necessary to establish relationships between the thermodynamic variables (ρ, p, T, e, h) and to relate the transport properties (μ, k) to them. For example, for a compressible flow without external heat addition, the relevant equations to be solved are Eq. (1.1) or Eq. (1.3) for the continuity equation, Eq. (1.11) for the three momentum equations, and Eq. (1.15) for the energy equation. These five equations contain seven unknowns ρ, p, e, T, V_1, V_2, V_3, assuming that the transport coefficients μ, k can be related to the thermodynamic variables. Clearly, two additional equations are required to close the system. These equations are provided by the relationships among the thermodynamic variables. Such relations are known as *equations of state*. From the *state principle* of thermodynamics, it is known that the local thermodynamic state is fixed by *any two independent* thermodynamic variables *provided* the chemical composition of the fluid does not change due to diffusion of finite-rate chemical reactions. Thus, for the present example, if we choose ρ and T as the two independent variables, equations of state of the form

$$p = p(\rho, T), \qquad e = e(\rho, T) \tag{1.22}$$

are required.

For a perfect gas, for example, the equation of state is

$$p = \rho R T \tag{1.23}$$

where R is the gas constant. Also for a perfect gas, we have

$$e = c_v T, \quad h = c_p T, \quad \gamma = \frac{c_p}{c_v}, \quad c_v = \frac{R}{\gamma - 1}, \quad c_p = \frac{\gamma R}{\gamma - 1} \tag{1.24}$$

where γ is the ratio of specific heats. For air at standard conditions, $R = 287$ J/kg-K and $\gamma = 1.4$. For a perfect gas then, the form of Eqs. (1.22) is obvious. For fluids that cannot be considered a perfect gas, the required state relations are generally given in terms of tables, charts, or curve fits.

Kinetic theory is used to relate the coefficients of viscosity and thermal conductivity to the thermodynamic variables. For example, Sutherland's law for dynamic viscosity (Schlichting, 1979) is

$$\mu = C_1 \frac{T^{3/2}}{T + C_2} \tag{1.25}$$

where C_1 and C_2 are constants for a given gas. For air at moderate temperatures, $C_1 = 1.458 \times 10^{-6}$ kg/(m-s-K$^{1/2}$) and $C_2 = 110.4$ K. Once μ is known, the coefficient of thermal conductivity k is determined from the definition of the Prandtl number,

$$Pr = \frac{c_p \mu}{k} \tag{1.26}$$

since the ratio c_p/Pr is approximately constant for most gases.

1.6   VECTOR FORM OF EQUATIONS

It is often convenient to combine the governing equations into a compact vector form before applying a numerical algorithm to them. For example, for a compressible flow without external heat addition or body forces, the governing equations in Cartesian coordinates can be written as

$$\frac{\partial U}{\partial t} + \frac{\partial E}{\partial x} + \frac{\partial F}{\partial y} + \frac{\partial G}{\partial z} = 0 \tag{1.27}$$

where U, E, F, and G are vectors given by

$$U = \begin{bmatrix} \rho & \rho u & \rho v & \rho w & E \end{bmatrix}^T$$

$$E = \begin{bmatrix} \rho u \\ \rho u^2 + p - \tau_{xx} \\ \rho u v - \tau_{xy} \\ \rho u w - \tau_{xz} \\ (E + p)u - u\tau_{xx} - v\tau_{xy} - w\tau_{xz} + q_x \end{bmatrix}$$

$$F = \begin{bmatrix} \rho v \\ \rho u v - \tau_{xy} \\ \rho v^2 + p - \tau_{yy} \\ \rho v w - \tau_{yz} \\ (E + p)v - u\tau_{xy} - v\tau_{yy} - w\tau_{yz} + q_y \end{bmatrix} \tag{1.28}$$

$$G = \begin{bmatrix} \rho w \\ \rho u w - \tau_{xz} \\ \rho v w - \tau_{yz} \\ \rho w^2 + p - \tau_{zz} \\ (E + p)w - u\tau_{xz} - v\tau_{yz} - w\tau_{zz} + q_z \end{bmatrix}$$

where u, v, w represent the three components of the velocity vector V, and q_x, q_y, q_z are the conductive heat fluxes in the x, y, and z directions, respectively. The first row of the vector equation (1.27) corresponds to the continuity equation (1.1) in Cartesian coordinates, the second, third, and fourth rows are the momentum equations (1.11), while the fifth row is the energy equation (1.15).

1.7 NONDIMENSIONAL FORM OF EQUATIONS

It is advisable to cast the governing equations in nondimensional form before carrying out a numerical solution. This enables the flow variables to be "normalized" so that their values fall between prescribed limits such as 0 and 1. Also, the characteristic parameters such as Reynolds number, Prandtl number, Mach number, etc., can be varied independently. The proper nondimensionalization is problem dependent. However, if L is the characteristic length, and other characteristic quantities are taken to be freestream values denoted by subscript ∞, we may define the dimensionless variables, denoted by an asterisk, as

$$x^* = \frac{x}{L}, \quad y^* = \frac{y}{L}, \qquad z^* = \frac{z}{L}, \quad t^* = \frac{t}{L/V_\infty}$$

$$u^* = \frac{u}{V_\infty}, \quad v^* = \frac{v}{V_\infty}, \qquad w^* = \frac{w}{V_\infty}, \quad \mu^* = \frac{\mu}{\mu_\infty} \qquad (1.29)$$

$$\rho^* = \frac{\rho}{\rho_\infty}, \quad p^* = \frac{p}{\rho_\infty V_\infty^2}, \quad T^* = \frac{T}{T_\infty}, \quad e^* = \frac{e}{V_\infty^2}$$

If this nondimensionalizing procedure is applied to the compressible Navier-Stokes equations given by Eqs. (1.27) and (1.28), the following dimensionless equations are obtained:

$$\frac{\partial U^*}{\partial t^*} + \frac{\partial E^*}{\partial x^*} + \frac{\partial F^*}{\partial y^*} + \frac{\partial G^*}{\partial z^*} = 0 \qquad (1.30)$$

where U^*, E^*, F^*, and G^* are the vectors given by Eqs. (1.28) *except* that each term is dimensionless, denoted by an asterisks, and the dimensionless total energy per unit volume is

$$E^* = \rho^* \left(e^* + \frac{u^{*2} + v^{*2} + w^{*2}}{2} \right) \qquad (1.31)$$

The components of the shear stress tensor and the heat flux vector in dimensionless form are

$$\tau_{xx}^* = \frac{2\mu^*}{3\,\mathrm{Re}_L}\left(2\frac{\partial u^*}{\partial x^*} - \frac{\partial v^*}{\partial y^*} - \frac{\partial w^*}{\partial z^*}\right)$$

$$\tau_{xy}^* = \frac{\mu^*}{\mathrm{Re}_L}\left(\frac{\partial u^*}{\partial y^*} + \frac{\partial v^*}{\partial x^*}\right)$$

$$\tau_{yy}^* = \frac{2\mu^*}{3\,\mathrm{Re}_L}\left(2\frac{\partial v^*}{\partial y^*} - \frac{\partial u^*}{\partial x^*} - \frac{\partial w^*}{\partial z^*}\right)$$

$$\tau_{yz}^* = \frac{\mu^*}{\mathrm{Re}_L}\left(\frac{\partial v^*}{\partial z^*} + \frac{\partial w^*}{\partial y^*}\right)$$

$$\tau_{zz}^* = \frac{2\mu^*}{3\,\mathrm{Re}_L}\left(2\frac{\partial w^*}{\partial z^*} - \frac{\partial u^*}{\partial x^*} - \frac{\partial v^*}{\partial y^*}\right) \qquad (1.32)$$

$$\tau_{xz}^* = \frac{\mu^*}{\mathrm{Re}_L}\left(\frac{\partial u^*}{\partial z^*} + \frac{\partial w^*}{\partial x^*}\right)$$

$$q_x^* = -\frac{\mu^*}{(\gamma - 1)M_\infty^2\,\mathrm{Re}_L\,\mathrm{Pr}}\frac{\partial T^*}{\partial x^*}$$

$$q_y^* = -\frac{\mu^*}{(\gamma - 1)M_\infty^2\,\mathrm{Re}_L\,\mathrm{Pr}}\frac{\partial T^*}{\partial y^*}$$

$$q_z^* = -\frac{\mu^*}{(\gamma - 1)M_\infty^2\,\mathrm{Re}_L\,\mathrm{Pr}}\frac{\partial T^*}{\partial z^*}$$

where M_∞ and Re_L are the freestream Mach number and Reynolds number, respectively, given by

$$\mathrm{Re}_L = \frac{\rho_\infty V_\infty L}{\mu_\infty}, \qquad M_\infty = \frac{V_\infty}{\sqrt{\gamma R T_\infty}}$$

and the perfect gas equations of state [Eqs. (1.22)] become

$$p^* = \frac{\rho^* T^*}{\gamma M_\infty^2}, \qquad e^* = \frac{T^*}{\gamma(\gamma - 1)M_\infty^2}$$

Note that the dimensionless forms of the governing equations in Eqs. (1.30) are identical (except for the asterisks) to the dimensional form in Eqs. (1.27). For convenience, the asterisks are usually dropped.

1.8 BOUNDARY LAYER EQUATIONS

Prandtl (1904) originated the concept of a boundary layer. From experimental evidence, he reasoned that when the Reynolds number is large, a thin region (called the *boundary layer*) existed near a solid boundary where viscous effects were just as important as inertia effects no matter how small the viscosity of the fluid might be. Outside this boundary layer, velocity (as well as temperature) gradients were small, and since the viscosity is small or the Reynolds number is high, viscous effects are negligible. Thus, the outside mean flow pattern, being determined primarily by the boundary form, is practically that of inviscid flow past the boundary. This concept helps to reduce the governing equations considerably for the boundary layer region. Since Prandtl it has been found that a similar reduction in the governing equations is possible for flows in which a primary flow direction can be identified. Such flows include jets, wakes, mixing layers, and developing flow in pipes and other internal passages. It is thus common to refer to these reduced equations as the *thin-shear-layer equations*.

1.8.1 Boundary Layer Equations for Incompressible, Two-Dimensional Flow

Based on an order-of-magnitude analysis of the Navier-Stokes equations, the following boundary layer equations (in Cartesian coordinates) for an unsteady, incompressible, constant property flow over a two-dimensional body can be derived (Anderson et al., 1984):

continuity:

$$\frac{\partial u}{\partial x} + \frac{\partial v}{\partial y} = 0 \tag{1.33}$$

momentum:

$$\frac{\partial u}{\partial t} + u\frac{\partial u}{\partial x} + v\frac{\partial u}{\partial y} = -\frac{1}{\rho}\frac{\partial p}{\partial x} + v\frac{\partial^2 u}{\partial y^2} \tag{1.34}$$

$$\frac{\partial p}{\partial y} \simeq 0 \tag{1.35}$$

energy:

$$\frac{\partial T}{\partial t} + u\frac{\partial T}{\partial x} + v\frac{\partial T}{\partial y} = \alpha\frac{\partial^2 T}{\partial y^2} + \frac{\beta Tu}{\rho c_p}\frac{\partial p}{\partial x} + \frac{\mu}{\rho c_p}\left(\frac{\partial u}{\partial y}\right)^2 \tag{1.36}$$

where v is the kinematic viscosity, μ/ρ; α is the thermal diffusivity, $k/\rho c_p$; and β is the coefficient of volumetric expansion

$$\beta = -\frac{1}{\rho}\frac{\partial \rho}{\partial T}\bigg|_p$$

For an ideal gas $\beta = 1/T$, where T is the absolute temperature. The last two terms in Eq. (1.36) are retained on the assumption that the Eckert number is of order unity, which is true only for high-speed flows.

The initial and boundary conditions are

Initial condition:
   $u(x, y, 0)$, $T(x, y, 0)$ known everywhere
No slip:
   $u(x, 0, t) = v(x, 0, t) = 0$

   $T(x, 0, t) = T_w(x, t)$ or $\left.\dfrac{\partial T}{\partial y}\right|_{y=0} = \dfrac{q(x, t)}{k}$

Inlet condition:
   $u(x_0, y, t)$, $T(x_0, y, t)$ known at some x_0
Patching to the outer layer:
   $u(x, y, t) \to U_e(x, t)$ as $y \to \infty$, $T(x, y, t) \to T_e(x, t)$ as $y \to \infty$

where the subscript e refers to conditions at the edge of the boundary layer. The pressure gradient term in Eqs. (1.34) and (1.36) is evaluated for the given boundary from

$$-\frac{1}{\rho}\frac{\partial p}{\partial x} = \frac{\partial U_e}{\partial t} + U_e \frac{\partial U_e}{\partial x} \tag{1.37}$$

This follows from the Euler equation for the inviscid outer flow.

These boundary layer equations hold for flow over a curved wall as long as the boundary layer thickness is much smaller than the radius of curvature of the wall. The Reynolds averaged form of boundary layer equations for turbulent flow will be presented in Chapter 3.

1.8.2   Boundary Layer Equations for Compressible Flow

In order to reduce the Navier-Stokes equations, the order-of-magnitude analysis can also be carried out for compressible flow. Again referring the reader elsewhere for details (Cebeci and Smith, 1974), we provide the unsteady boundary layer equations for two-dimensional or axisymmetric laminar, compressible flow for the coordinate system in Fig. 1.1:

continuity:

$$\frac{\partial \rho}{\partial t} + \frac{\partial}{\partial x}(r^n \rho u) + \frac{\partial}{\partial y}(r^n \rho v) = 0 \tag{1.38}$$

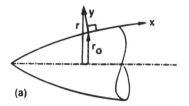

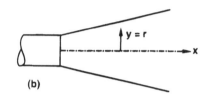

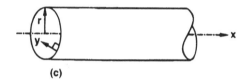

Figure 1.1 Coordinate system for axisymmetric boundary layer equations: (a) external boundary layer; (b) axisymmetric free shear layer flow; (c) confined axisymmetric flow.

momentum:

$$\rho \frac{\partial u}{\partial t} + \rho u \frac{\partial u}{\partial x} + \rho v \frac{\partial u}{\partial y} = -\frac{\partial p}{\partial x} + \frac{1}{r^n} \frac{\partial}{\partial y} \left(r^n \mu \frac{\partial u}{\partial y} \right) \tag{1.39}$$

$$\frac{\partial p}{\partial y} \simeq 0 \tag{1.40}$$

energy:

$$\rho \frac{\partial H}{\partial t} + \rho u \frac{\partial H}{\partial x} + \rho v \frac{\partial H}{\partial y}$$
$$= \frac{1}{r^n} \frac{\partial}{\partial y} \left[r^n \mu \left\{ \frac{1}{\text{Pr}} \frac{\partial H}{\partial y} + \left(1 - \frac{1}{\text{Pr}} \right) u \frac{\partial u}{\partial y} \right\} \right] + \frac{\partial p}{\partial t} \tag{1.41}$$

state:

$$\rho = \rho(p, T) \tag{1.42}$$

The enthalpy H, following boundary layer approximation, is

$$H = c_p T + \frac{u^2}{2}$$

and n is an index equal to zero for two-dimensional flow ($r^n = 1$) and equal to unity for axisymmetric flow ($r^n = r$).

The boundary layer equations for compressible flow are not significantly more complex than those for incompressible flow. The main difference is in the property variations of μ, k, and ρ.

The boundary layer approximation can also be made for a three-dimensional flow as long as velocity derivatives with respect to only one coordinate direction are large. Thus, the three-dimensional boundary layer flow remains "thin" with respect to only one coordinate direction. With y-direction normal to the wall, the three-dimensional boundary layer equations in Cartesian coordinates, applicable to a compressible flow, are

continuity:

$$\frac{\partial \rho}{\partial t} + \frac{\partial(\rho u)}{\partial x} + \frac{\partial(\rho v)}{\partial y} + \frac{\partial(\rho w)}{\partial z} = 0 \tag{1.43}$$

momentum:

$$\rho \frac{\partial u}{\partial t} + \rho u \frac{\partial u}{\partial x} + \rho v \frac{\partial u}{\partial y} + \rho w \frac{\partial u}{\partial z} = -\frac{\partial p}{\partial x} + \frac{\partial}{\partial y}\left(\mu \frac{\partial u}{\partial y}\right) \tag{1.44}$$

$$\rho \frac{\partial w}{\partial t} + \rho u \frac{\partial w}{\partial x} + \rho v \frac{\partial w}{\partial y} + \rho w \frac{\partial w}{\partial z} = -\frac{\partial p}{\partial z} + \frac{\partial}{\partial y}\left(\mu \frac{\partial w}{\partial y}\right) \tag{1.45}$$

energy:

$$\rho \frac{\partial H}{\partial t} + \rho u \frac{\partial H}{\partial x} + \rho v \frac{\partial H}{\partial y} + \rho w \frac{\partial H}{\partial z}$$

$$= \frac{\partial}{\partial y}\left[\mu \left\{ \frac{1}{\Pr} \frac{\partial H}{\partial y} + \left(1 - \frac{1}{\Pr}\right)\left(u \frac{\partial u}{\partial y} + w \frac{\partial w}{\partial y}\right)\right\}\right] + \frac{\partial p}{\partial t} \tag{1.46}$$

The equation of state is the same as Eq. (1.42), and the enthalpy H is

$$H = c_p T + \frac{u^2}{2} + \frac{w^2}{2}$$

The three-dimensional boundary layer equations are used primarily for external flows, for which the pressure gradient terms can be evaluated from the solution of the Euler equations. Three-dimensional internal flows are generally computed from somewhat different equations, to be discussed in section 1.10.

Also, unlike the Navier-Stokes equations, the boundary layer equations are *parabolic* in the main flow direction, and thus the solution can be "marched" in that direction. This makes boundary layer equations relatively easier to solve than the *elliptic* Navier-Stokes equations.

1.9   EULER EQUATIONS

We observed in the preceding section that for solution of the boundary layer equations, we must first solve for the inviscid flow outside the boundary layer. Note that the inviscid part of the flow can be solved independently of the boundary layer part only if the boundary layer is very thin compared to a characteristic length of the flow field, so the interaction between the two parts is negligible. For flows in which this interaction is not negligible, it is still possible to use separate sets of equations for the two regions, but the equations must be solved iteratively. This iterative procedure can be computationally inefficient. It is then desirable to use a single set of equations that remain valid throughout the flow field. Such equations will be discussed in section 1.10. In this section, we discuss equations that are valid only in the *inviscid* (*nonviscous, nonconducting*) portion of the flow field. These equations are obtained simply by dropping the viscous and heat transfer terms from the Navier-Stokes equations. As a consequence of this reduction, inviscid flow equations are much simpler to solve than the Navier-Stokes equations. Some of these simplifications will be observed here. Strictly speaking, Euler equation refers only to the inviscid momentum equation. However, we will refer to the set of inviscid flow equations as the *Euler equations*.

1.9.1   Continuity Equation

The continuity equation does not contain any viscous or heat transfer terms. Thus simplification of the continuity equation for an inviscid flow is not possible. However, for a two-dimensional or axisymmetric, steady flow, it is possible to satisfy the continuity equation exactly by introducing a *stream function* ψ. This is true irrespective of the flow being viscous or inviscid. For example, the continuity equation for a two-dimensional, steady, compressible flow in Cartesian coordinates is

$$\frac{\partial}{\partial x}(\rho u) + \frac{\partial}{\partial y}(\rho v) = 0 \qquad (1.47)$$

If we define a stream function ψ such that

$$\rho u = \frac{\partial \psi}{\partial y}, \qquad \rho v = -\frac{\partial \psi}{\partial x} \tag{1.48}$$

it is clear that Eq. (1.47) is satisfied. Thus introduction of the stream function reduces the number of dependent variables by one. The price for this reduction is that the velocity derivatives in the remaining equations have to be replaced using Eq. (1.48). These remaining equations will therefore contain derivatives which are one order higher. The physical significance of the stream function is obvious from

$$d\psi = \frac{\partial \psi}{\partial x} dx + \frac{\partial \psi}{\partial y} dy = -\rho v \, dx + \rho u \, dy$$
$$= \rho V \cdot dA = d\dot{m} \tag{1.49}$$

Thus there is no mass flow ($d\dot{m} = 0$) across lines of constant ψ ($d\psi = 0$). Lines of constant ψ are called *streamlines*, and the difference between the values of ψ for any two streamlines represents the mass flow rate per unit width between those streamlines. By definition, a *streamline* is a line in the flow field whose tangent at any point is in the direction of the flow velocity at that point.

For a two-dimensional, incompressible flow the continuity equation in Cartesian coordinates is

$$\frac{\partial u}{\partial x} + \frac{\partial v}{\partial y} = 0 \tag{1.50}$$

and the stream function is defined by

$$u = \frac{\partial \psi}{\partial y}, \qquad v = -\frac{\partial \psi}{\partial x} \tag{1.51}$$

For a steady, axisymmetric, compressible flow the continuity equation in cylindrical coordinates is (see Appendix A)

$$\frac{1}{r} \frac{\partial}{\partial r} (r\rho V_r) + \frac{\partial}{\partial z} (\rho V_z) = 0 \tag{1.52}$$

and the stream function is defined by

$$\rho V_r = \frac{1}{r} \frac{\partial \psi}{\partial z}, \qquad \rho V_z = -\frac{1}{r} \frac{\partial \psi}{\partial r} \tag{1.53}$$

It is possible to replace the continuity equation for three-dimensional flows by *two* stream functions. However, it is more complex than using the continuity equation in its original form and is therefore seldom used.

1.9.2   Inviscid Momentum Equations

The inviscid momentum equation, also known as the *Euler equation*, is given by Eq. (1.14). If we assume steady flow and neglect body forces, it reduces to

$$V \cdot \nabla V = -\frac{1}{\rho} \nabla p \tag{1.54}$$

Integrating this equation along a line in the flow field gives

$$\int (V \cdot \nabla V) \cdot dr = -\int \frac{1}{\rho} \nabla p \cdot dr \tag{1.55}$$

where dr is the differential length along the line. Let us further assume that the line is a streamline. Thus, V has the same direction as dr and we can simplify the integrand on the left side of Eq. (1.55) to get

$$(V \cdot \nabla V) \cdot dr = V \frac{\partial V}{\partial r} \cdot dr = V \frac{\partial V}{\partial r} dr = V \, dV = d\left(\frac{V^2}{2}\right)$$

Similarly, the integrand on the right side of Eq. (1.55) yields

$$\frac{1}{\rho} \nabla p \cdot d\mathbf{r} = \frac{dp}{r}$$

and Eq. (1.55) reduces to

$$\frac{V^2}{2} + \int \frac{dp}{r} = \text{constant} \tag{1.56}$$

The integral in Eq. (1.56) can be evaluated if the flow is assumed *barotropic*, i.e, a fluid for which ρ is either constant or a function only of p; the former being an incompressible flow, and the latter an isentropic flow for which

$$\rho = (\text{constant}) \, p^{1/\gamma} \tag{1.57}$$

Thus for a steady incompressible flow, Eq. (1.56) reduces to

$$p + \frac{1}{2} \rho V^2 = \text{constant} \tag{1.58}$$

which is called the *Bernoulli equation*. For a steady isentropic, compressible flow, Eq. (1.56) yields

$$\frac{V^2}{2} + \frac{\gamma}{\gamma - 1} \frac{p}{\rho} = \text{constant} \tag{1.59}$$

which is sometimes referred to as the *compressible Bernoulli equation*. Note that Eqs. (1.58) and (1.59) are valid *only along a given streamline*.

It can be easily shown that Eqs. (1.58) and (1.59) are valid everywhere in the flow field if the flow is *irrotational* as well. For an irrotational flow the vorticity is zero; i.e., the fluid particles do not rotate about their axis. The vorticity ζ is defined by

$$\zeta = \nabla \times V \tag{1.60}$$

Thus, for an irrotational flow, V can be expressed as the gradient of a single-valued scalar ϕ since

$$\zeta = \nabla \times V = \nabla \times (\nabla \phi) = 0 \tag{1.61}$$

The scalar ϕ is called the *velocity potential*. Also, we can express the acceleration of a fluid particle as

$$\frac{DV}{Dt} = \frac{\partial V}{\partial t} + V \cdot \nabla V = \frac{\partial V}{\partial t} + \nabla\left(\frac{V^2}{2}\right) - V \times \zeta \tag{1.62}$$

For an irrotational flow this equation reduces to

$$\frac{DV}{Dt} = \frac{\partial V}{\partial t} + \nabla\left(\frac{V^2}{2}\right)$$

which can be substituted into Euler's equation [Eq. (1.14)] to yield

$$\frac{\partial V}{\partial t} + \nabla\left(\frac{V^2}{2}\right) = \frac{1}{\rho} F - \frac{1}{\rho} \nabla p \tag{1.63}$$

If we again assume steady flow and neglect body forces, Eq. (1.63) can be written as

$$\nabla\left(\frac{V^2}{2} + \int \frac{dp}{\rho}\right) = 0 \tag{1.64}$$

since

$$\nabla \int \frac{dp}{\rho} = \frac{\nabla p}{\rho}$$

Integrating Eq. (1.64) along any *arbitrary* line in the flow field yields Eq. (1.56) again, but the constant now has the same value *everywhere* in the flow field. The incompressible Bernoulli equation [Eq. (1.58)] and the compressible Bernoulli equation [Eq. (1.59)] follow in the same manner as before, the only difference being that the equations are valid everywhere in the flow field because of the additional assumption of irrotationality.

For the special case of an inviscid, incompressible, irrotational flow, the continuity equation

$$\nabla \cdot V = 0$$

can be combined with

$$V = \nabla \phi$$

to yield the *Laplace* equation

$$\nabla^2 \phi = 0$$

1.9.3   Inviscid Energy Equations

We have already seen that for an *adiabatic* flow of an inviscid fluid, Eq. (1.21) holds. This can also be written in terms of specific enthalpy as

$$\rho \frac{Dh}{Dt} = \frac{Dp}{Dt} \tag{1.66}$$

Defining the specific stagnation enthalpy H as

$$H = h + \frac{V^2}{2}$$

we can rewrite Eq. (1.66) as

$$\rho \frac{DH}{Dt} = \frac{\partial p}{\partial t} + V \cdot \left(\rho \frac{DV}{Dt} + \nabla p \right) \tag{1.67}$$

If the body forces are neglected, the term within parentheses on the right side vanishes due to the Euler equation [Eq. (1.14)]. Equation (1.67) thus reduces to

$$\frac{DH}{Dt} = \frac{1}{\rho} \frac{\partial p}{\partial t} \tag{1.68}$$

which for a steady flow becomes

$$V \cdot \nabla H = 0 \tag{1.69}$$

This equation can be integrated along a streamline to yield H = constant.

For an incompressible flow, Eq. (1.21) reduces to

$$\frac{De}{Dt} = 0 \tag{1.70}$$

which implies that the internal energy is constant along a streamline in steady flow.

1.9.4   Simplified Form of Euler Equations

It is possible to simplify the Euler equations by making additional assumptions. If we assume the flow to be steady, irrotational, and isentropic, the Euler equations can be combined into a single equation called the *velocity potential equation*. In the Cartesian coordinate system, replacing the velocity components by

$$u = \frac{\partial \phi}{\partial x}, \qquad v = \frac{\partial \phi}{\partial y}, \qquad w = \frac{\partial \phi}{\partial z}$$

the continuity equation can be written as

$$\frac{\partial}{\partial x}(\rho \phi_x) + \frac{\partial}{\partial y}(\rho \phi_y) + \frac{\partial}{\partial z}(\rho \phi_z) = 0 \qquad (1.71)$$

where the subscript denotes partial differentiation with respect to the variable. The momentum and energy equations reduce to Eq. (1.56) under the present assumptions. In differential form this equation can be written as

$$dp = -\rho d\left(\frac{V^2}{2}\right) = -\rho d\left(\frac{\phi_x^2 + \phi_y^2 + \phi_z^2}{2}\right) \qquad (1.72)$$

With the speed of sound defined by

$$a = \sqrt{\left(\frac{\partial p}{\partial \rho}\right)_s}$$

Eq. (1.72) can be written as

$$d\rho = -\frac{\rho}{a^2} d\left(\frac{\phi_x^2 + \phi_y^2 + \phi_z^2}{2}\right) \qquad (1.73)$$

which can be used to find the derivatives of ρ in each direction. Substituting these expressions for ρ_x, ρ_y, and ρ_z into Eq. (1.71) and simplifying, the velocity potential equation is obtained:

$$\left(1 - \frac{\phi_x^2}{a^2}\right)\phi_{xx} + \left(1 - \frac{\phi_y^2}{a^2}\right)\phi_{yy} + \left(1 - \frac{\phi_z^2}{a^2}\right)\phi_{zz}$$
$$- \frac{2\phi_x \phi_y}{a^2}\phi_{xy} - \frac{2\phi_x \phi_z}{a^2}\phi_{xz} - \frac{2\phi_y \phi_z}{a^2}\phi_{yz} = 0 \qquad (1.74)$$

For an incompressible flow, the velocity potential equation reduces to the Laplace equation as $a \to \infty$.

1.10   PARABOLIZED NAVIER-STOKES EQUATIONS

The boundary layer equations can be used to solve many, but not all, viscous flow problems since the boundary layer assumptions are invalid for some viscous flow problems. For example, if the inviscid flow is fully merged with the viscous flow, the two flows cannot be solved independent of each other as required by boundary layer theory. It then becomes necessary to solve a set of equations which are valid in both the inviscid and viscous flow regions. Some examples of viscous flow fields where the boundary layer equations are not appropriate include (1) a supersonic flow around a blunt body at high altitude, and (2) a flow along a corner formed by two intersecting surfaces, among others. In the first example, there is a strong interaction between the boundary layer and inviscid flow in the region between the shock wave and the blunt body, while in the second example, viscous derivatives with respect to *two* "normal" directions are important very near the corner. As we pointed out earlier, the boundary layer equations include viscous derivatives with respect to a single "normal" coordinate direction only. Obviously, the complete Navier-Stokes equations can be used to solve such flow fields. However, they are very difficult to solve. In general, a large amount of computer time and storage is required to solve the complete Navier-Stokes equations. Fortunately, for some of the viscous flow problems for which the boundary layer equations are not appropriate, it is possible to solve a reduced set of equations that fall between the complete Navier-Stokes equations and the boundary layer equations in terms of complexity. These reduced equations are often referred to as the "parabolized" Navier-Stokes equations. They contain a nonzero normal pressure gradient which is a necessary condition for solving the viscous and inviscid flow regions simultaneously.

The most important advantage in using the parabolized Navier-Stokes (PNS) equations instead of the complete Navier-Stokes equations is that for a steady flow the former are a mixed set of hyperbolic-parabolic equations in the streamwise direction, provided certain conditions are met. In other words, the Navier-Stokes are "parabolized" in the streamwise direction, leading to a boundary-layer-type marching technique for the solution. This results in a substantial saving in computer time and storage. Another saving in computation time results from the fact that the PNS equations have fewer terms compared to those in the complete equations. The conditions under which the PNS equations are a set of hyperbolic-parabolic equations are that the inviscid outer region of the flow be supersonic, and that the streamwise velocity component be positive everywhere. The last condition excludes streamwise flow separation but crossflow separation is permitted. An additional complication is caused by the presence of streamwise pressure gradient in the streamwise momentum equation. With this term present everywhere in the flow field, upstream influence can occur in the subsonic part of the boundary layer, and a space-marching technique is not well-posed. The reader is referred to Anderson et al. (1984) for techniques to overcome this difficulty.

Different versions of the PNS equations are available in the literature since their derivation from the complete Navier-Stokes equations is not as rigorous as that of the boundary layer equations. These versions differ sometimes based on the type of flow problem being solved. In all cases, however, the normal pressure gradient term is retained, and the second derivative terms with respect to the streamwise direction are omitted. Rudman and Rubin (1968) were perhaps the first to use the PNS equations to study the hypersonic laminar flow near the leading edge of a flat plate. They used a series expansion technique to reduce the complete Navier-Stokes equations to the PNS equations. The set of PNS equations derived by Rudman and Rubin does not contain a streamwise pressure gradient term. Thus, no upstream influence through the subsonic part of the boundary layer is allowed, and the equations behave in a strictly parabolic manner in the boundary layer region. These PNS equations have been used to solve leading-edge flows about both two- and three-dimensional geometries including flat plates, rectangular corners, cones, and wing tips (see Lin and Rubin, 1973, for references). Leaving the details to the reader (see, for example, Anderson et al., 1984), the three-dimensional PNS equations of Rudman and Rubin in Cartesian coordinates are

continuity:

$$\frac{\partial \rho}{\partial t} + \frac{\partial(\rho u)}{\partial x} + \frac{\partial(\rho v)}{\partial y} + \frac{\partial(\rho w)}{\partial z} = 0 \tag{1.75}$$

x-momentum:

$$\rho \frac{\partial u}{\partial t} + \rho u \frac{\partial u}{\partial x} + \rho v \frac{\partial u}{\partial y} + \rho w \frac{\partial u}{\partial z} = \frac{\partial}{\partial y}\left(\mu \frac{\partial u}{\partial y}\right) + \frac{\partial}{\partial z}\left(\mu \frac{\partial u}{\partial z}\right) \tag{1.76}$$

y-momentum:

$$\rho \frac{\partial v}{\partial t} + \rho u \frac{\partial v}{\partial x} + \rho v \frac{\partial v}{\partial y} + \rho w \frac{\partial v}{\partial z}$$

$$= -\frac{\partial p}{\partial y} + \frac{\partial}{\partial z}\left(\mu \frac{\partial v}{\partial z}\right) + \frac{4}{3}\frac{\partial}{\partial y}\left(\mu \frac{\partial v}{\partial y}\right)$$

$$+ \frac{\partial}{\partial x}\left(\mu \frac{\partial u}{\partial y}\right) + \frac{\partial}{\partial z}\left(\mu \frac{\partial w}{\partial y}\right) - \frac{2}{3}\frac{\partial}{\partial y}\left(\mu \frac{\partial u}{\partial x} + \mu \frac{\partial w}{\partial z}\right) \tag{1.77}$$

z-momentum:

$$\rho \frac{\partial w}{\partial t} + \rho u \frac{\partial w}{\partial x} + \rho v \frac{\partial w}{\partial y} + \rho w \frac{\partial w}{\partial z}$$

$$= -\frac{\partial p}{\partial z} + \frac{\partial}{\partial y}\left(\mu \frac{\partial w}{\partial y}\right) + \frac{4}{3}\frac{\partial}{\partial z}\left(\mu \frac{\partial w}{\partial z}\right)$$

$$+ \frac{\partial}{\partial x}\left(\mu \frac{\partial u}{\partial z}\right) + \frac{\partial}{\partial y}\left(\mu \frac{\partial v}{\partial z}\right) - \frac{2}{3}\frac{\partial}{\partial z}\left(\mu \frac{\partial v}{\partial y} + \mu \frac{\partial u}{\partial x}\right) \tag{1.78}$$

energy:

$$\rho c_v \frac{\partial T}{\partial t} + \rho u c_v \frac{\partial T}{\partial x} + \rho v c_v \frac{\partial T}{\partial y} + \rho w c_v \frac{\partial T}{\partial z}$$

$$= -p\left(\frac{\partial u}{\partial x} + \frac{\partial v}{\partial y} + \frac{\partial w}{\partial z}\right) + \frac{\partial}{\partial y}\left(k\frac{\partial T}{\partial y}\right) + \frac{\partial}{\partial z}\left(k\frac{\partial T}{\partial z}\right)$$

$$+ \mu\left[\left(\frac{\partial u}{\partial y}\right)^2 + \left(\frac{\partial u}{\partial z}\right)^2 + \left(\frac{\partial w}{\partial y} + \frac{\partial v}{\partial z}\right)^2\right]$$

$$+ \frac{4}{3}\mu\left[\left(\frac{\partial v}{\partial y}\right)^2 + \left(\frac{\partial w}{\partial z}\right)^2 - \frac{\partial v}{\partial y}\frac{\partial w}{\partial z}\right] \tag{1.79}$$

The most common form, perhaps, of the PNS equations (Lubard and Helliwell, 1973, 1974) is obtained by assuming that the streamwise viscous derivative terms (including the heat flux terms) are negligible [assumed to be of $O(1)$] compared to the normal and transverse viscous derivative terms [of $O(\text{Re}_L^{1/2})$]. The resulting set of equations for a Cartesian coordinate system (x is the streamwise direction) is

continuity:

$$\frac{\partial \rho}{\partial t} + \frac{\partial(\rho u)}{\partial x} + \frac{\partial(\rho v)}{\partial y} + \frac{\partial(\rho w)}{\partial z} = 0 \tag{1.80}$$

x-momentum:

$$\rho\frac{\partial u}{\partial t} + \rho u\frac{\partial u}{\partial x} + \rho v\frac{\partial u}{\partial y} + \rho w\frac{\partial u}{\partial z} = -\frac{\partial p}{\partial x} + \frac{\partial}{\partial y}\left(\mu\frac{\partial u}{\partial y}\right) + \frac{\partial}{\partial z}\left(\mu\frac{\partial u}{\partial z}\right) \tag{1.81}$$

y-momentum:

$$\rho\frac{\partial v}{\partial t} + \rho u\frac{\partial v}{\partial x} + \rho v\frac{\partial v}{\partial y} + \rho w\frac{\partial v}{\partial z} = -\frac{\partial p}{\partial y} + \frac{\partial}{\partial z}\left(\mu\frac{\partial v}{\partial z}\right) + \frac{4}{3}\frac{\partial}{\partial y}\left(\mu\frac{\partial v}{\partial y}\right)$$

$$+ \frac{\partial}{\partial z}\left(\mu\frac{\partial w}{\partial y}\right) - \frac{2}{3}\frac{\partial}{\partial y}\left(\mu\frac{\partial w}{\partial z}\right) \tag{1.82}$$

z-momentum:

$$\rho\frac{\partial w}{\partial t} + \rho u\frac{\partial w}{\partial x} + \rho v\frac{\partial w}{\partial y} + \rho w\frac{\partial w}{\partial z} = -\frac{\partial p}{\partial z} + \frac{\partial}{\partial y}\left(\mu\frac{\partial w}{\partial y}\right) + \frac{4}{3}\frac{\partial}{\partial z}\left(\mu\frac{\partial w}{\partial z}\right)$$

$$+ \frac{\partial}{\partial y}\left(\mu\frac{\partial v}{\partial z}\right) - \frac{2}{3}\frac{\partial}{\partial z}\left(\mu\frac{\partial v}{\partial y}\right) \tag{1.83}$$

energy:

$$
\begin{aligned}
\rho c_v \frac{\partial T}{\partial t} &+ \rho u c_v \frac{\partial T}{\partial x} + \rho v c_v \frac{\partial T}{\partial y} + \rho w c_v \frac{\partial T}{\partial z} \\
&= -p\left(\frac{\partial u}{\partial x} + \frac{\partial v}{\partial y} + \frac{\partial w}{\partial z}\right) + \frac{\partial}{\partial y}\left(k\frac{\partial T}{\partial y}\right) + \frac{\partial}{\partial z}\left(k\frac{\partial T}{\partial z}\right) \\
&\quad + \mu\left[\left(\frac{\partial u}{\partial y}\right)^2 + \left(\frac{\partial u}{\partial z}\right)^2 + \left(\frac{\partial w}{\partial y} + \frac{\partial v}{\partial z}\right)^2\right] \\
&\quad + \frac{4}{3}\mu\left[\left(\frac{\partial v}{\partial y}\right)^2 + \left(\frac{\partial w}{\partial z}\right)^2 - \frac{\partial v}{\partial y}\frac{\partial w}{\partial z}\right]
\end{aligned}
\tag{1.84}
$$

A comparison of this set of PNS equation with that of Rudman and Rubin [Eqs. (1.75) to (1.79)] reveals that while the continuity and energy equations are identical, the momentum equations are different. In particular, the present x-momentum equation contains the streamwise pressure gradient term.

1.11   GOVERNING EQUATIONS IN GENERALIZED COORDINATES

For many problems, a transformation from physical space to computational space is performed in order to simplify implementation of the boundary conditions and to enhance the efficiency and accuracy of the numerical scheme. This transformation allows clustering of grid points in regions where the flow variables undergo high gradients, and grid point motion when required. The computational domain is rectangular with a uniformly spaced grid. While grid generation is discussed in Chapter 4, it is clear that a transformation of the governing equations from the physical space into the computational space is also required before any solution can be obtained. In this section we will see how the governing equations can be transformed from a Cartesian coordinate system (x, y, z) in the physical space to any general nonorthogonal (or orthogonal) coordinate system (ξ, η, ζ) in the computational space. The governing equations are written in *strong conservative* form to include the capability for shock capturing (Anderson, 1992a).

Let us consider a completely general transformation of the form

$$
\tau = t, \quad \xi = \xi(t, x, y, z), \quad \eta = \eta(t, x, y, z), \quad \zeta = \zeta(t, x, y, z)
\tag{1.85}
$$

The chain rule of partial differentiation yields the following for the Cartesian derivatives:

$$\frac{\partial}{\partial t} = \frac{\partial}{\partial \tau} + \xi_t \frac{\partial}{\partial \xi} + \eta_t \frac{\partial}{\partial \eta} + \zeta_t \frac{\partial}{\partial \zeta}$$

$$\frac{\partial}{\partial x} = \xi_x \frac{\partial}{\partial \xi} + \eta_x \frac{\partial}{\partial \eta} + \zeta_x \frac{\partial}{\partial \zeta} \tag{1.86}$$

$$\frac{\partial}{\partial y} = \xi_y \frac{\partial}{\partial \xi} + \eta_y \frac{\partial}{\partial \eta} + \zeta_y \frac{\partial}{\partial \zeta}$$

$$\frac{\partial}{\partial z} = \xi_z \frac{\partial}{\partial \xi} + \eta_z \frac{\partial}{\partial \eta} + \zeta_z \frac{\partial}{\partial \zeta}$$

The metrics $(\xi_t, \eta_t, \zeta_t, \xi_x, \eta_x, \zeta_x, \xi_y, \eta_y, \zeta_y, \xi_z, \eta_z, \zeta_z)$ appearing in these equations must be evaluated. In most cases analytical determination of the metrics is not possible; therefore they must be computed numerically. Since the grid size in the computational space is uniform, x_ξ, x_η, x_ζ, etc., can be computed readily. Thus, if the metrics appearing in Eqs. (1.86) can be expressed in terms of these derivatives, the numerical computation of metrics is easy. In order to obtain such relations, we first write the differential expression

$$dt = \frac{\partial t}{\partial \tau} d\tau + \frac{\partial t}{\partial \xi} d\xi + \frac{\partial t}{\partial \eta} d\eta + \frac{\partial t}{\partial \zeta} d\zeta$$

But according to (1.85),

$$\frac{\partial t}{\partial \tau} = 1, \qquad \frac{\partial t}{\partial \xi} = \frac{\partial t}{\partial \eta} = \frac{\partial t}{\partial \zeta} = 0$$

Thus,

$$dt = d\tau \tag{1.87}$$

Similarly

$$dx = x_\tau \, d\tau + x_\xi \, d\xi + x_\eta \, d\eta + x_\zeta \, d\zeta$$
$$dy = y_\tau \, d\tau + y_\xi \, d\xi + y_\eta \, d\eta + y_\zeta \, d\zeta \tag{1.88}$$
$$dz = z_\tau \, d\tau + z_\xi \, d\xi + z_\eta \, d\eta + z_\zeta \, d\zeta$$

Equations (1.87) and (1.88) can be expressed in matrix form as

$$\begin{bmatrix} dt \\ dx \\ dy \\ dz \end{bmatrix} = \begin{bmatrix} 1 & 0 & 0 & 0 \\ x_\tau & x_\xi & x_\eta & x_\zeta \\ y_\tau & y_\xi & y_\eta & y_\zeta \\ z_\tau & z_\xi & z_\eta & z_\zeta \end{bmatrix} \begin{bmatrix} d\tau \\ d\xi \\ d\eta \\ d\zeta \end{bmatrix} \tag{1.89}$$

Reversing the role of the independent variables, we can write

$$
\begin{bmatrix} d\tau \\ d\xi \\ d\eta \\ d\zeta \end{bmatrix} = \begin{bmatrix} 1 & 0 & 0 & 0 \\ \xi_t & \xi_x & \xi_y & \xi_z \\ \eta_t & \eta_x & \eta_y & \eta_z \\ \zeta_t & \zeta_x & \zeta_y & \zeta_z \end{bmatrix} \begin{bmatrix} dt \\ dx \\ dy \\ dz \end{bmatrix}
\tag{1.90}
$$

Comparing Eqs. (1.89) and (1.90), we get

$$
\begin{bmatrix} 1 & 0 & 0 & 0 \\ \xi_t & \xi_x & \xi_y & \xi_z \\ \eta_t & \eta_x & \eta_y & \eta_z \\ \zeta_t & \zeta_x & \zeta_y & \zeta_z \end{bmatrix} = \begin{bmatrix} 1 & 0 & 0 & 0 \\ x_\tau & x_\xi & x_\eta & x_\zeta \\ y_\tau & y_\xi & y_\eta & y_\zeta \\ z_\tau & z_\xi & z_\eta & z_\zeta \end{bmatrix}^{-1}
\tag{1.91}
$$

Thus, the metrics are

$$
\begin{aligned}
\xi_x &= J(y_\eta z_\zeta - y_\zeta z_\eta), & \zeta_y &= J(x_\zeta z_\eta - x_\eta z_\zeta), & \xi_z &= J(x_\eta y_\zeta - x_\zeta y_\eta) \\
\eta_x &= J(y_\zeta z_\xi - y_\xi z_\zeta), & \eta_y &= J(x_\xi z_\zeta - x_\zeta z_\xi), & \eta_z &= J(x_\zeta y_\xi - x_\xi y_\zeta) \\
\zeta_x &= J(y_\xi z_\eta - y_\eta z_\xi), & \zeta_y &= J(x_\eta z_\xi - x_\xi z_\eta), & \zeta_z &= J(x_\xi y_\eta - x_\eta y_\xi)
\end{aligned}
\tag{1.92}
$$

and

$$
\xi_t = -(x_\tau \xi_x + y_\tau \xi_y + z_\tau \xi_z), \qquad \eta_t = -(x_\tau \eta_x + y_\tau \eta_y + z_\tau \eta_z)
$$

$$
\zeta_t = -(x_\tau \zeta_x + y_\tau \zeta_y + z_\tau \zeta_z)
\tag{1.93}
$$

Substituting Eqs. (1.92) into (1.93) yields

$$
\begin{aligned}
\xi_t &= J[x_\tau(y_\zeta z_\eta - y_\eta z_\zeta) + y_\tau(x_\eta z_\zeta - x_\zeta z_\eta) + z_\tau(x_\zeta y_\eta - x_\eta y_\zeta)] \\
\eta_t &= J[x_\tau(y_\xi z_\zeta - y_\zeta z_\xi) + y_\tau(x_\zeta z_\xi - x_\xi z_\zeta) + z_\tau(x_\xi y_\zeta - x_\zeta y_\xi)] \\
\zeta_t &= J[x_\tau(y_\eta z_\xi - y_\xi z_\eta) + y_\tau(x_\xi z_\eta - x_\eta z_\xi) + z_\tau(x_\eta y_\xi - x_\xi y_\eta)]
\end{aligned}
\tag{1.94}
$$

where J is the Jacobian of the transformation:

$$
J = \frac{\partial(\xi, \eta, \zeta)}{\partial(x, y, z)} = \begin{vmatrix} \xi_x & \xi_y & \xi_z \\ \eta_x & \eta_y & \eta_z \\ \zeta_x & \zeta_y & \zeta_z \end{vmatrix}
\tag{1.95}
$$

which can be evaluated as follows:

$$
J = \frac{1}{J^{-1}} = 1 \Big/ \frac{\partial(x, y, z)}{\partial(\xi, \eta, \zeta)} = \begin{vmatrix} x_\xi & x_\eta & x_\zeta \\ y_\xi & y_\eta & y_\zeta \\ z_\xi & z_\eta & z_\zeta \end{vmatrix}^{-1}
$$

$$
= [x_\xi(y_\eta z_\zeta - y_\zeta z_\eta) - x_\eta(y_\xi z_\zeta - y_\zeta z_\xi) + x_\zeta(y_\xi z_\eta - y_\eta z_\xi)]^{-1}
\tag{1.96}
$$

For a discussion of the proper way to compute metrics, see Anderson et al. (1984, Chap. 10). Applying the generalized transformation to the compressible Navier-Stokes equations written in vector form (Eqs. (1.27)], we obtain the transformed equation

$$U_\tau + \xi_t U_\xi + \eta_t U_\eta + \zeta_t U_\zeta + \xi_x E_\xi + \eta_x E_\eta + \zeta_x E_\zeta$$
$$+ \xi_y F_\xi + \eta_y F_\eta + \zeta_y F_\zeta + \xi_z G_\xi + \eta_z G_\eta + \zeta_z G_\zeta = 0 \tag{1.97}$$

This equation is no longer in conservative form. Following Viviand (1974) and Vinokur (1974), it can be cast into *strong conservative* form to yield

$$\frac{\partial}{\partial \tau}\left(\frac{U}{J}\right) + \frac{\partial}{\partial \xi}\left(\frac{\xi_t U + \xi_x E + \xi_y F + \xi_z G}{J}\right)$$
$$+ \frac{\partial}{\partial \eta}\left(\frac{\eta_t U + \eta_x E + \eta_y F + \eta_z G}{J}\right)$$
$$+ \frac{\partial}{\partial \zeta}\left(\frac{\zeta_t U + \zeta_x E + \zeta_y F + \zeta_z G}{J}\right) = 0 \tag{1.98}$$

We can redefine the terms in this equation to write it in the form

$$\frac{\partial U_1}{\partial \tau} + \frac{\partial E_1}{\partial \xi} + \frac{\partial F_1}{\partial \eta} + \frac{\partial G_1}{\partial \zeta} = 0 \tag{1.99}$$

where

$$U_1 = \frac{U}{J}$$

$$E_1 = \frac{\xi_t U + \xi_x E + \xi_y F + \xi_z G}{J} \tag{1.100}$$

$$F_1 = \frac{\eta_t U + \eta_x E + \eta_y F + \eta_z G}{J}$$

$$G_1 = \frac{\zeta_t U + \zeta_x E + \zeta_y F + \zeta_z G}{J}$$

Note that the vectors E_1, F_1, and G_1 contain viscous and heat flux terms [cf. Eqs. (1.28)] that involve partial derivatives. These partial derivatives are also to be transformed using Eqs. (1.86).

The viscous stresses, using Stokes' hypothesis ($\mu' = -2\mu/3$), in the transformed computational space are

$$\tau_{xx} = \mu\left[\frac{4}{3}(\xi_x u_\xi + \eta_x u_\eta + \zeta_x u_\zeta)\right.$$
$$\left. - \frac{2}{3}(\xi_y v_\xi + \eta_y v_\eta + \zeta_y v_\zeta) - \frac{2}{3}(\xi_z w_\xi + \eta_z w_\eta + \zeta_z w_\zeta)\right]$$

$$\tau_{yy} = \mu\left[\frac{4}{3}(\xi_y v_\xi + \eta_y v_\eta + \zeta_y v_\zeta)\right.$$

$$\left. -\frac{2}{3}(\xi_x u_\xi + \eta_x u_\eta + \zeta_x u_\zeta) - \frac{2}{3}(\xi_z w_\xi + \eta_z w_\eta + \zeta_z w_\zeta)\right]$$

$$\tau_{zz} = \mu\left[\frac{4}{3}(\xi_z w_\xi + \eta_z w_\eta + \zeta_z w_\zeta)\right.$$

$$\left. -\frac{2}{3}(\xi_x u_\xi + \eta_x u_\eta + \zeta_x u_\zeta) - \frac{2}{3}(\xi_y v_\xi + \eta_y v_\eta + \zeta_y v_\zeta)\right] \qquad (1.101)$$

$$\tau_{xy} = \tau_{yx} = \mu(\xi_y u_\xi + \eta_y u_\eta + \zeta_y u_\zeta + \xi_x v_\xi + \eta_x v_\eta + \zeta_x v_\zeta)$$

$$\tau_{xz} = \tau_{zx} = \mu(\xi_z u_\xi + \eta_z u_\eta + \zeta_z u_\zeta + \xi_x w_\xi + \eta_x w_\eta + \zeta_x w_\zeta)$$

$$\tau_{yz} = \tau_{zy} = \mu(\xi_z v_\xi + \eta_z v_\eta + \zeta_z v_\zeta + \xi_y w_\xi + \eta_y w_\eta + \zeta_y w_\zeta)$$

and the heat conduction terms in the computational space are

$$q_x = -k(\xi_x T_\xi + \eta_x T_\eta + \zeta_x T_\zeta)$$
$$q_y = -k(\xi_y T_\xi + \eta_y T_\eta + \zeta_y T_\zeta) \qquad (1.102)$$
$$q_z = -k(\xi_z T_\xi + \eta_z T_\eta + \zeta_z T_\zeta)$$

The *conservative* form of the governing equations is convenient for applying finite-difference schemes (Anderson, 1992a). For many applications, the grid is independent of time, and thus time gradients of the metrics are zero. However, when the grid is changing with time, a constraint on the way the metrics are differenced, called the *geometric conservation* law (Thomas and Lombard, 1978), must be satisfied so as to prevent the introduction of additional errors into the solution. Details are available in Anderson et al. (1984).

1.12 MATHEMATICAL PROPERTIES OF THE GOVERNING EQUATIONS

It is useful to examine some mathematical properties of the Navier-Stokes equations since any valid solution of the equations should obey these properties. Depending upon the flow situation, the Navier-Stokes equations can be classified as hyperbolic, elliptic, or parabolic. For the rules governing this classification, the reader may refer to any text on partial differential equations or to Anderson et al. (1984), Fletcher (1991, Chap. 2), and Anderson (1992b). We make some relevant observations here.

For *hyperbolic* equations, information at a given point influences only those regions between the advancing characteristics. The *method of characteristics* takes advantage of this property during solution; one can march along the characteristic. For *parabolic* equations, information at a point P in, say, the x-y plane influences the entire region of the plane to one side of P only. The solution to parabolic equations can thus be marched in the main flow direction. For *elliptic* equations, information at a point P in the x-y plane influences *all* other regions of the domain. Therefore, the solution at point P must be carried out *simultaneously* with the solution at all other points in the domain. This is in stark contrast to the 'marching' solutions germane to parabolic and hyperbolic equations.

For *inviscid compressible* flows, the system of equations is always *hyperbolic* if the motion is *unsteady*, no matter whether the flow is locally subsonic or supersonic. If the motion is *steady*, the classification of the system depends upon the fluid speed, being *hyperbolic* if it is *supersonic* and *elliptic* if *subsonic*. Thus, inviscid compressible flows can be of mixed type when the flow is steady and subsonic in one region while being supersonic elsewhere. Since incompressible flow (which theoretically implies that the Mach number is zero) is a subcase of the subsonic flow, *steady inviscid incompressible* flows are also *elliptic*. For such flows, physical boundary conditions must be applied over a closed boundary that totally surrounds the flow domain. Examples of *parabolic* flows are the *boundary layer* flows governed by the boundary layer equations of section 1.8 and flows governed by the PNS equations of section 1.10.

Owing to the totally different mathematical behavior of elliptic and hyperbolic equations, we can appreciate the difficulties early researchers encountered in trying to solve mixed problems, a prime example of which is the supersonic flow over a blunt body such as an atmospheric entry vehicle. The sudden change in the nature of the governing equations across the sonic line precluded any practical solution of the steady flow blunt body problem involving a uniform treatment of both the subsonic and supersonic regions. However, since *unsteady* inviscid flow is governed by hyperbolic equations, no matter whether the flow is locally subsonic or supersonic, we can solve the *unsteady* inviscid flow equations, marching forward in time. At large times, the solution approaches steady state, which is the desired result. Thus, we get the steady-state solution for the *entire* flow field, *including* both the subsonic and supersonic regions, using the same, uniform method throughout the flow field. This is the basic philosophy behind the *time-dependent technique* for the solution of flow problems.

It may be helpful to the reader to examine the closed-form solution to some linear partial differential equations of the elliptic, parabolic, and hyperbolic types. Numerous classical solutions can be found in Hildebrand (1976) and Anderson et al. (1984). We move on to numerical solutions in the remainder of the book.

NOMENCLATURE

C	specific heat
c_p	specific heat at constant pressure
c_v	specific heat at constant volume
e	specific internal energy
E	total energy per unit volume
$\mathbf{F}$	body force per unit volume
h	specific enthalpy
H	enthalpy
J	Jacobian of the coordinate transformation
k	thermal conductivity
L	characteristic length
m	mass flow
M	Mach number
p	pressure
$\mathbf{P}$	surface force per unit volume
Pr	Prandtl number
q	conductive heat flux
R	gas constant
Re	Reynolds number
t	time
T	temperature
u, v, w	Cartesian velocity components
$\mathbf{V}$	velocity vector
V_r, V_z	radial and axial components of velocity in cylindrical coordinates
x, y, z	Cartesian coordinates

Greek

α	thermal diffusivity
β	coefficient of volumetric expansion
γ	ratio of specific heats ($= c_p/c_v$)
δ_{ij}	Kronecker delta ($= 0$ for $i \neq j$, and $= 1$ for $i = j$)
ζ	vorticity
κ	coefficient of bulk viscosity
μ	dynamic viscosity
μ'	second coefficient of viscosity
ν	kinematic viscosity
ξ, η, ζ	generalized coordinates

ρ density
σ_{ij} stress on a plane normal to axis i in the direction of axis j
ϕ velocity potential
Φ dissipation function
τ time
τ_{ij} viscous (shear) stress tensor
ψ stream function

Subscripts

e refers to value at the edge of the boundary layer
∞ refers to value in the freestream

REFERENCES

Anderson DA, Tannehill JC, Pletcher RH Computational Fluid Mechanics and Heat Transfer. Washington DC: Hemisphere: 1984.

Anderson JD Jr. Governing equations of fluid dynamics. In: Wendt JF, ed. Computational Fluid Dynamics—An Introduction. Berlin: Springer, 1992a pp 15–51.

Anderson JD Jr. Mathematical properties of the fluid dynamic equations. In: Wendt JF, ed. Computational Fluid Dynamics—An Introduction, Berlin: Springer, 1992b, pp 75–84.

Cebeci T, Smith AMO. Analysis of Turbulent Boundary Layers. New York: Academic Press; 1974.

Fletcher CAJ Computational Techniques for Fluid Dynamics. 2nd ed. Berlin: Springer, 1991.

Hildebrand FB. Advanced Calculus for Applications. Englewood Cliffs, NJ: Prentice-Hall, 1976.

Lin TC, Rubin SG Viscous flow over a cone at moderate incidence. I: Hypersonic tip region. Comput Fluids 1: 37–57, 1973.

Lubard SC, Helliwell WS. Calculation of the flow on a cone at high angle of attack. R&D Associates Technical Report, RDA-TR-150, Santa Monica, CA, 1973.

Lubard SC, Helliwell WS. Calculation of the flow on a cone at high angle of attack. AIAA J 12: 965–974, 1974.

Prandtl L. Verh. 3rd Intl. Math. Kongr. Heidel, p 484 (translated as NACA TM 452), 1984.

Round GF, Garg VK. Applications of Fluid Dynamics. London: Edward Arnold, 1986.

Rudman S, Rubin SG Hypersonic viscous flow over slender bodies with sharp leading edges AIAA J 6: 1883–1889, 1968.

Schlichting H. Boundary Layer Theory. 7th ed. New York: McGraw-Hill, 1979.

Thomas PD, Lombard CK The geometric conservation law—a link between finite-difference and finite-volume methods of flow computation on moving grids. AIAA Paper 78-1208, Seattle, WA, 1978.

Vinokur M. Conservation equations of gas-dynamics in curvilinear coordinate systems. J Comp Phys 14: 105–125, 1974.
Viviand H. Conservation forms of gas dynamic equations. La Recherche Aérospatiale, No 1974-1, pp 65–68, 1974.

Numerical Techniques

Vijay K. Garg

AYT Corporation/NASA Lewis Research Center, Cleveland, Ohio

2.1 INTRODUCTION

We notice from Chapter 1 that the equations governing fluid flow and heat transfer are complex partial differential equations for which no analytical solution can be found except in rather simple situations. For many practical problems, in general, obtaining a numerical solution to the Navier-Stokes equations is the only possibility, short of actual experimentation. With the easy availability of high-speed digital computers, growth of computational fluid dynamics (CFD) has mushroomed in the last three decades. While analytical solutions, if possible, yield closed-form expressions which give the variation of dependent variables *continuously* throughout the domain, numerical solutions can give answers only at *discrete points* in the domain, called *grid points*. The end product of CFD is thus a collection of numbers, in contrast to a closed-form analytical solution. In the long run, however, the objective of most engineering analyses, closed-form or otherwise, is a quantitative description of the problem, i.e., numbers (Anderson, 1976).

For incompressible flows there are several possibilities for the formulation of the problem. These include primitive variables, stream-function vorticity, and vorticity velocity formulations. The primitive variable approach offers the fewest complications in extending two-dimensional schemes to three dimensions. The primary difficulty with this approach is the specification of boundary conditions on pressure. See Peyret and Taylor (1983) and Anderson et al. (1984) for ways to overcome this problem. The stream-function vorticity formulation is best suited for two-dimensional flows, though it has been applied to three-dimensional incompressible flows as well [see, for example, Aziz and Hellums (1967), and Mallinson and De Vahl Davis (1973, 1977) for application details, and Hirasaki and Hellums (1970) and Richardson and Cornish (1977) for boundary condition considerations]. The difficulty with such methods is primarily associated with determination of vorticity at a boundary. A number of ways to overcome this difficulty are available (Peyret and Taylor, 1983; Anderson et al., 1984). An inconvenience of this formulation is that the pressure is not directly available, and additional computation is required for its determination. We may point out that for two-dimensional flows, the stream-function-only formulation can also be utilized. The interested reader is referred to Bourcier and François (1969), Roache and Ellis (1975), Morchoisne (1979), Cebeci et al. (1981), and Jaluria and Torrance (1986) for the details. The vorticity velocity formulation requires the vorticity equation, the continuity equation, and the equations that define vorticity in terms of velocity gradients. A combination of the continuity equation and the definition of vorticity yields elliptic equations for the velocity components. The interested reader is referred to Fasel (1976) and Dennis et al. (1979) for details.

Today there are a number of numerical techniques available for solving the fluid flow and heat transfer problems. Some of these are specific to the type of flow under investigation, as mentioned above, and for example, the *panel* methods for inviscid incompressible flows, *Green's function* methods for incompressible flows (generally two-dimensional), the *method of characteristics* for inviscid supersonic or inviscid transient compressible flow problems, etc. In this chapter, however, we will describe briefly three techniques that, in principle, encompass all the methods. These are the finite difference method (FDM), the finite element method (FEM), and the finite volume method (FVM). Peyret and Taylor (1983) point out that the *spectral* method can be considered as a variant of the FEM. For some relationships between the FDM, FEM and spectral methods, see Patankar (1980), Peyret and Taylor (1983), and Fletcher (1991). The relative merits of the FDM, FEM, and spectral methods are given in Fletcher (1984, Chap. 6).

2.2   FINITE DIFFERENCE METHOD

The finite difference method is widely used and is perhaps the oldest method. The essence of a finite difference method is to replace the partial derivatives appearing in the governing equations with algebraic difference quotients, thus yielding a system of *algebraic equations* which can be solved for the flow-field variables at the specific, *discrete grid points* in the flow domain. The nature of the resulting algebraic system depends upon the character of the problem defined by the original partial differential equation (PDE) or system of PDEs.

For convenience, let us consider a two-dimensional problem, and let Fig. 2.1 show a section of the discrete grid in the x-y plane. Let us assume that the spacing of the grid points in the x-direction is uniform, and given by Δx, and that the spacing of grid points in the y-direction is also uniform, and given by Δy, as shown in Fig. 2.1. In general, Δx and Δy are different. Indeed, it is not necessary that Δx and Δy be uniform; we can have totally unequal spacing in both directions so that Δx is a different value between each successive pair of grid points, and similarly for Δy. The vast majority of CFD applications, however, involve numerical solutions on a grid which involves uniform spacing in each direction, since this greatly simplifies the

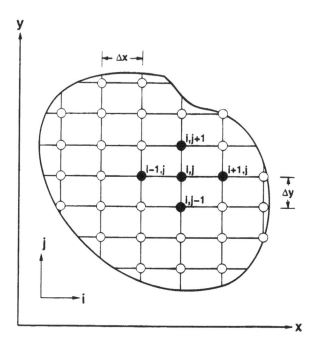

Figure 2.1 Finite difference grid in a two-dimensional region.

programming, saves storage space, and usually results in greater accuracy. This uniform spacing does not occur in the physical x-y space. Generally, in CFD, numerical calculations are carried out in a transformed computational space which has uniform spacing in the transformed independent variables, but which corresponds to nonuniform spacing in the physical space. The grid lines are indexed by integers i and j which increase monotonically along the x and y coordinates, respectively. Let us now derive some common finite difference expressions used to replace the partial derivatives in the PDEs.

2.2.1   Finite Difference Representation of Derivatives

Finite difference representation of the partial derivatives can be derived from Taylor series expansion. Let $f_{i,j}$ denote the value of dependent variable f at the point (i, j) with coordinates (x_i, y_j), where f may be a component of fluid velocity, temperature or any other dependent variable. Then the value $f_{i+1,j}$ at point $(i + 1, j)$ can be expressed in terms of the Taylor series expanded about point (i, j) as follows:

$$f_{i+1, j} = f_{i, j} + \left(\frac{\partial f}{\partial x}\right)_{i, j} \Delta x + \left(\frac{\partial^2 f}{\partial x^2}\right)_{i, j} \frac{(\Delta x)^2}{2} + \left(\frac{\partial^3 f}{\partial x^3}\right)_{i, j} \frac{(\Delta x)^3}{6} + \cdots \quad (2.1)$$

Equation (2.1) is a mathematically exact expression for $f_{i+1, j}$ if (a) the number of terms is infinite and the series converges, and/or (b) $\Delta x \to 0$. Clearly, only a finite number of terms in Eq. (2.1) can be carried in numerical computations. Thus, Eq. (2.1) is *truncated*. For example, if terms of order $(\Delta x)^2$ and higher are neglected, Eq. (2.1) reduces to

$$f_{i+1, j} \approx f_{i, j} + \left(\frac{\partial f}{\partial x}\right)_{i, j} \Delta x \quad (2.2)$$

Equation (2.2) is said to be *first-order accurate*. If terms of order $(\Delta x)^3$ and higher are neglected, we get from Eq. (2.1)

$$f_{i+1, j} \approx f_{i, j} + \left(\frac{\partial f}{\partial x}\right)_{i, j} \Delta x + \left(\frac{\partial^2 f}{\partial x^2}\right)_{i, j} \frac{(\Delta x)^2}{2} \quad (2.3)$$

where Eq. (2.3) is *second-order accurate*. In Eqs. (2.2) and (2.3), the neglected higher-order terms represent the *truncation error* in the finite series representation. The truncation error can be reduced by (a) carrying more terms in the Taylor series expansion, Eq. (2.1), leading to higher-order accuracy in the representation of $f_{i, j}$, and (b) reducing the magnitude of Δx. We can solve Eq. (2.1) for $(\partial f/\partial x)_{i, j}$ to get

$$\left(\frac{\partial f}{\partial x}\right)_{i, j} = \frac{f_{i+1, j} - f_{i, j}}{\Delta x} - \left(\frac{\partial^2 f}{\partial x^2}\right)_{i, j} \frac{\Delta x}{2} - \left(\frac{\partial^3 f}{\partial x^3}\right)_{i, j} \frac{(\Delta x)^2}{6} - \cdots$$

or

$$\left(\frac{\partial f}{\partial x}\right)_{i, j} = \frac{f_{i+1, j} - f_{i, j}}{\Delta x} + O(\Delta x) \quad (2.4)$$

The symbol $O(\Delta x)$ in Eq. (2.4) is a formal mathematical notation which represents "terms of order of Δx." Details on the O notation can be found in Whittaker and Watson (1927). Equation (2.4) is a more precise notation than Eq. (2.2), which involves the "approximately equal" sign. The first-order-accurate difference representation for the derivative $(\partial f/\partial x)_{i, j}$, expressed in Eq. (2.4), is known as the *first-order forward difference*.

Similarly, the *first-order backward difference* expression for the derivative $(\partial f/\partial x)_{i, j}$ can be written as

$$\left(\frac{\partial f}{\partial x}\right)_{i, j} = \frac{f_{i, j} - f_{i-1, j}}{\Delta x} + O(\Delta x) \quad (2.5)$$

which follows from the Taylor series expansion for $f_{i-1,j}$ about the point (i, j), obtained by replacing Δx by $-\Delta x$ on the right side of Eq. (2.1) to yield

$$f_{i-1,j} = f_{i,j} + \left(\frac{\partial f}{\partial x}\right)_{i,j}(-\Delta x) + \left(\frac{\partial^2 f}{\partial x^2}\right)_{i,j}$$
$$\times \frac{(-\Delta x)^2}{2} + \left(\frac{\partial^3 f}{\partial x^3}\right)_{i,j}\frac{(-\Delta x)^3}{6} + \cdots \tag{2.6}$$

Subtracting Eq. (2.6) from Eq. (2.1), and rearranging, we get

$$\left(\frac{\partial f}{\partial x}\right)_{i,j} = \frac{f_{i+1,j} - f_{i-1,j}}{2\,\Delta x} + O(\Delta x)^2 \tag{2.7}$$

which is the *second-order central difference* for the derivative $(\partial f/\partial x)_{i,j}$.

To obtain a finite difference expression for the second partial derivative $(\partial^2 f/\partial x^2)_{i,j}$, we add Eqs. (2.1) and (2.6) and rearrange to yield

$$\left(\frac{\partial^2 f}{\partial x^2}\right)_{i,j} = \frac{f_{i+1,j} - 2f_{i,j} + f_{i-1,j}}{(\Delta x)^2} + O(\Delta x)^2 \tag{2.8}$$

Equation (2.8) is a *second-order central second difference* for the derivative $(\partial^2 f/\partial x^2)$ at grid point (i, j).

Difference expressions for the y-derivatives are obtained in the same manner. The results, analogous to the previous equations for the x-derivatives, are

$$\left(\frac{\partial f}{\partial y}\right)_{i,j} = \frac{f_{i,j+1} - f_{i,j}}{\Delta y} + O(\Delta y) \qquad \text{Forward difference}$$

$$\left(\frac{\partial f}{\partial y}\right)_{i,j} = \frac{f_{i,j} - f_{i,j-1}}{\Delta y} + O(\Delta y) \qquad \text{Backward difference}$$

$$\left(\frac{\partial f}{\partial y}\right)_{i,j} = \frac{f_{i,j+1} - f_{i,j-1}}{2\Delta y} + O(\Delta y)^2 \qquad \text{Central difference}$$

$$\left(\frac{\partial^2 f}{\partial y^2}\right)_{i,j} = \frac{f_{i,j+1} - 2f_{i,j} + f_{i,j-1}}{(\Delta y)^2} + O(\Delta y)^2 \qquad \text{Central second difference}$$

We leave it as an exercise for the reader to show that the central second difference in Eq. (2.8) can be interpreted as a forward difference of the first derivatives, with backward differences used for each of the first derivatives. A similar philosophy can be used to generate a finite difference expression for the mixed derivative $(\partial^2 f/\partial x\partial y)$ at grid point (i, j). Clearly,

$$\frac{\partial^2 f}{\partial x\,\partial y} = \frac{\partial}{\partial x}\left(\frac{\partial f}{\partial y}\right) \tag{2.9}$$

In Eq. (2.9), we write the x-derivative as a central difference of the y-derivatives, and then write the y-derivatives also in terms of central differences to get

$$\frac{\partial^2 f}{\partial x\, \partial y} = \frac{\partial}{\partial x}\left(\frac{\partial f}{\partial y}\right) \approx \frac{(\partial f/\partial y)_{i+1,\,j} - (\partial f/\partial y)_{i-1,\,j}}{2\,\Delta x}$$

$$\frac{\partial^2 f}{\partial x\, \partial y} \approx \left[\left(\frac{f_{i+1,\,j+1} - f_{i+1,\,j-1}}{2\,\Delta y}\right) - \left(\frac{f_{i-1,\,j+1} - f_{i-1,\,j-1}}{2\,\Delta y}\right)\right]\frac{1}{2\,\Delta x}$$

or

$$\left(\frac{\partial^2 f}{\partial x\, \partial y}\right)_{i,\,j} = \frac{1}{4\,\Delta x\,\Delta y}\left(f_{i+1,\,j+1} + f_{i-1,\,j-1} - f_{i+1,\,j-1} - f_{i-1,\,j+1}\right)$$
$$+\ O[(\Delta x)^2, (\Delta y)^2] \tag{2.10}$$

Many other difference approximations can be written for the above derivatives, as well as for derivatives of higher order. For a tabulation of many forms of difference approximations, see Hyman and Larrouturou (1982).

2.2.2   Finite Difference Approximations at a Boundary

At a boundary, only one direction, away from the boundary, is available. For example, Fig. 2.2 illustrates a portion of the boundary with a normal in the y-direction. Let grid point 1 be on the boundary, with points 2 and 3 a

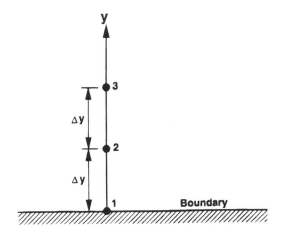

Figure 2.2   Grid points at a boundary.

distance Δy and $2 \Delta y$ above the boundary, respectively. It is easy to write a forward difference approximation for $\partial f / \partial y$ at the boundary as

$$\left(\frac{\partial f}{\partial y}\right)_1 = \frac{f_2 - f_1}{\Delta y} + O(\Delta y) \tag{2.11}$$

which is first-order accurate. The second-order accurate approximation for $\partial f / \partial y$ at the boundary can be found by fitting a parabola

$$f = a + by + cy^2 \tag{2.12}$$

through the three points 1, 2, and 3 in Fig. 2.2. This yields

$$\left(\frac{\partial f}{\partial y}\right)_1 = b = \frac{-3f_1 + 4f_2 - f_3}{2 \Delta y} \tag{2.13}$$

Using a Taylor series expansion about point 1, it can be shown that Eq. (2.13) is second-order accurate. Thus, the second-order-accurate approximation for $\partial f / \partial y$ at the boundary is

$$\left(\frac{\partial f}{\partial y}\right)_1 = \frac{-3f_1 + 4f_2 - f_3}{2 \Delta y} + O(\Delta y)^2 \tag{2.14}$$

Both Eqs. (2.11) and (2.14) are called *one-sided differences*, since they express a derivative at a point in terms of dependent variables on *only one side* of the point. Many other one-sided differences can be formed, with higher degrees of accuracy, using additional grid points (Hyman and Larrouturou, 1982).

2.2.3 Solution of Finite Difference Equations—Some Considerations

As stated earlier, the FDM replaces the partial derivatives in the governing equations by the difference quotients, leading to a system of algebraic equations for the dependent variable(s) at each grid point. Let us examine some aspects in the solution of these equations by considering the model equation

$$\frac{\partial f}{\partial t} = \frac{\partial^2 f}{\partial x^2} \tag{2.15}$$

Here, we have assumed, for convenience, that the dependent variable f is a function of x and t only. There is no advantage at this stage in considering a more complex example. It may be noted that Eq. (2.15) is parabolic, and its solution can therefore be "marched" (cf. section 1.12). Replacing the time derivative in Eq. (2.15) with a forward difference, and the spatial derivative with a central difference at time t, we get

$$\frac{f_i^{n+1} - f_i^n}{\Delta t} = \frac{f_{i+1}^n - 2f_i^n + f_{i-1}^n}{(\Delta x)^2} \qquad (2.16)$$

In Eq. (2.16), the index for time appears as a superscript, where n denotes values at time t, $n + 1$ denotes values at time $t + \Delta t$, and so on. The subscript denotes the grid point location, as usual. Since forward difference is used for the time derivative, and central difference for the spatial derivative, the *truncation error for the complete equation* is $O[\Delta t, (\Delta x)^2]$.

If truncation error approaches zero as the grid is refined (in general, Δx, Δy, Δz, and $\Delta t \to 0$), the finite difference approximation is said to be *consistent* with the governing differential equation. Consistency is an essential requirement for a valid numerical simulation of a partial differential equation. In our example, we note that the truncation error approaches zero as $\Delta x \to 0$ and $\Delta t \to 0$. Hence the difference Eq. (2.16) is consistent with the differential Eq. (2.15).

Let us now examine the solution of Eq. (2.16). From the given initial condition, we know the dependent variable at all x at the initial instant. Examining Eq. (2.16), we find that it contains only one unknown, namely f_i^{n+1}. Thus, the solution of Eq. (2.16) can be marched in time such that the value of f at time $t + \Delta t$ can be obtained *explicitly* from the *known* values at time t; i.e., f_i^{n+1} is obtained directly from the known values f_{i+1}^n, f_i^n, and f_{i-1}^n. This is an example of an *explicit finite difference solution*.

As a counterexample, let us replace the spatial derivative in Eq. (2.15) by the central difference at time $t + \Delta t$, instead of at time t as earlier, to get

$$\frac{f_i^{n+1} - f_i^n}{\Delta t} = \frac{f_{i+1}^{n+1} - 2f_i^{n+1} + f_{i-1}^{n+1}}{(\Delta x)^2} \qquad (2.17)$$

The differencing given in Eq. (2.17) is the *fully implicit* form. If the spatial difference on the right side of Eq. (2.17) is written in terms of the *average* values between time t and $t + \Delta t$, we get the well-known *Crank-Nicolson form*. Examining Eq. (2.17), we find that the unknown f_i^{n+1} is not only expressed in terms of the known values at time index n, namely f_i^n, but also in terms of unknown values at time index $n + 1$, namely f_{i+1}^{n+1} and f_{i-1}^{n+1}. Hence, Eq. (2.17) applied at a given grid point i cannot by itself result in the solution for f_i^{n+1}. Rather, Eq. (2.17) must be written at all grid points, resulting in a system of algebraic equations, which must be solved simultaneously to yield the unknown f_i^{n+1} for all i. This is an example of an *implicit finite difference solution*. Note that the system of algebraic equations that results in *sparse*. For the present example, we get a *tridiagonal system* of equations, which can be easily solved using an algorithm based on Gaussian elimination, often known as the Thomas algorithm (Hornbeck, 1973).

The number of arithmetic operations required for solving a tridiagonal system of n equations by this algorithm is $O(n)$, as compared to $O(n^3)$ needed for the solution of a general algebraic system by Gaussian elimination. This leads to a very efficient algorithm and small *roundoff error*.

The relative major advantages and disadvantages of the explicit and implicit methods are:

1. *Explicit Approach*
   a. Advantage: relatively simple to set up and program.
   b. Disadvantage: For a given spatial grid, Δt must be less than some limit imposed by stability constraints. In many cases, Δt must be very small to maintain stability, resulting in a long computational time to obtain results over a given interval of time.
2. *Implicit Approach*
   a. Advantage: stability can be maintained over much larger values of Δt, thereby requiring fewer time steps to obtain results over a given interval of time. For the *fully* implicit approach, there is no stability constraint on Δt. Note, however, that the *Crank-Nicolson* scheme, contrary to popular belief, can lead to physically unrealistic solutions (Patankar and Baliga, 1978; Patankar, 1980).
   b. Disadvantage: more complicated to set up and program.
   c. Disadvantage: since matrix manipulations are usually required at each time step, computer time per time step is much larger than that in the explicit approach.
   d. Disadvantage: since large Δt can be taken, truncation error is larger. Thus, use of the implicit method to follow the exact transients (time variations of the dependent variable) may not be as accurate as an explicit approach. However, if the steady state is the desired result, the relative timewise inaccuracy for a time-dependent solution is unimportant.

2.2.4 Errors and Stability Analysis

While the FDM is a philosophically straightforward technique, there is no guarantee that the solution of the algebraic equations will be accurate, or even stable, under all conditions for a given problem. For *linear* PDEs, however, there is a formal way of examining the accuracy and stability, and these ideas provide some guidance for understanding the behavior of more complex nonlinear PDEs that govern fluid flow and heat transfer, in general. While more details are available in Hornbeck (1973) and in Anderson et al. (1984), we present some basic ideas here. The numerical solution of a PDE is influenced by two sources of error:

1. *Discretization Error*: The difference between the exact analytical solution of the PDE [for example, Eq. (2.15)] and the exact solution (to infinite precision) of the corresponding difference equation [for example, Eq. (2.16)]. Based on the previous discussion, the discretization error is simply the truncation error for the difference equation plus any errors introduced by the numerical treatment of the boundary conditions. This is an *error of approximation.*

2. *Roundoff Error*: The difference between the exact and machine solutions of the finite difference approximation. This is a consequence of the finite word length in any computer. This is an *error of calculation.*

If we let

E = exact (analytical) solution of the PDE

D = exact solution of the finite difference approximation

N = numerical solution from a computer with finite accuracy

then

Discretization error $= E - D$

Roundoff error $= \varepsilon = N - D$ \hfill (2.18)

Equation (2.18) can be written as

$$N = D + \varepsilon \tag{2.19}$$

where ε is the roundoff error, which we will simply call "error" for brevity in the remainder of this section. The numerical solution N must satisfy the difference equation. Hence from Eq. (2.16), we have

$$\frac{D_i^{n+1} + \varepsilon_i^{n+1} - D_i^n - \varepsilon_i^n}{\Delta t} = \frac{D_{i+1}^n + \varepsilon_{i+1}^n - 2D_i^n - 2\varepsilon_i^n + D_{i-1}^n + \varepsilon_{i-1}^n}{(\Delta x)^2}$$

$$\tag{2.20}$$

By definition, D is the *exact* solution of the difference equation, hence

$$\frac{D_i^{n+1} - D_i^n}{\Delta t} = \frac{D_{i+1}^n - 2D_i^n + D_{i-1}^n}{(\Delta x)^2} \tag{2.21}$$

Subtracting Eq. (2.21) from (2.20), we get

$$\frac{\varepsilon_i^{n+1} - \varepsilon_i^n}{\Delta t} = \frac{\varepsilon_{i+1}^n - 2\varepsilon_i^n + \varepsilon_{i-1}^n}{(\Delta x)^2} \tag{2.22}$$

Thus the error ε also satisfies the difference equation.

Let us now consider the *stability* of the difference Eq. (2.16). The solution of Eq. (2.16) will be *stable* if the errors ε_i decay, or at best stay the same, as the solution progresses from time step n to $n + 1$. If the ε_i's grow larger during marching of the solution from step n to $n + 1$, the solution is *unstable*. Thus, for the solution to be *stable*,

$$\frac{\varepsilon_i^{n+1}}{\varepsilon_i^n} \le 1 \tag{2.23}$$

In application to the difference Eq. (2.16), we follow the Fourier series of von Neumann stability analysis (Smith, 1978). This method is applicable to homogeneous equations with homogeneous boundary conditions.

Let the error distribution $\varepsilon(x, t)$ be represented by a Fourier series in x as

$$\varepsilon(x, t) = e^{\alpha t} \sum_m e^{ik_m x} \tag{2.24}$$

where k_m is the real wave number but α may be complex. When Eq. (2.24) is substituted into Eq. (2.22), the behavior of each term of the series is the same as the series itself, since the difference equation is *linear*. We can thus consider just one term of the series, say

$$\varepsilon_m(x, t) = e^{\alpha t} e^{ik_m x} \tag{2.25}$$

Substituting Eq. (2.25) into (2.22), we get

$$\frac{e^{\alpha(t + \Delta t)} e^{ik_m x} - e^{\alpha t} e^{ik_m x}}{\Delta t} = \frac{e^{\alpha t} e^{ik_m(x + \Delta x)} - 2e^{\alpha t} e^{ik_m x} + e^{\alpha t} e^{ik_m(x - \Delta x)}}{(\Delta x)^2} \tag{2.26}$$

Dividing Eq. (2.26) by $e^{\alpha t} e^{ik_m x}$, rearranging, and using some trigonometric identities, we get

$$e^{\alpha \Delta t} = 1 - \frac{4 \Delta t}{(\Delta x)^2} \sin^2 \frac{k_m \Delta x}{2} \tag{2.27}$$

From Eq. (2.25), we get

$$\frac{\varepsilon_i^{n+1}}{\varepsilon_i^n} = \frac{e^{\alpha(t + \Delta t)} e^{ik_m x}}{e^{\alpha t} e^{ik_m x}} = e^{\alpha \Delta t} \tag{2.28}$$

Combining Eqs. (2.28), (2.27), and (2.23), we get

$$\left| \frac{\varepsilon_i^{n+1}}{\varepsilon_i^n} \right| = |e^{\alpha \Delta t}| = \left| 1 - \frac{4 \Delta t}{(\Delta x)^2} \sin^2[(k_m \Delta x)/2] \right| \le 1 \tag{2.29}$$

Equation (2.29) *must* be satisfied to have a *stable* solution, as dictated by Eq. (2.23). If we let $r = \Delta t/(\Delta x)^2$ and $\beta = k_m \Delta x/2$, we can rewrite Eq. (2.29) as

$$|1 - 4r \sin^2 \beta| \le 1$$

The left side of this inequality is called the *amplification factor*. For the above inequality to hold, we have two possible situations:

1. $1 - 4r \sin^2 \beta \le 1$ or $4r \sin^2 \beta \ge 0$
2. $1 - 4r \sin^2 \beta \ge -1$ or $4r \sin^2 \beta \le 2$

The first condition is always satisfied since r is positive. The second condition is satisfied only if $r \le 1/2$, which is the *stability requirement* for the solution of difference Eq. (2.16) to be *stable*. This places a constraint on the size of the time step relative to the mesh size. The method also provides information on the growth rate of various harmonics. In fact, the first harmonic to amplify when r exceeds $1/2$ is the harmonic with wavelength $2 \Delta x$. This is the sawtooth wave which often appears in explicit calculations on the verge of instability.

The von Neumann stability analysis can also be carried out for hyperbolic equations. Consider the first-order wave equation in one dimension,

$$\frac{\partial f}{\partial t} + c \frac{\partial f}{\partial x} = 0 \tag{2.30}$$

where c is the wave speed. The analytical solution of Eq. (2.30) is

$$f(x - ct) = \text{constant}$$

Lax (1954) proposed the following first-order differencing for solving equations of this form:

$$f_i^{n+1} = \frac{f_{i+1}^n + f_{i-1}^n}{2} - c \frac{\Delta t}{\Delta x} \left(\frac{f_{i+1}^n - f_{i-1}^n}{2} \right) \tag{2.31}$$

The first term on the right represents an average value at the previous time level, while the second term contains the central difference for the spatial derivative. If we now assume an error of the same form as in Eq. (2.25) and substitute it into Eq. (2.31) after replacing f by ε (recall that ε satisfies the difference equation for linear problems), the amplification factor becomes

$$e^{\alpha \Delta t} = \cos \gamma - iC \sin \gamma \tag{2.32}$$

where $\gamma = k_m \Delta x$ and $C = c \Delta t / \Delta x$ is called the *Courant number*. The stability requirement is

$$|e^{\alpha \Delta t}| = |\cos \gamma - iC \sin \gamma| \le 1$$

which when applied to Eq. (2.32) yields

$$C = c \frac{\Delta t}{\Delta x} \leq 1 \tag{2.33}$$

This is again the *stability constraint* on the time step relative to the mesh size. This is called the Courant-Friedrichs-Lewy condition, generally written as the CFL condition (Courant et al., 1967). It is an important stability criterion for hyperbolic equations.

The above discussion illustrates how von Neumann analysis can be applied to study the stability for a single equation. The basic ideas can also be extended to a system of equations. The reader is referred to Anderson et al. (1984) for details.

2.2.5 Convergence

A successful numerical solution is one in which all error components are bounded. A numerical solution is said to be *convergent* if the discretization error approaches zero as the mesh is refined, i.e., as Δx, Δy, Δz, and Δt, as appropriate, $\rightarrow 0$. Convergence in this sense implies that the exact analytical solution is approached numerically by mesh refinement.

If an iterative procedure is used to solve the difference equations, another type of convergence comes into play. An iterative procedure is said to converge if the error in the solution of difference equations approaches zero as the number of iterations $\rightarrow \infty$. In practice, the iteration is considered to have converged when the error becomes reasonably small or when it drops by several orders of magnitude from the original value.

For transient problems, there is an important theorem that relates consistency, stability, and convergence. This is the Lax's equivalence theorem (Lax and Richtmyer, 1956; Richtmyer and Morton, 1967). It states, "Given a properly posed, linear initial-value problem and a linear finite difference approximation to it that satisfies the consistency condition, stability is the necessary and sufficient condition for convergence. Thus, *well posedness + consistency + stability* implies *convergence* for linear problems."

The equivalence theorem allows the convergence of a solution to be established indirectly, thus obviating the need for a direct proof. This theorem is of considerable value in analyzing finite difference approximations with time-dependent coefficient matrices or for systems in which coefficients are quasi-linearized. Stability of the finite difference approximation, along with the equivalence theorem, is sufficient to ensure that both discretization and roundoff errors are bounded.

2.3 FINITE ELEMENT METHOD

The finite element method was originally conceived in the 1950s to analyze the stresses in complex airframe structures. It has since been developed and applied to a wide variety of problems in continuum mechanics. Unlike the FDM, which envisions the solution region as a collection of discrete grid points, the FEM envisions the solution region as built up of many small, interconnected subregions or *elements* which form a grid. The elements can be rectilinear or curved. The grid itself need not be structured. Due to this *unstructured* form, very complex geometries can be handled with ease. This is clearly the most important advantage of the FEM, and is not shared by the FDM, which needs a structured grid. A finite element model of a problem gives a *piecewise* approximation to the governing equations. The unknown field variable is expressed in terms of assumed approximating functions within each element. The approximating functions, also known as *shape functions* or *interpolation functions*, are defined in terms of the values of the field variables at specified points called *nodes* or *nodal points*. Nodes usually lie on the element boundaries where adjacent elements are supposed to be connected. In addition to boundary nodes, an element may also have a few interior nodes. For the finite element representation of a problem, the nodal values of the field variable become the new unknowns. Once these unknowns are found, the interpolation functions define the field variable throughout the assemblage of elements. Certain compatibility conditions must be satisfied by the interpolation functions. Often functions are chosen so that the field variable or its derivatives are continuous across adjoining element boundaries. An essential characteristic of the FEM is the modular way in which the discretization is obtained. The discrete equations are constructed from contributions at the element level which are then *assembled*.

Another characteristic of the FEM is that it looks for a solution of some integral form of the PDE rather than a solution of the PDE itself. There are basically four different approaches. The first approach is called the *direct approach* because its origin is traceable to the direct stiffness method of structural analysis. This approach can, however, be used only for relatively simple problems. A more versatile and advanced approach is the *variational approach*. This relies on the calculus of variations and involves extremizing a *functional*. For problems in solid mechanics the functional turns out to be the potential energy, but in fluid dynamics a functional is not available in general. For such problems, an even more versatile approach is obtained from a *weighted residual formulation*. This renders the method an ability to naturally incorporate *differential type boundary conditions*. This property constitutes another important advantage of the FEM that is not shared by the FDM. This approach also imparts a strong and rigorous mathematical foundation to the FEM. This allows, for instance, a precise definition of accuracy, a concept only loosely defined in the FDM. A fourth approach relies on thermal and/or mechanical energy of a system. The *energy balance approach*, like the weighted residuals approach, requires no variational statement, thereby broadening the range of possible applications of the FEM.

There are many books available these days for the FEM and its application to various problems in continuum mechanics. Some of these are listed in the references, for example, Zienkiewicz (1977), Huebner and Thornton (1982), Zienkiewicz and Morgan (1983), Reddy (1984), Baker and Pepper (1991), and Pepper and Heinrich (1992). In this section, only the basic ingredients of the FEM are explained. Also, of the four approaches listed above, only the *weighted residual approach* is described in some detail. The reader is referred to the above-mentioned books for other approaches as well as for other details.

2.3.1 Weighted Residual Formulation

The method of weighted residuals is a general technique, not necessarily associated with the FEM, for obtaining approximate solutions to linear and nonlinear PDEs (Crandall, 1956; Finlayson, 1972). Applying this method involves two steps. The *first* step is to assume some functional behavior of the dependent variable in order to approximately satisfy the differential equation and boundary conditions. Substitution of this approximation into the original differential equation and boundary conditions results in an error called a *residual*. This residual is required to vanish in some average sense over the entire solution domain. The *second* step is to solve the equation(s) resulting from the first step, thereby obtaining the functional form which yields the approximate solution sought.

Let us consider a problem defined by the following differential equation in general:

$$L(\phi) - f = 0 \qquad\qquad\qquad (2.34)$$

in the domain Ω bounded by the surface Σ. Here, ϕ is the dependent variable whose approximate solution is sought, and f is a known function of the independent variables. The proper boundary conditions are prescribed on Σ. The method of weighted residuals is applied in two steps as follows.

First, the unknown exact solution ϕ is approximated by $\tilde{\phi}$, where either the functional behavior of $\tilde{\phi}$ is completely specified in terms of unknown parameters, or the functional behavior on all but one of the independent variables is specified while the functional dependence on the remaining independent variable is left unspecified. Thus, the dependent variable is approximated by

$$\phi \approx \tilde{\phi} = \sum_{i=1}^{m} N_i u_i \qquad (2.35)$$

where N_i are the assumed functions and u_i are either the unknown parameters or unknown functions of one of the independent variables. The upper limit on the summation, m, is the number of unknowns u_i. The m functions N_i are usually chosen to satisfy the global boundary conditions.

When $\tilde{\phi}$ is substituted into Eq. (2.34), an error will result, that is,

$$L(\tilde{\phi}) - f = R$$

where R is the error or residual, as a result of approximating ϕ by $\tilde{\phi}$. The method of weighted residuals seeks to determine the m unknowns u_i in such a way that the residual R over the entire solution domain is small. This is accomplished by forming a weighted average of the residual, and specifying that this weighted average vanish over the solution domain. Thus m linearly independent weighting functions W_i are chosen, and the condition

$$\int_{\Omega} [L(\tilde{\phi}) - f] W_i \, d\Omega = \int_{\Omega} R W_i \, d\Omega = 0, \qquad i = 1, 2, \ldots, m \qquad (2.36)$$

is assumed to imply $R \approx 0$ in some sense.

Once the weighting functions W_i are specified, Eq. (2.36) represents a set of m equations, either algebraic or ordinary differential equations, to be solved for the u_i. Ordinary differential equations result when the dependent variable is a function of spatial coordinates and time. In this case the u_i are functions of time, $u_i(t)$. The second step is to solve Eq. (2.36) for the u_i, thus obtaining an approximate solution for the unknown dependent variable via Eq. (2.35). For many linear and some nonlinear problems it can be shown that as $m \to \infty$, $\tilde{\phi} \to \phi$, thereby implying convergence. For nonlinear problems, in general, convergence studies are scarce.

2.3.2 Galerkin Formulation

There are a variety of weighted residual techniques due to the broad choice of weighting functions. Among the possible choices for the set of weighting functions, some are the most obvious. For example, the weighting functions can be taken to be Dirac delta functions in m points. This choice leads to vanishing of the residual at the selected m points. This method is called the *point collocation method*. Clearly, it has much in common with the finite difference philosophy.

The most popular choice for the weighting functions in the FEM is the shape functions themselves; i.e., $W_i = N_i$ for $i = 1, 2, \ldots, m$. This is called the *Galerkin method*. It implies that the residual is made orthogonal to the space of shape functions. Thus Galerkin's method requires that

$$\int_\Omega [L(\tilde{\phi}) - f]N_i \, d\Omega = \int_\Omega R N_i \, d\Omega = 0, \qquad i = 1, 2, \ldots, m \qquad (2.37)$$

2.3.3   Application to an Element

In the above two sections, we considered the entire solution domain. However, Eq. (2.34) holds for any point in the solution domain or for any collection of points defining an arbitrary subdomain or *element* of the whole domain. Thus, we may focus our attention on an individual element and define a local approximation analogous to Eq. (2.35) that is valid for the individual element. The functions N_i are recognized as the shape functions $N_i^{(e)}$ defined over the element, and the u_i are the undetermined parameters, which may be the nodal values of the dependent variable or its derivatives. Then Galerkin's method leads to the following equations governing the behavior of an element:

$$\int_{\Omega^{(e)}} [L(\phi^{(e)}) - f^{(e)}]N_i^{(e)} \, d\Omega^{(e)} = 0, \qquad i = 1, 2, \ldots, r \qquad (2.38)$$

where the superscript (e) restricts the range to one element, and

$$\phi^{(e)} = \lfloor N^{(e)} \rfloor \{\phi\}^{(e)}$$

$f^{(e)}$ is the forcing function defined over element (e), and r is the number of unknown parameters assigned to the element.

We have a set of equations like Eqs. (2.38) for each element of the solution domain. In order to be able to *assemble* the system equations from the element equations, it is essential that our choice of shape functions N_i satisfy certain conditions in terms of interelement continuity described in section 2.3.5. In general, the higher the order of interelement continuity required of the shape functions, the narrower our choice of functions becomes. Often there is a way to escape this dilemma by applying integration by parts to the integral expression of Eq. (2.38). This leads to expressions containing lower-order derivatives, thus enabling the use of shape functions with lower-order interelement continuity. When integration by parts is possible, it also offers a convenient way to introduce the natural boundary conditions that must be satisfied on some portion of the boundary. Let us summarize the integration by parts equations used in finite element formulations with Galerkin's method.

In one dimension, where the domain of interest is $a \le x \le b$, integration by parts is

$$\int_a^b u \, dv = uv \vert_a^b - \int_a^b v \, du \tag{2.39}$$

where we identify u as the weighting function W_i, and dv with derivatives in the differential operator L in Eq. (2.36). Integration by parts in two dimensions is known as Green's theorem, while integration by parts in three dimensions is known as Gauss's theorem. For a two- or three-dimensional domain Ω with boundary Σ integration by parts is

$$\int_\Omega u(\nabla \cdot v) \, d\Omega = \int_\Sigma u(v \cdot n) \, d\Sigma - \int_\Omega v \cdot \nabla u \, d\Omega \tag{2.40}$$

where, as in one dimension, we identify u as the weighting function W_i, and $\nabla \cdot v$ with derivatives in the differential operator L. In Eq. (2.40) ∇ is the gradient operator, and n is the unit normal vector to the boundary.

2.3.4 Assembly

Once the element matrices (or equations) have been determined, it is necessary to assemble them for the entire solution domain. Following Huebner and Thornton (1982), the general procedure for this assembly is summarized below. Any special considerations for improving computing efficiency have been omitted from the following procedure.

1. Set up $n \times n$ and $n \times 1$ null matrices (all zero entries), where $n = $ number of system nodal variables.
2. Starting with one element, transform the element equations from local to global coordinates if the two coordinate systems are not coincident.
3. Perform any necessary matrix operations on the element matrices if the element has one or more nodes without connectivity. In such a case, it is necessary to eliminate the nodal unknowns or degrees of freedom associated with these nodes.
4. Using the correspondence between local and global numbering schemes, change to the global indices.
5. Insert these terms into the corresponding $n \times n$ and $n \times 1$ master matrices in the locations denoted by their indices. Each time a term is placed in a location where another term has already been placed, it is added to the value already there.
6. Return to step 2 and repeat this procedure for the other elements in the domain.

The above steps can be easily programmed for the digital computer and used for all problems solved by the FEM.

2.3.5 Elements

The selection of the type of element is of utmost importance in the FEM. Many types of finite elements have been developed. Here we outline the general principles governing elements and shape functions; details are available in Huebner and Thornton (1982). The following notation is used to express the degree of continuity of a field variable at element interfaces. If the field variable is continuous at element interfaces, it is said to have C^0 continuity. If, in addition, the first derivatives are continuous, the variable is said to have C^1 continuity, and so on. Suppose that the functions appearing within the integrals of the element equations possess derivatives up to the nth order. Then, to have rigorous assurance of convergence as the element size decreases, the following requirements must be met:

Compatibility: At element interfaces the field variable must have C^{n-1} continuity.
Completeness: Within an element the field variable must have C^n continuity.

Elements whose shape functions satisfy the first requirement are called *compatible* or *conforming* elements. Those that satisfy the second requirement are called *complete* elements. While rigorous proof of convergence may not be possible in all cases, compatible and complete elements are generally used.

The complete specification of an element requires four pieces of information: the element shape, the number and type (exterior or interior) of nodes, the type of nodal variables (e.g., ϕ, derivatives of ϕ, and so on), and the type of shape function. *Exterior nodes* lie on the boundary of the element and represent the points of connection between bordering elements. Nodes positioned at the corners of the elements, along the edges, or on the surfaces are all exterior nodes. *Interior nodes* are those that do not connect with neighboring elements. Often the nodal variables or the parameters assigned to the element are called the *degrees of freedom* of the element. Several common finite element shapes are shown in Fig. 2.3. Elements are shown for one, two, and three space dimensions, for axisymmetric geometries, and for curved boundaries (the *isoparametric* elements).

A *natural coordinate system* is generally used to represent the geometry of the element. It is essentially a local coordinate system that relies on the element geometry for its definition, and whose coordinates range between zero and unity within the element. Such a system has the property that one

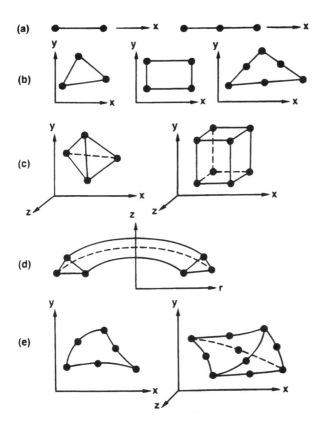

Figure 2.3 Finite elements. (a) One-dimensional elements; (b) two-dimensional elements: three-node triangle, rectangle, six-node triangle; (c) three-dimensional elements: tetrahedron, right prism; (d) two-dimensional axisymmetric ring element; (e) isoparametric elements: triangle, tetrahedron.

particular coordinate has unit value at one node of the element and zero value at the other node(s); its variation between nodes is linear. We can construct a natural coordinate system for any element; see Huebner and Thornton (1982) for details. The use of natural coordinates in deriving shape functions is particularly advantageous since special closed-form integration formulas can often be used to evaluate the integrals in the element equations. Natural coordinates also play a crucial role in the development of isoparametric elements.

The number of available elements possessing C^0 continuity is essentially infinite. Only the simplest of such elements are generally used. The number of elements possessing higher levels of continuity, C^1, C^2, and so on, decreases rapidly. For the two-node linear element, the three-node triangle, and the four-node tetrahedron in Fig. 2.3, C^0 continuity exists when the nodal variable is the unknown function and linear polynomials are used for interpolation. C^0 continuity also exists for the six-node planar triangle when the nodal variable is the unknown function and a quadratic shape function is used. For the same six-node planar triangle, however, C^1 continuity exists when the dependent variable, say ϕ, and $\partial\phi/\partial x$, $\partial\phi/\partial y$, $\partial^2\phi/\partial x^2$, $\partial^2\phi/\partial x\partial y$, and $\partial^2\phi/\partial y^2$ are prescribed at each vertex, and the outward normal derivative of ϕ is prescribed at the middle nodes. This implies that $\tilde{\phi}$ is described in the element by a quintic polynomial. Thus, C^1 continuity requires much more complex representation than C^0 continuity. The choice and allocation of elements involve a trade-off between convenience and complexity. The type of element that yields good accuracy and low computing time is problem dependent. The general trend is to use many simple elements (linear or quadratic) rather than a few complex ones.

2.3.6   Condensation and Substructuring

Condensation and substructuring are concepts uniquely associated with the FEM. *Condensation* is used to handle elements that have internal nodes. Internal nodes do not connect with the nodes of other elements during the assembly process; consequently the degrees of freedom associated with internal nodes do not affect interelement continuity. Internal nodes are sometimes used to improve the field variable representation *within* an element. To reduce the overall size of the assembled system of equations, we may eliminate internal nodal degrees of freedom at the element level before assembly. This is done by a process called condensation.

Substructuring is the extension of the process of condensation to groups of finite elements in a complex physical continuum. It is a method whereby a large complex domain such as an aircraft, ship, bridge, or automobile is viewed as an assemblage of a small number of very complex elements. For such large domains, the global matrix system may be too large even for present-day computers. If so, the internal nodes in the substructure can be removed by condensation. The condensed substructures may then be assembled to form a matrix system that can be solved. Substructuring is readily extended to diffusion problems, such as conduction heat transfer.

2.4   FINITE VOLUME METHOD

The *finite volume method* is a technique by which the integral formulation of the conservation laws are discretized directly in the physical space. Thus, this approach employs numerical balances of a conserved variable over small control volumes, thereby ensuring that the basic quantities like mass, momentum, and energy remain conserved at the discrete level. This is the fundamental advantage of the FVM. If appropriately formulated, the numerical solution satisfies the conservation principle globally as well. This method is particularly useful for discontinuous flows such as transonic flows with shock waves. Another example of a flow with a discontinuity is a slipline that occurs behind an airfoil or blade if the entropy production is different on streamlines on both sides of the airfoil. In this case, a tangential discontinuity occurs. Another typical situation is an incompressible flow where the imposition of incompressibility, as a conservation law for mass, determines the pressure field (Dick, 1992).

Like the FEM, the flow field or *domain* is subdivided into a set of *nonoverlapping cells* that cover the whole domain. In the FVM, the term *cell* is used in place of the term *element* used in the FEM. On each cell the conservation laws are applied to determine the flow-field variables at some discrete points of the cells, called *nodes*. The shape and location of the cells associated with a given node, as well as the rules and accuracy for the evaluation of the fluxes through the cell surfaces, can be varied. As in the FEM, the nodes are at typical locations of the cells, such as cell centers, cell vertices, or mid sides. Cells can be *triangular, quadrilaterial*, etc. They can be elements of a *structured grid* or an *unstructured grid*. Only the coordinates of the cell vertices are really necessary, and curvilinear coordinates, not necessarily orthogonal, can be used to define the set of cells. Figure 2.4 shows some typical grids used in the FVM.

A typical choice for the nodes is *cell centers* when the solution over the cell can be represented as a piecewise constant function or *cell vertices* when

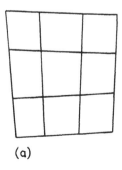 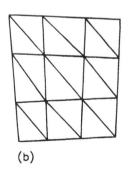 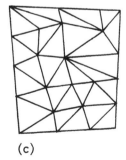

(a)   (b)   (c)

Figure 2.4 Various grids for the FVM: (a) structured quadrilateral grid; (b) structured triangular grid; (c) unstructured triangular grid.

this representation is piecewise linear or bilinear. However, in the FVM, unlike the FEM, a function space for the solution need not be defined, and nodes can be chosen in a way that does not imply an interpolation structure. Figure 2.5 shows some typical examples of choices of nodes with the associated definition of variables. The first two choices in Fig. 2.5 imply an interpolation structure; the last two do not. In Fig. 2.5d, *all* function values are not defined at *all* nodes. The nodes on which pressure and density are defined are different from those on which x- and y-components of velocity are defined. This approach is known as the *staggered grid approach*. Figure 2.5c shows that the *control volumes* on which the conservation laws are applied *need not coincide* with the cells of the grid. Control volumes can even be overlapping but not the cells. For the *cell-centered* FVM, the cells coincide with the control volumes, but for the *cell vertex* FVM, they may not (Hirsch, 1990).

Owing to the decoupling of volumes and cells, the freedom in the representation of the solution over the flow field becomes much larger in the

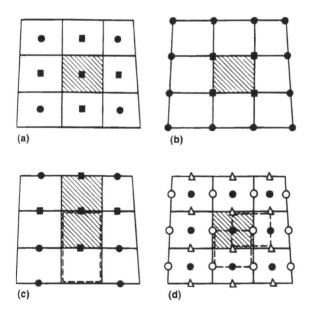

Figure 2.5 Choices for nodes in the FVM. Nodes marked ■ participate in the flux balance of the control volume: (a) piecewise constant interpolation structure; (b) piecewise linear interpolation structure; (c) no interpolation structure with all variables defined at each node; (d) no interpolation structure with ρ and p defined at ●, u defined at ○, and v at △ nodes.

FVM than in both the FEM and the FDM. Clearly, the FVM tries to combine the best from the FEM, i.e., the *geometric flexibility*, with the best from the FDM, i.e., the flexibility in defining the *discrete flow field*. This makes the FVM extremely popular in engineering applications. There is one disadvantage, however. The FVM has difficulty in the definition of derivatives. Since the computational grid is not necessarily orthogonal and equally spaced, as in the FDM, a definition of a derivative based on the Taylor series expansion is not possible. Also, there is no mechanism like the weak formulation in the FEM to convert higher-order derivatives into lower ones. Despite these difficulties, the FVM is applied to a whole range of applications.

2.4.1 Basic Considerations

The integral form of the conservation law for a variable ϕ can be written as

$$\frac{\partial}{\partial t} \int_\Omega \phi \, d\Omega + \oint_S \mathbf{F} \cdot d\mathbf{S} = \int_\Omega Q \, d\Omega \tag{2.41}$$

where $\mathbf{F}$ is the flux of ϕ crossing the control surface S that encloses the control volume Ω, and Q is the source of ϕ within the control volume. This equation can be written for the discretized control volume Ω_a with the associated variable ϕ_a as

$$\frac{\partial}{\partial t} (\phi_a \Omega_a) + \sum_{\text{sides}} (\mathbf{F} \cdot \mathbf{S}) = Q_a \Omega_a \tag{2.42}$$

where the sum of the flux terms refers to all the external sides of the control volume Ω_a. Referring to Fig. 2.6a and to cell 1 (i, j), we identify ϕ_a with $\phi_{i,j}$, Ω_a with the area of $ABCD$, and the flux terms are summed over the four sides AB, BC, CD, DA. For the mesh of Fig. 2.6c, Ω_a is the shaded area of the triangles having node a in common, and the flux summation extends over the five sides 12, 23, 34, 45, 51. This is the general formulation of the FVM. For a selected Ω_a, we need to define how to estimate the volume and face areas of the control volume Ω_a and how to estimate the fluxes at the faces. We discuss these issues briefly here; more details are available in Hirsch (1990) and Dick (1992).

Considering Eq. (2.42), we observe some features that distinguish the FVM from the FEM and the FDM:

1. The coordinates of point a, and thus the precise location of the variable ϕ inside the control volume Ω_a, do not appear explicitly. Consequently, ϕ_a is not necessarily attached to a fixed point inside the control volume and can be considered as an *average value* of the flow variable ϕ over the control volume. The first term of Eq.

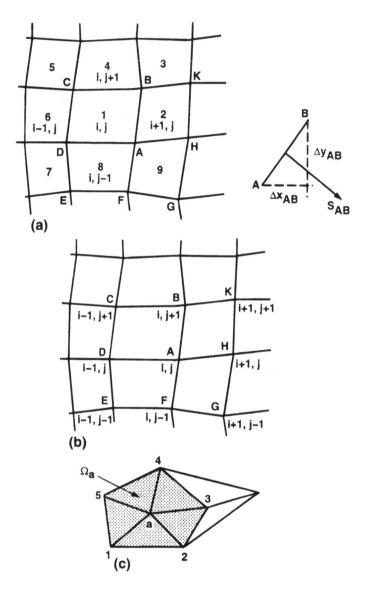

Figure 2.6 Two-dimensional finite volume mesh systems: (a) cell-centered structured mesh; (b) cell vertex structured mesh; (c) cell vertex unstructured mesh.

(2.42) therefore represents the time rate of change of the averaged flow variable over the selected control volume.

2. The cell coordinates appear only in the determination of the cell volume and face areas. Referring to Fig. 2.6a, and considering the control volume $ABCD$, for example, only the coordinates of A, B, C, D are needed.

3. In the absence of source terms, the finite volume formulation states that the variation of the average value ϕ over a time interval Δt is equal to the sum of fluxes exchanged between the neighboring cells. For steady flows it implies that the sum of fluxes entering the control volume is zero; that is,

$$\sum_{sides} (\boldsymbol{F} \cdot \boldsymbol{S}) = 0 \qquad (2.43)$$

When adjacent cells are considered, for example, cells 1 and 2 in Fig. 2.6a, the flux through the common face AB contributes *equally* to the two cells but with *opposite* signs. Thus, during computation, the flux through face AB is added to the flux balance of cell 1 and subtracted from the flux balance of cell 2. This automatically guarantees global conservation.

4. Like the FEM, the FVM also allows a natural introduction of boundary conditions such as the vanishing of certain normal components at the solid walls. For the mass conservation equation, $\boldsymbol{F} = \rho \boldsymbol{V}$ and at a solid boundary $\boldsymbol{F} \cdot d\boldsymbol{S} = 0$. Hence the corresponding contributions to Eqs. (2.42) and (2.43) would vanish.

For a conservative FVM, the choice of discrete control volumes Ω_a must satisfy the following constraints:

1. Their sum should span the whole domain Ω.
2. Adjacent Ω_a may overlap if each internal surface is common to two volumes.
3. Fluxes along a cell surface must be computed *independent* of the cell in which they are considered.

The last constraint ensures the conservative property of the FVM, since the flux contributions of internal faces will cancel when contributions of the associated control volumes are added.

2.4.2 Two-Dimensional Finite Volume Method

For the control volume $ABCD$ of Fig. 2.6a, Eq. (2.41) can be written as

$$\frac{\partial}{\partial t} \int_{\Omega_{ij}} \phi \, d\Omega + \oint_{ABCD} (f \, dy - g \, dx) = \int_{\Omega_{ij}} Q \, d\Omega \qquad (2.44)$$

where f and g are the Cartesian components of the flux vector $\boldsymbol{F}$. Equation (2.44) can be easily discretized. The surface vector for side AB is

$$\boldsymbol{S}_{AB} = \Delta y_{AB}\, \boldsymbol{n}_x - \Delta x_{AB}\, \boldsymbol{n}_y = (y_B - y_A)\boldsymbol{n}_x - (x_B - x_A)\boldsymbol{n}_y \qquad (2.45)$$

where $\boldsymbol{n}_x$ and $\boldsymbol{n}_y$ denote unit vectors in x and y directions, respectively. With this, the discretized equation for the cell Ω_{ij} can be written as

$$\frac{\partial}{\partial t}(\phi\Omega)_{ij} + \sum_{ABCD} [f_{AB}(y_B - y_A) - g_{AB}(x_B - x_A)] = (Q\Omega)_{ij} \qquad (2.46)$$

The sum Σ_{ABCD} extends over the four sides of the quadrilateral $ABCD$. For a general quadrilateral $ABCD$, the area Ω can be evaluated from the vector product of the diagonals. With $\boldsymbol{r}_{AB} = \boldsymbol{r}_B - \boldsymbol{r}_A$, where $\boldsymbol{r}_A$ is the position vector of point A, we get

$$\begin{aligned}
\Omega_{ABCD} &= \frac{1}{2}\,|\boldsymbol{r}_{AC} \times \boldsymbol{r}_{BD}| \\
&= \frac{1}{2}\,[(x_C - x_A)(y_D - y_B) - (y_C - y_A)(x_D - x_B)] \\
&\equiv \frac{1}{2}\,(\Delta x_{AC}\,\Delta y_{BD} - \Delta x_{BD}\,\Delta y_{AC})
\end{aligned} \qquad (2.47)$$

The right side of Eq. (2.47) is positive for the cell $ABCD$ when A, B, C, D are positioned counterclockwise.

Flux Evaluation through Cell Faces

The evaluation of flux components through the faces, such as f_{AB}, g_{AB}, depends upon the scheme selected and upon the location of the flow variables with respect to the mesh. There are two popular schemes, the *central* and *upwind* schemes. Central schemes are based on local flux approximation, while upwind schemes determine the cell face fluxes according to the direction of propagation of the convection speed.

For *central schemes* and *cell-centered* FVM, the following alternatives may be used:

(1) Take f as the average of the fluxes in A and B:

$$f_{AB} = \frac{1}{2}(f_A + f_B) \qquad (2.48)$$

where either the variables are evaluated in A and B,

$$\phi_A = \frac{1}{4}(\phi_{ij} + \phi_{i+1,j} + \phi_{i,j-1} + \phi_{i+1,j-1}) \quad \text{and} \quad f_A = f(\phi_A) \qquad (2.49)$$

or the fluxes are averaged, as

$$f_A = \frac{1}{4}(f_{ij} + f_{i+1,j} + f_{i,j-1} + f_{i+1,j-1}) \tag{2.50}$$

(2) Average the fluxes:

$$f_{AB} = \frac{1}{2}(f_{ij} + f_{i+1,j}) \qquad \text{where} \qquad f_{ij} = f(\phi_{ij}) \tag{2.51}$$

(3) Another choice given below is not identical to Eq. (2.51) since the flux components are generally nonlinear functions of ϕ:

$$f_{AB} = f\left(\frac{\phi_{ij} + \phi_{i+1,j}}{2}\right) \tag{2.52}$$

Similar choices hold for the component g of the flux vector F. Equations (2.49) and (2.52) generally require a lower number of flux evaluations than Eqs. (2.50) and (2.51).

For *central schemes* and *cell vertex* FVM, Eqs. (2.48) or (2.52) may be used. Equation (2.48) corresponds to the trapezoidal rule for evaluation of the integral $\int_{AB} f\, dy = (f_A + f_B)(y_B - y_A)/2$. For *upwind schemes* and *cell-centred* FVM, a convective flux is evaluated as a function of the direction of propagation of the associated convection speed. The latter is defined in terms of the flux Jacobian:

$$A(\phi) = \frac{\partial F}{\partial \phi} = a\mathbf{n}_x + b\mathbf{n}_y \tag{2.53}$$

with $a(\phi) = \partial f/\partial \phi$ and $b(\phi) = \partial g/\partial \phi$. The simplest upwind scheme takes the cell-side flux equal to the flux generated in the upstream cell. This implies that the cell face flux is determined by contributions transported in the direction of the convection velocity. Considering Fig. 2.6a, we get

$$\begin{aligned}(F \cdot S)_{AB} &= (F \cdot S)_{ij} & \text{if} \quad (A \cdot S)_{AB} > 0 \\ (F \cdot S)_{AB} &= (F \cdot S)_{i+1,j} & \text{if} \quad (A \cdot S)_{AB} < 0 \end{aligned} \tag{2.54}$$

For *upwind schemes* and *cell vertex* FVM, we may define, in reference to Fig. 2.6b,

$$\begin{aligned}(F \cdot S)_{AB} &= (F \cdot S)_{CD} & \text{if} \quad (A \cdot S)_{AB} > 0 \\ (F \cdot S)_{AB} &= (F \cdot S)_{HK} & \text{if} \quad (A \cdot S)_{AB} < 0 \end{aligned} \tag{2.55}$$

When applied to the control volume $GHKBCDEF$ of Fig. 2.6b, we get contributions from such points as $(i-2, j)$ and $(i, j-2)$ for positive convection speeds. This leads to schemes with an unnecessary large support for the same accuracy and is seldom used.

The above equations for the evaluation of fluxes imply some regularity of the mesh even though the FVM can be applied to arbitrary grids. For example, considering Eqs. (2.51) and (2.52) for the cell-centered FVM, and interpreting the cell-averaged values ϕ_{ij} in Fig. 2.6a as mid-cell values, it is clear that these equations perform an arithmetic average of the fluxes (or the variables) on both sides of the cell face AB. This leads to a second-order approximation on a Cartesian mesh if AB is at mid-distance from the cell centers 1 and 2. For a nonuniform mesh, however, this will not be the case, and a loss of accuracy will result. Similar considerations apply to Eqs. (2.49) and (2.50). In such a case, a linear interpolation of f_{AB} (or ϕ_{AB}) between the cell values f_{ij} and $f_{i+1,j}$ (or ϕ_{ij} and $\phi_{i+1,j}$) is taken. An analysis of the truncation errors for certain finite volume discretizations on nonuniform meshes is available in Turkel (1985) and Roe (1987).

2.4.3   Integration Formulas for Finite Volumes

For the Navier-Stokes equations, the viscous flux components are functions of the velocity gradients. Thus we need to evaluate these gradients at the cell faces. This is done by application of the Gauss divergence theorem. According to this theorem for the gradient of a scalar ϕ,

$$\int_{\Omega} \nabla\phi \, d\Omega = \oint_{S} \phi \, dS \tag{2.56}$$

where S is the closed surface bounding the arbitrary volume Ω. Thus the gradients averaged over the control volume Ω may be defined as

$$\overline{\left(\frac{\partial\phi}{\partial x}\right)_{\Omega}} \equiv \frac{1}{\Omega} \int_{\Omega} \frac{\partial\phi}{\partial x} \, d\Omega = \frac{1}{\Omega} \oint_{S} \phi n_x \cdot dS \tag{2.57a}$$

and

$$\overline{\left(\frac{\partial\phi}{\partial y}\right)_{\Omega}} \equiv \frac{1}{\Omega} \int_{\Omega} \frac{\partial\phi}{\partial y} \, d\Omega = \frac{1}{\Omega} \oint_{S} \phi n_y \cdot dS \tag{2.57b}$$

For two-dimensional control volume Ω, we have

$$\overline{\left(\frac{\partial\phi}{\partial x}\right)_{\Omega}} \equiv \frac{1}{\Omega} \oint_{S} \phi \, dy = -\frac{1}{\Omega} \oint_{S} y \, d\phi \tag{2.58a}$$

following partial integration. Similarly, the averaged y-derivatives are obtained from

$$\overline{\left(\frac{\partial \phi}{\partial y}\right)}_\Omega \equiv -\frac{1}{\Omega} \oint_S \phi \, dx = \frac{1}{\Omega} \oint_S x \, d\phi \qquad (2.58b)$$

For an arbitrary vector a, the divergence theorem can also be written as

$$\int_\Omega \nabla \cdot a \, d\Omega = \oint_S a \cdot dS \qquad (2.59)$$

This is particularly useful for determining the cell face areas and volumes. For a two-dimensional cell, for example, taking $a = r$ with $\Delta \cdot r = 2$, we have

$$2\Omega = \oint_S r \cdot dS = \oint_S (x \, dy - y \, dx) \qquad (2.60)$$

Application to three-dimensional control volumes is discussed next.

2.4.4   Three-Dimensional Finite Volume Method

The three-dimensional domain is generally divided into tetrahedral control volumes or hexahedral volumes (Fig. 2.7), where the four points forming a cell face are not necessarily coplanar. Equation (2.42) remains unchanged,

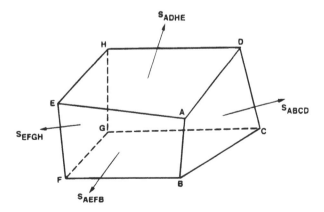

Figure 2.7   Three-dimensional hexahedral control volume.

but care is to be exercised in the evaluation of cell face areas and volumes so as to ensure that the sum of the computed volumes of adjacent cells is indeed equal to the total volume of the combined cells.

Evaluation of Cell Face Areas

The divergence theorem yields an important property of the area vector S attached to a cell face, since Eq. (2.56) with $\phi = 1$ becomes

$$\oint_S dS = 0 \tag{2.61}$$

Thus the outward surface vector of a given face contained in the closed surface S,

$$S_{\text{face}} = \int_{\text{face}} dS \tag{2.62}$$

is dependent only upon the boundaries of the face. Thus for face $ABCD$ of Fig. 2.7, we can use Eq. (2.47) to get

$$S_{ABCD} = \frac{1}{2} (r_{AC} \times r_{BD}) \tag{2.63}$$

It may be noted that other alternatives, based on the circumstances that the cell face $ABCD$ may not be coplanar, are identical to the simple Eq. (2.63).

Evaluation of Control Volumes

There are several alternatives to determine the volume of a hexahedral cell, the most straightforward being based on the subdivision into tetrahedra or pyramids (Fig. 2.8). The volume of the tetrahedron Ω_{PABC} is obtained by applying Eq. (2.59) for a vector a equal to the position vector r. Since $\nabla \cdot r = 3$, we get

$$\Omega_{PQBC} = \frac{1}{3} \oint_{PABC} r \cdot dS = \frac{1}{3} \sum_{\text{faces}} r \cdot S_{\text{faces}} \tag{2.64}$$

or

$$\Omega_{PABC} = \frac{1}{3} r_{(P)} \cdot S_{ABC} \tag{2.65}$$

when $r_{(P)}$ represents a vector originating in P. This follows from the fact that when $r_{(P)}$ lies in the faces containing P, it is orthogonal to the associated S vector, and the contribution comes only from the face ABC opposite

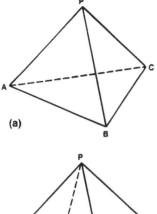

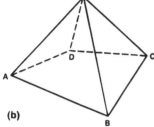

Figure 2.8   Three-dimensional control volumes: (a) tetrahedron; (b) pyramid.

P. Thus with $r_{(P)} = r_{PA}$, we get

$$\Omega_{PABC} = \frac{1}{6} r_{PA} \cdot (r_{AB} \times r_{BC}) = \frac{1}{6} r_{PA} \cdot (r_{BC} \times r_{CA}) \qquad (2.66)$$

Equation (2.66) can also be expressed as a determinant:

$$\Omega_{PABC} = \frac{1}{6} \begin{vmatrix} x_P & y_P & z_P & 1 \\ x_A & y_A & z_A & 1 \\ x_B & y_B & z_B & 1 \\ x_C & y_C & z_C & 1 \end{vmatrix} \qquad (2.67)$$

Care must be exercised while evaluating the tetrahedra volumes, since the sign of the volume Ω_{PABC} in Eqs. (2.65) through (2.67) depends on the orientation of the triangular decompositions. In addition, when the cell surfaces are not coplanar, the same diagonal must be used in the evaluation of the tetrahedra in the two cells which share this surface. A useful guideline, in order to avoid sign errors, is to apply a right-hand rotation rule from the base toward the summit of each tetrahedron.

In a similar way, for a pyramid $PABCD$, we have

$$\Omega_{PABCD} = \frac{1}{3} \oint_{PABCD} \boldsymbol{r} \cdot d\boldsymbol{S} = \frac{1}{3} \boldsymbol{r}_{(P)} \cdot \boldsymbol{S}_{ABCD} \tag{2.68}$$

Since $ABCD$ is not necessarily coplanar, $\boldsymbol{r}_{(P)}$ may be approximated by

$$\boldsymbol{r}_{(P)} = \frac{1}{4} \left(\boldsymbol{r}_{PA} + \boldsymbol{r}_{PB} + \boldsymbol{r}_{PC} + \boldsymbol{r}_{PD} \right) \tag{2.69}$$

and with Eq. (2.63) for S_{ABCD} we get

$$\begin{aligned}
\Omega_{PABCD} &= \frac{1}{24} \left(\boldsymbol{r}_{PA} + \boldsymbol{r}_{PB} + \boldsymbol{r}_{PC} + \boldsymbol{r}_{PD} \right) \cdot \left(\boldsymbol{r}_{AC} \times \boldsymbol{r}_{BD} \right) \\
&= \frac{1}{12} \left(\boldsymbol{r}_{PA} + \boldsymbol{r}_{PB} \right) \cdot \left(\boldsymbol{r}_{AC} \times \boldsymbol{r}_{BD} \right)
\end{aligned} \tag{2.70}$$

If the face $ABCD$ is coplanar, Eq. (2.70) reduces to

$$\Omega_{PABCD} = \frac{1}{6} \boldsymbol{r}_{PA} \cdot \left(\boldsymbol{r}_{AC} \times \boldsymbol{r}_{BD} \right) \tag{2.71}$$

More elaborate decompositions of hexahedral volumes in pyramids can be found in Davies and Salmond (1985).

2.4.5   Time Integration

Having discussed the evaluation of fluxes, and the associated surface area and volume calculation, we now turn to the integration in time for Eq. (2.42). As for the FDM, the Lax-Wendroff time-stepping scheme can also be applied to a finite volume formulation. In the FDM, the Lax-Wendroff scheme is a one-step method. In the FVM, a one-step method is not used since it involves a flux balance of terms containing derivatives and since the definition of derivatives is not simple in the FVM. However, two-step schemes, such as the Richtmyer scheme and the MacCormack scheme, can be used without any problem in the FVM. In the terminology of ordinary differential equations, the Richtmyer scheme is a *two-stage (Runge-Kutta)* scheme, while the MacCormack method is a *predictor-corrector method*. We will not provide any details about these methods here; the interested reader is referred to Dick (1992) and Hirsch (1990) for details. Here we briefly describe the four-step Runge-Kutta time stepping scheme devised by Jameson et al. (1981) that is very popular. This four-stage method gives the best ratio of allowable time step to computational work per time step. The four-stage scheme can be written as

$$\phi_{i,j}^0 = \phi_{i,j}^n$$

$$\phi_{i,j}^1 = \phi_{i,j}^0 - \alpha_1 \frac{\Delta t}{\Omega_{i,j}} R^0$$

$$\phi_{i,j}^2 = \phi_{i,j}^0 - \alpha_2 \frac{\Delta t}{\Omega_{i,j}} R^1 \tag{2.72}$$

$$\phi_{i,j}^3 = \phi_{i,j}^0 - \alpha_3 \frac{\Delta t}{\Omega_{i,j}} R^2$$

$$\phi_{i,j}^{n+1} = \phi_{i,j}^0 - \alpha_4 \frac{\Delta t}{\Omega_{i,j}} R^3$$

where $\alpha_1 = 1/4$, $\alpha_2 = 1/3$, $\alpha_3 = 1/2$, and $\alpha_4 = 1$, and the superscript denotes the (intermediate) time level. In reference to Eq. (2.44), the residual R is given by

$$R = \oint_{ABCD} (f\, dy - g\, dx) - \int_{\Omega_{ij}} Q\, d\Omega \tag{2.73}$$

This integration scheme is explicit with a CFL condition of CFL $\leq 2\sqrt{2}$. Note that Eq. (2.72) is *not* the classical fourth-order Runge-Kutta scheme. It is second-order accurate in time, which is sufficient, since the accuracy of the space discretization is also second-order. Jameson also introduced an artificial viscosity as a blend of a second-order and a fourth-order term, and used it in all steps of Eq. (2.72). In order to keep the calculation conservative, the added diffusive term, for a structured quadrilateral grid, is

$$d_{i+1/2,j} - d_{i-1/2,j} + d_{i,j+1/2} - d_{i,j-1/2} \tag{2.74}$$

where

$$d_{i+1/2,j} = \varepsilon_{i+1/2,j}^{(2)}(\phi_{i+1,j} - \phi_{i,j}) \\ - \varepsilon_{i+1/2,j}^{(4)}(\phi_{i+2,j} - 3\phi_{i+1,j} + 3\phi_{i,j} - \phi_{i-1,j}) \tag{2.75}$$

with similar definitions for other terms in Eq. (2.74). The coefficients of the second-order term $\varepsilon^{(2)}$ and the fourth-order term $\varepsilon^{(4)}$ are chosen in a self-adaptive way by Jameson as follows:

$$\varepsilon_{i+1/2,j}^{(2)} = k_2 \max(v_{i+1,j}, v_{i,j}) \\ \varepsilon_{i+1/2,j}^{(4)} = \max(0, k_4 - \varepsilon_{i+1/2,j}^{(2)}) \tag{2.76}$$

with $k_2 = 1/4$, $k_4 = 1/256$, and the pressure switch $v_{i,j}$ is taken to be

$$v_{i,j} = \frac{|p_{i+1,j} - 2p_{i,j} + p_{i-1,j}|}{p_{i+1,j} + 2p_{i,j} + p_{i-1,j}} \tag{2.77}$$

where p is the pressure. Other values for k_2 and k_4 are also in use. With this definition, the second-order term is significant only in shock regions. Also, at solid boundaries, the diffusive terms in Eq. (2.74) are to be set equal to zero in the direction normal to the boundary.

Several variations of Jameson's artificial diffusive term are in use. Also, the diffusive correction in the second to fourth step is often taken to be the same as in the first step. More details along with examples are available in Schmidt and Jameson (1982). The most commonly used formulation for the artificial viscosity applicable to unstructured grids can be found in Pike and Roe (1985).

Jameson introduced several techniques in order to accelerate the convergence toward the steady-state solution. Within each cell the solution is allowed to advance in time at the maximum rate compatible with a fixed CFL number and the size of the cell. Also, the residual R is generally smoothed. This gives the scheme an implicit character, thereby increasing the allowable CFL number. With residual smoothing, a CFL number of 5 is generally used. Moreover, it has the effect of smoothing the high-frequency variations of the residual and is essential to multigrid convergence acceleration. These details are available in Hirsch (1990). For application of Jameson's technique to three-dimensional viscous flows in turbomachines using the multigrid approach, the reader is referred to Arnone (1994). This technique has been used for computation of heat transfer to gas turbine blades that are film-cooled (Garg, 1997; Garg and Gaugler, 1996, 1997). Jameson's technique has also been applied to many other problems.

NOMENCLATURE

A	flux Jacobian [Eq. (2.53)]
c	wave speed
C	Courant number
d	artificial dissipation in Jameson's scheme
D	exact solution of the finite difference approximation
E	exact (analytical) solution of the partial differential equation
f	dependent variable or known function of independent variables or component of flux $\mathbf{F}$
$\mathbf{F}$	flux of ϕ across the control surface
g	component of flux vector $\mathbf{F}$
i, j	index for grid lines or nodes
k	wave number
$\mathbf{n}$	unit vector
N	numerical solution of the finite difference approximation

N_i	shape functions for the FEM
Q	source of ϕ within the control volume
$\mathbf{r}$	position vector
R	residue
$\mathbf{S}$	control surface
t	time
u_i	unknown parameters or functions of one of the independent variables for the FEM
$\mathbf{V}$	velocity vector
W_i	linearly independent weighting functions
x, y, z	Cartesian coordinates

Greek

α	complex number (section 2.2.4) or four-stage Runge-Kutta scheme coefficients (section 2.4.5)
$\Delta x, \Delta y, \Delta z$	mesh sizes in the x, y, z directions, respectively
Δt	time step
ε	roundoff error (section 2.2.4) or terms contributing to artificial dissipation (section 2.4.5)
ρ	density
Σ	control surface
ϕ	dependent variable
$\tilde{\phi}$	approximation to ϕ
Ω	control volume

Subscripts

i, j	refer to the grid node or cell (i, j)

Superscripts

(e)	refers to individual element
n	refers to value at time t

REFERENCES

Anderson DA, Tannehill JC, Pletcher RH. Computational Fluid Mechanics and Heat Transfer. Washington DC: Hemisphere, 1984.

Anderson JD Jr. Computational fluid dynamics—an engineering tool? In: Pouring AA, ed. Numerical/Laboratory Computer Methods in Fluid Dynamics. New York: ASME, 1976, pp 1–12.

Arnone A. Viscous analysis of three-dimensional rotor flow using a multigrid method. J Turbomachinery 116: 435–445, 1994.

Aziz K, Hellums, JD. Numerical solution of the three-dimensional equations of motion for laminar natural convection. Phys. Fluids 10: 314–324, 1967.

Baker AJ, Pepper DW. Finite Elements 1-2-3. New York: McGraw-Hill, 1991.

Bourcier M, François C. Numerical integration of the Navier-Stokes equations in a square domain. Rech Aérosp 131: 23–33, 1969.

Cebeci T, Hirsh RS, Keller HB, Williams PG. Studies of numerical methods for the plane Navier-Stokes equations. Comput Meth Appl Mech Eng 27: pp 13–44, 1981.

Crandall SH. Engineering Analysis. New York: McGraw-Hill, 1956.

Courant R, Friedrichs KO, Lewy H. On the partial differential equations of mathematical physics. IBM J Res Dev 11: 215–234, 1967 (Translated from Math Ann 100: 32–74, 1928).

Davies DE, Salmond DJ. Calculation of the volume of a general hexahedron for flow predictions. AIAA J 23: 954–956, 1985.

Dennis, SCR, Ingham DB, Cook RN. Finite-difference methods for calculating steady incompressible flows in three dimensions. J Comput Phys 33: pp 325–339, 1979.

Dick E. Introduction to finite element techniques in computational fluid dynamics. In: Wendt JF, ed. Computational Fluid Dynamics—An Introduction. Berlin: Springer, 1992, pp 261–288.

Fasel H. Investigation of the stability of boundary layers by a finite-difference model of the Navier-Stokes equations. J Fluid Mech 78: 355–383, 1976.

Finlayson BA The Method of Weighted Residuals and Variational Principles. New York: Academic, 1972.

Fletcher CAJ. Computational Galerkin Methods. Berlin: Springer, 1984.

Fletcher CAJ. Computational Techniques for Fluid Dynamics. 2nd ed. Berlin: Springer, 1991.

Garg VK. Adiabatic effectiveness and heat transfer coefficient on a film-cooled rotating blade. Numer Heat Transfer, part A, 32: 811–830, 1997.

Garg VK, Gaugler RE. Effect of coolant temperature and mass flow on film cooling of turbine blades. Int J Heat Mass Transfer 40: 435–445, 1997.

Garg VK, Gaugler RE. Leading edge film-cooling effects on turbine blade heat transfer. Numer Heat Transfer, part A, 30: 165–187, 1996.

Hirasaki GJ, Hellums JD. Boundary conditions on the vector and scalar potentials in viscous three-dimensional hydrodynamics. Q Appl Math 28: 293–296, 1970.

Hirsch C. Numerical Computation of Internal and External Flows. New York: Wiley, 1990.

Hornbeck RW. Numerical marching techniques for fluid flows with heat transfer. NASA SP-297, 1973.

Huebner KH, Thornton EA. The Finite Element Method for Engineers. 2nd ed. New York: Wiley, 1982.

Hyman JM, Larrouturou B. The numerical differentiation of discrete functions using polynomial interpolation methods. In: Thompson JF, ed. Numerical Grid Generation. New York: Elsevier, 1982, 487–506.

Jaluria Y, Torrance KE. Computational Heat Transfer. Washington DC: Hemisphere, 1986.

Jameson A, Schmidt W, Turkel E. Numerical solution of the Euler equations by finite volume methods using Runge-Kutta time stepping schemes. AIAA Paper 81-1259, 1981.

Lax, PD. Weak solutions of nonlinear hyperbolic equations and their numerical computation. Comm Pure Appl Math 7: 159–193, 1954.

Lax PD, Richtmyer RD. Survey of the stability of linear finite difference equations. Comm Pure Appl Math 9: 267–293, 1956.

Mallinson GD, De Vahl Davis G. The method of the false transient for the solution of coupled elliptic equations. J Comput Phys 12: 435–461, 1973.

Mallinson GD, De Vahl Davis G. Three-dimensional natural convection in a box: a numerical study. J Fluid Mech 83: 1–31, 1977.

Morchoisne Y. Resolution of the Navier-Stokes equations by a space-time pseudospectral method. Rech Aérosp 1979-5: 293–306, 1979.

Patankar SV. Numerical Heat Transfer and Fluid Flow. Washington DC: Hemisphere, 1980.

Patankar SV, Baliga BR. A new finite-difference scheme for parabolic differential equations. Numer Heat Transfer 1: 27, 1978.

Pepper DW, Heinrich JC. The Finite Element Method: Basic Concepts and Applications. Washington DC: Hemisphere, 1992.

Peyret R, Taylor TD. Computational Methods for Fluid Flow. Berlin: Springer, 1983.

Pike J, Roe PL. Accelerated convergence of Jameson's finite volume Euler scheme using Van der Houwen integrators. Computers & Fluids 13: 223–236, 1985.

Reddy JN. An Introduction to the Finite Element Method. New York: McGraw-Hill, 1984.

Richardson SM, Cornish ARH. Solution of three-dimensional incompressible flow problems. J Fluid Mech 82: 309–319, 1977.

Richtmyer RD, Morton KW. Difference Methods in Initial-Value Problems. 2nd ed. New York: Interscience, 1967.

Roache PJ, Ellis MA. The biharmonic driver method for the steady-state Navier-Stokes equations. Comput Fluids 3: 305–320, 1975.

Roe PL. Error estimates for cell-vertex solutions of the compressible Euler equations. ICASE Rep. No. 87-6, NASA Langley Research Center, 1987.

Schmidt W, Jameson A. Recent Developments in finite-volume time-dependent techniques for two- and three-dimensional transonic flows. VKI-Lecture Ser. 1982-04, 1982.

Smith GD. Numerical Solution of Partial Differential Equations. 2nd ed. London: Oxford University Press, 1978.

Turkel E. Accuracy of schemes with non-uniform meshes for compressible fluid flows. ICASE Rep. No. 85-43, NASA Langley Research Center, 1985.

Whittaker ET, Watson GN. A Course in Modern Analysis. 4th ed. (reprinted 1962, Cambridge University Press), 1927.

Zienkiewicz OC. The Finite Element Method. New York: McGraw-Hill, 1977.

Zienkiewicz OC, Morgan K. Finite Elements and Approximation. New York: Wiley, 1983.

3

Turbulence Modeling

John R. Schwab

Fluid and Thermal Analysis Consultant, North Olmsted, Ohio

As stated in Chapter 1, the unaveraged unsteady form of the Navier-Stokes equations remains valid for the entire flow regime from laminar to turbulent, as long as appropriate values are specified for the transport coefficients v and α. However, the extremely small length and time scales associated with turbulent motion require impractically dense computational grids and small time steps when attempting to solve the unaveraged unsteady governing equations for engineering applications at realistic Reynolds numbers. Turbulence modeling is the method of incorporating the critical effects of turbulent motion on the mean unsteady flow and heat transfer without having to resolve the actual small scales of turbulent motion.

This chapter describes the fundamental aspects of turbulence modeling for incompressible and compressible single-phase nonreacting flows, including thermal turbulence modeling. Although they are extremely interesting and important, transitional flows are not discussed, since their complexity requires a separate detailed treatment.

3.1 BASIC CONCEPTS

3.1.1 Reynolds Averaging

Reynolds averaging is the standard method of determining the influence of the turbulent motion on the mean unsteady flow and heat transfer. The instantaneous flow variables appearing in the governing equations are first decomposed into mean and fluctuating components. The mean component may still be unsteady, since it is defined as

$$\bar{q} = \frac{1}{T} \int_{t}^{t+T} q \, dt \tag{3.1}$$

where T is an arbitrary time scale much larger than the characteristic turbulent time scale but much smaller than the gross unsteady time scale. The turbulent time scale will be several orders of magnitude smaller than the gross unsteady time scale for flows at realistic Reynolds numbers. The fluctuating component is defined as the difference between the instantaneous value and the mean component:

$$q' = q - \bar{q} \tag{3.2}$$

Since the averaging process is linear, it is applied to each term in the governing equations after making the substitution

$$q = \bar{q} + q' \tag{3.3}$$

and exploiting the definition

$$\overline{q'} = \frac{1}{T} \int_t^{t+T} q' \, dt = 0 \tag{3.4}$$

The conservation law form of the continuity equation (1.3) becomes

$$\frac{\partial \bar{\rho}}{\partial t} + \frac{\partial \overline{\rho u_i}}{\partial x_i} + \frac{\partial \bar{\rho}}{\partial t} + \frac{\partial \overline{\rho u_i}}{\partial x_i} + \frac{\partial \overline{\rho' u_i'}}{\partial x_i} = 0 \tag{3.5}$$

The conservation law form of the momentum equation (1.11) becomes

$$\frac{\partial}{\partial t} \left(\overline{\rho u_i} + \overline{\rho' u_i'} \right) + \frac{\partial}{\partial x_j} \left(\overline{\rho u_i} \, \bar{u}_j + \bar{u}_i \overline{\rho' u_j'} \right)$$

$$= -\frac{\partial \bar{p}}{\partial x_i} + \frac{\partial}{\partial x_j} \left(\overline{\tau_{ij}} - \bar{u}_j \overline{\rho' u_i'} - \overline{\rho} \, \overline{u_i' u_j'} - \overline{\rho' u_i' u_j'} \right) \tag{3.6}$$

where

$$\overline{\tau_{ij}} = \mu \left[\left(\frac{\partial \bar{u}_i}{\partial x_j} + \frac{\partial \bar{u}_j}{\partial x_i} \right) - \frac{2}{3} \delta_{ij} \frac{\partial \bar{u}_k}{\partial x_k} \right] \tag{3.7}$$

The conservation law form of the energy equation (1.17) becomes

$$\frac{\partial}{\partial t} \left(c_p \bar{\rho} \bar{\theta} + c_p \overline{\rho' \theta'} \right) + \frac{\partial}{\partial x_j} \left(c_p \overline{\rho} \bar{u}_j \bar{\theta} \right)$$

$$= \frac{\partial \bar{p}}{\partial t} + \bar{u}_j \frac{\partial \bar{p}}{\partial x_j} + \overline{u_j' \frac{\partial p'}{\partial x_j}}$$

$$+ \frac{\partial}{\partial x_j} \left(k \frac{\partial \bar{\theta}}{\partial x_j} - c_p \bar{\rho} \overline{u_j' \theta'} - c_p \overline{\rho' u_j' \theta'} \right) + \bar{\Phi} \tag{3.8}$$

where

$$\bar{\Phi} = \overline{\tau_{ij} \frac{\partial u_i}{\partial x_j}} = \overline{\tau_{ij}} \frac{\partial \bar{u}_i}{\partial x_j} + \overline{\tau_{ij}' \frac{\partial u_i'}{\partial x_j}} \tag{3.9}$$

The terms involving the fluctuating pressure gradient and the fluctuating stress tensor are generally neglected. For incompressible flows, the fluctuating density vanishes and the mean density is constant, producing the following set of Reynolds-averaged governing equations:

$$\frac{\partial \bar{u}_i}{\partial x_i} = 0 \tag{3.10}$$

$$\rho \frac{\partial}{\partial t}(\bar{u}_i) + \rho \frac{\partial}{\partial x_j}(\bar{u}_i \bar{u}_j) = -\frac{\partial \bar{p}}{\partial x_i} + \frac{\partial}{\partial x_j}(\bar{\tau}_{ij} - \rho \overline{u_i' u_j'}) \tag{3.11}$$

$$\rho \frac{\partial}{\partial t}(c_p \bar{\theta}) + \rho \frac{\partial}{\partial x_j}(c_p \bar{u}_j \bar{\theta})$$

$$= \frac{\partial \bar{p}}{\partial t} + \bar{u}_j \frac{\partial \bar{p}}{\partial x_j} + \frac{\partial}{\partial x_j}\left(k \frac{\partial \bar{\theta}}{\partial x_j} - c_p \rho \overline{u_j' \theta'}\right) + \bar{\tau}_{ij} \frac{\partial \bar{u}_i}{\partial x_j} \tag{3.12}$$

The crux of incompressible turbulence modeling is to find rational methods for computing or approximating the two additional terms introduced by Reynolds averaging, $-\overline{u_i' u_j'}$ and $-\overline{u_i' \theta'}$, which are referred to as the Reynolds stress and Reynolds flux terms to recognize their apparent similarity to the laminar stress and flux terms.

3.1.2   Favre Averaging

For compressible flows, additional models must be developed for the correlation terms involving the fluctuating density that vanished for incompressible flow. Since they have no analogous terms in the laminar equations, these terms are difficult to model without involving additional transport equations.

Favre averaging offers a useful mathematical simplification to eliminate the appearance of correlations with density fluctuations in the governing equations. The averaging is density-weighted as follows

$$\tilde{q} = \frac{1}{\bar{\rho} T} \int_t^{t+T} \rho q \, dt \tag{3.13}$$

where the density remains Reynolds-averaged, as does the pressure. The fluctuating component is defined as the difference between the instantaneous value and the density-averaged mean component

$$q'' = q - \tilde{q} \tag{3.14}$$

Favre averaging is applied to each term in the governing equations after making the substitution

$$q = \tilde{q} + q'' \tag{3.15}$$

and exploiting the definition

$$\overline{\rho q''} = 0 \tag{3.16}$$

The conservation law form of the continuity equation (1.3) becomes

$$\frac{\partial \overline{\rho}}{\partial t} + \frac{\partial (\overline{\rho} \tilde{u})}{\partial x_i} = 0 \tag{3.17}$$

The conservation law form of the momentum equation (1.11) becomes

$$\frac{\partial}{\partial t} (\overline{\rho} \tilde{u}_i) + \frac{\partial}{\partial x_j} (\overline{\rho} \tilde{u}_i \tilde{u}_j) = -\frac{\partial \overline{p}}{\partial x_i} + \frac{\partial}{\partial x_j} (\overline{\tau}_{ij} - \overline{\rho u_i'' u_j''}) \tag{3.18}$$

where

$$\overline{\tau}_{ij} = \mu \left[\left(\frac{\partial \tilde{u}_i}{\partial x_j} + \frac{\partial \tilde{u}_j}{\partial x_i} \right) - \frac{2}{3} \delta_{ij} \frac{\partial \tilde{u}_k}{\partial x_k} \right] + \mu \left[\left(\frac{\overline{\partial u_i''}}{\partial x_j} + \frac{\overline{\partial u_j''}}{\partial x_i} \right) - \frac{2}{3} \delta_{ij} \frac{\overline{\partial u_k''}}{\partial x_k} \right] \tag{3.19}$$

The second bracketed term is expected to be relatively small and conveniently neglected (Anderson, Tannehill, and Pletcher, 1984, p. 203).

The conservation law form of the energy equation (1.17) becomes

$$\frac{\partial}{\partial t} (c_p \overline{\rho} \tilde{\theta}) + \frac{\partial}{\partial x_j} (c_p \overline{\rho} \tilde{u}_j \tilde{\theta}) = \frac{\partial \overline{p}}{\partial t} + \tilde{u}_j \frac{\partial \overline{p}}{\partial x_j} + \overline{u_j'' \frac{\partial p}{\partial x_j}}$$
$$+ \frac{\partial}{\partial x_j} \left(k \frac{\partial \tilde{\theta}}{\partial x_j} + k \frac{\overline{\partial \theta''}}{\partial x_j} - c_p \overline{\rho u_j'' \theta''} \right) + \overline{\Phi} \tag{3.20}$$

where

$$\overline{\Phi} = \overline{\tau_{ij} \frac{\partial u_i}{\partial x_j}} = \overline{\tau}_{ij} \frac{\partial \tilde{u}_i}{\partial x_j} + \overline{\tau_{ij} \frac{\partial u_i''}{\partial x_j}} \tag{3.21}$$

The governing equations now implicitly incorporate the fluctuating density, while the turbulent stress and flux become $-\overline{\rho u_i'' u_j''}$ and $-\overline{\rho u_i'' \theta''}$, respectively. It is important to remember that Favre averaging is a mathematical process that changes only the symbolic representation of the averaged equations. The effects of the fluctuating density must still either be modeled or neglected on the basis of rational physical arguments.

Reynolds-averaged variables may be evaluated from Favre-averaged variables using

$$\overline{\rho q} = \overline{\rho} \tilde{q} + \overline{\rho' q'} \tag{3.22}$$

where the covariance term should be computed through a transport equation. For direct comparison of computed results with experimental data, the physical averaging process of the measurement or the data-reduction averaging method should dictate which formulation is most appropriate.

3.1.3 Boussinesq Assumption

Since the main effect of turbulence is to dramatically augment diffusion and dissipation governed by molecular viscosity, it seems natural to attempt to define a turbulent or eddy viscosity to augment the molecular viscosity, so that the turbulent terms have the same form as the molecular terms. The Boussinesq assumption postulates a turbulent analogy to the fundamental Newtonian stress-strain relation for a simple shear layer

$$\overline{\tau_{12}} = \nu \frac{\partial \overline{u}_1}{\partial x_2} \tag{3.23}$$

$$-\overline{u_1' u_2'} = \nu_T \frac{\partial \overline{u}_1}{\partial x_2} \tag{3.24}$$

The parallel turbulent analogy to the fundamental Fourier heat flux relation for a simple shear layer is

$$\overline{q}_2 = \alpha \frac{\partial \overline{\theta}}{\partial x_2} \tag{3.25}$$

$$-\overline{u_2' \theta'} = \alpha_T \frac{\partial \overline{\theta}}{\partial x_2} \tag{3.26}$$

The generalization of these simple relations yields

$$-\overline{u_i' u_j'} = \nu_T \left(\frac{\partial \overline{u}_i}{\partial x_j} + \frac{\partial \overline{u}_j}{\partial x_i} - \frac{2}{3} \delta_{ij} \frac{\partial \overline{u}_k}{\partial x_k} \right) - \frac{2}{3} \delta_{ij} k \tag{3.27}$$

$$-\overline{u_i' \theta'} = \alpha_T \frac{\partial \overline{\theta}}{\partial x_i} \tag{3.28}$$

While the concept of turbulent viscosity and diffusivity is extremely useful, it is important to realize that it is merely a phenomenological concept without any theoretical basis. Molecular viscosity and diffusivity are fundamentally defined fluid properties, while turbulence must remain a flow property. Models for ν_T and α_T attempt to produce reasonable turbulence effects within the governing equations as functions of the mean flow quantities.

A turbulent Prandtl number may now be defined analogous to the molecular Prandtl number:

$$\Pr_T = \frac{\nu_T}{\alpha_T} = \frac{\overline{u_i' u_j'}}{\overline{u_i' \theta'}} \frac{\partial \overline{\theta}/\partial x_i}{\partial \overline{u}_i/\partial x_j} \tag{3.29}$$

While the molecular Prandtl number is a physical property, just like the molecular diffusivities, the turbulent Prandtl number is not. Even though a constant value for Pr_T is very useful as a good approximation for simple boundary layers and fully developed duct flows, there is absolutely no physical basis to assume that Pr_T remains constant or even isotropic in complex flows. However, the concept of rationalizing some sort of model for Pr_T remains extremely attractive for many practical situations (Kays, 1994).

3.1.4   Modeling Hierarchy

The most fundamental division of turbulence modeling distinguishes between models that employ the basic eddy viscosity/diffusivity concept and those that solve for the Reynolds stresses and fluxes directly. The former category includes algebraic models for the eddy viscosity, as well as one-equation and two-equation models that compute the eddy viscosity as a function of one or two turbulent quantities governed by partial differential transport equations. The latter category includes models with coupled partial differential equations for each Reynolds stress and flux, as well as simplified or truncated models using algebraic relations to compute the individual stresses and fluxes.

The algebraic eddy viscosity models generally are based upon the Prandtl mixing-length concept. They offer the distinct advantage of computational simplicity and often produce acceptable results for simple flows. For complex flows involving multiple strain rates, the mixing-length concept derived for a simple shear layer becomes inadequate.

One-equation models usually involve a transport equation for the turbulent kinetic energy, and require the specification of a turbulent length scale to complete the model. The transport equation allows the computed eddy viscosity to include some nonlocal effects, but the required specification of a length scale inhibits the generality of these models.

Two-equation models involve a second transport equation for a dissipation-related turbulence quantity, usually the dissipation rate of the turbulent kinetic energy. They can be considered complete turbulence models, since they require no further specification of any turbulent quantity (Wilcox, 1993a, p. 73).

Stress transport models solve coupled partial differential equations governing each Reynolds stress. This is clearly the highest level of turbulence modeling, where the transport equation balances specific terms for the convection, diffusion, production, dissipation, and body forces affecting each stress. The anisotropic effects of multiple strain rates for curvature and rotation on each stress can be directly captured. However, the potential fidelity of these models is offset by their complexity, especially for the near-wall region.

Algebraic stress transport models fall into two categories: those starting from a nonlinear constitutive relation designed to remedy the inadequacies of the Bousinessq assumption at the level of the two-equation models, and those resulting from truncation or simplification of the stress transport equations to reduce the set of coupled partial differential equations to a set of coupled algebraic equations to be solved with a two-equation model. Both methods result in similar models representing relatively complex anisotropic extensions to the two-equation eddy viscosity models.

Thermal turbulence modeling follows a hierarchy parallel to that for the velocity field after distinguishing between models computing thermal turbulence and models specifying thermal turbulence effects through a turbulent Prandtl number. The transport of any passive scalar can be modeled with the same techniques as those specifically described for temperature later in this chapter.

The vast majority of turbulence models have been developed for incompressible flows. When density fluctuations become significant in supersonic and hypersonic flows, their effects on the turbulence must be incorporated. In the absence of strong shocks and substantial heat transfer, compressible boundary layers below Mach 5 and mixing layers below Mach 1 can be adequately computed by simply including the variation of the mean density in the turbulence model governing equations. Special formulations of some turbulence models have been developed for hypersonic flows and supersonic flows with large pressure and temperature gradients; these will be discussed in separate sections later in this chapter.

Traditional turbulence models at all levels must employ empirically determined constants and functions that detract from any claim of universal applicability. Renormalization group (RNG) theory offers a method for creating a hierarchy of models (Yakhot and Orszag, 1986) with analytically determined constants and functions that reduce the empiricism of traditional modeling. The RNG method involves the systematic elimination of the small scales of turbulence while maintaining their effects on the large scales through a renormalized viscosity. While the current RNG models have not been as extensively tested as the traditional models, they remain a promising alternative and are worthy of future development and validation.

The final frontier of turbulence modeling removes all modeling assumptions and solves the unaveraged unsteady form of the governing equations with grid sizes and time steps small enough to resolve the Kolmogorov scales. This direct numerical simulation (DNS) approach will remain impractical for flows at realistic Reynolds number for many years due to its enormous computer time and storage requirements, although the low-

Reynolds-number solutions computed to data have provided invaluable data for developing and validating near-wall turbulence models (Mansour, Kim, and Moin, 1989; Rodi and Mansour, 1993).

Large eddy simulation (LES) is an alternative to DNS where only the largest turbulent eddies are resolved while the remaining small eddies are modeled. LES requires an order-of-magnitude less computer time and storage than does DNS to compute a flow at the same Reynolds number, but the subgrid-scale model remains crucial for accurate predictions using LES. Both DNS and LES are under intense development at this time; an adequate treatment of either technique is beyond the scope of this chapter. Rogallo and Moin (1984) provide an introduction to this area, while Rai and Moin (1991) and Germano et al. (1991) present some examples of recent approaches.

3.2 EDDY VISCOSITY MODELS

3.2.1 Algebraic Models

Algebraic eddy viscosity models are based upon the Prandtl mixing-length concept, which postulates the following form for the eddy viscosity in a boundary layer:

$$v_T = v_{mix} l_{mix} = l_{mix}^2 \left| \frac{\partial \overline{u}}{\partial y} \right| \tag{3.30}$$

In the inner region of the boundary layer, the mixing length, l, is assumed to be proportional to the distance from the wall:

$$l_{inner} = \kappa y \tag{3.31}$$

where $\kappa = 0.41$ is the von Karman constant. In the outer region, the mixing length is assumed to be proportional to the boundary layer thickness:

$$l_{outer} = \lambda \delta_{99} \tag{3.32}$$

where $\lambda = 0.085$ is Escudier's constant. Van Driest (1956) proposed a widely used modification to improve the prediction through the viscous sublayer immediately next to the wall:

$$l_{inner} = \kappa y (1 - e^{-y^+/A^+})$$

where $A^+ = 26$. Kays and Moffat (1975) developed an empirical function to account for the effects of pressure gradient and transpiration on a boundary layer:

$$A^+ = \frac{25}{a\{v_0^+ + b[p^+/(1 + cv_0^+)]\} + 1} \tag{3.33}$$

where

$$p^+ = \frac{\mu(\partial \overline{p}/\partial y)}{\sqrt{\overline{\rho}\tau_0^3}} \tag{3.34}$$

and

$$v_0^+ = \frac{v_0}{\sqrt{\tau_0/\overline{\rho}}} \tag{3.35}$$

with $a = 7.1$, $b = 4.25$, $c = 10.0$ for $p^+ < 0$, and $a = 7.1$, $b = 2.9$, $c = 0.0$ for $p^+ > 0$.

Cebeci and Smith (1974) developed a widely used version of the two-layer mixing-length model with

$$v_{Ti} = [\kappa y(1 - e^{-y^+/A^+})]^2 \sqrt{\left(\frac{\partial \overline{u}}{\partial y}\right)^2 + \left(\frac{\partial \overline{v}}{\partial x}\right)^2} \tag{3.36}$$

$$v_{To} = 0.0168\overline{u}_\infty \, \delta_v^* \, F_{\text{Kleb}} \tag{3.37}$$

$$\delta_v^* = \int_0^\delta \frac{1 - \overline{u}}{\overline{u}_\infty} \, dy \tag{3.38}$$

Pressure gradient effects are incorporated through

$$A^+ = \frac{26}{\sqrt{1 + y \dfrac{\partial \overline{p}/\partial x}{\overline{\rho}u_\tau^2}}} \tag{3.39}$$

while intermittency effects at the boundary layer edge are described by

$$F_{\text{Kleb}} = \left[1 + \left(\frac{y}{\delta}\right)^6\right]^{-1} \tag{3.40}$$

The eddy viscosity computation switches from the inner function to the outer function at the point where they are equal, which can be estimated (Wilcox, 1993a, p. 51) as

$$y_{\text{match}} = 0.04 \, \text{Re}_{\delta^*} \tag{3.41}$$

Baldwin and Lomax (1978) formulated another widely used two-layer eddy viscosity model that does not require an expensive calculation of the boundary layer thickness in order to determine a length scale in the outer region. Their model uses the distance away from the wall corresponding to the location of the maximum vorticity as the outer region length scale, making it especially attractive for use in a Navier-Stokes code.

$$v_{Ti} = [\kappa y(1 - e^{-y^+/A^+})]^2 |\omega| \tag{3.42}$$

$$v_{To} = 0.0168 C_{cp} F_{\text{wake}} F_{\text{Kleb}} \tag{3.43}$$

$$F_{\text{wake}} = \min\left[y_{\text{max}} F_{\text{max}}, \frac{C_{\text{wake}} y_{\text{max}} \bar{u}_{\text{max}}^2}{F_{\text{max}}} \right] \tag{3.44}$$

$$F_{\text{max}} = \max[y(1 - e^{-y^+/A^+})|\omega|] \tag{3.45}$$

$$F_{\text{Kleb}} = \left[1 + \left(\frac{C_{\text{Kleb}} y}{y_{\text{max}}}\right)^6\right]^{-1} \tag{3.46}$$

where $C_{cp} = 1.6$, $C_{\text{Kleb}} = 0.3$, $C_{\text{wake}} = 1.0$, y_{max} is the location of F_{max}, and $\bar{u}_{\text{max}}$ is the maximum velocity. Granville (1987) has proposed modifications for these constants and factors to improve prediction of boundary layers under pressure gradients.

Despite their obvious theoretical shortcomings, these algebraic models have been widely used due to their simplicity and robustness. Wilcox (1993a, pp. 53–64) presents a detailed comparison of results for channel and pipe flows and attached and separated boundary layers. Cebeci and Smith (1974) and Kays and Crawford (1993) describe a variety of modifications to incorporate blowing, suction, curvature, roughness, pressure gradients, and near-wall effects.

Martinelli and Yakhot (1989) used an RNG algebraic model to predict the transonic flow over an airfoil with good results. Kirtley (1992) applied a similar model to the computation of the complex flow within a compressor rotor and showed improved results compared to the Baldwin-Lomax model. Despite having their constants derived analytically as part of the RNG procedure, both of these models are still dependent on the empirical specification of an appropriate length scale.

3.2.2   One-Equation Models

The most obvious shortcoming of the algebraic eddy viscosity models discussed above is that they unrealistically compute $v_T = 0$ and $\partial \bar{u}/\partial y = 0$ at the centerline of a pipe or channel. Prandtl postulated the use of a turbulent velocity scale $V_T = \sqrt{k}$ to replace $v_{\text{mix}} = l_{\text{mix}} \partial \bar{u}/\partial y$ in (3.30) where $k = (1/2)(\overline{u'u'} + \overline{v'v'} + \overline{w'w'})$ is the turbulent kinetic energy. The eddy viscosity is then computed as

$$v_T = l\sqrt{k} \tag{3.47}$$

An exact governing equation for the transport of k can be derived by subtracting the Reynolds-averaged momentum equation from the instantaneous momentum equation and then Reynolds-averaging the result (Kays and

Crawford, 1993, p. 56):

$$\frac{\partial k}{\partial t} + \bar{u}_j \frac{\partial k}{\partial x_j} = -\overline{u_i' u_j'} \frac{\partial \bar{u}_i}{\partial x_j} - \varepsilon + \frac{\partial}{\partial x_j} \left[v \frac{\partial k}{\partial x_j} - \frac{1}{2} \overline{u_i' u_i' u_j'} - \frac{1}{\rho} \overline{p' u_j'} \right] \quad (3.48)$$

From left to right, the terms are unsteady rate, convection, production, dissipation, molecular diffusion, turbulent diffusion, and pressure diffusion. The last two terms are generally combined and modeled as a turbulent gradient-diffusion term using the eddy viscosity, while the dissipation is modeled as $\varepsilon = C_D k^{3/2}/l$ to yield

$$\frac{\partial k}{\partial t} + \bar{u}_j \frac{\partial k}{\partial x_j} = v_T \left(\frac{\partial \bar{u}_i}{\partial x_j} \right)^2 - C_D \frac{k^{3/2}}{l} + \frac{\partial}{\partial x_j} \left[\left(v + \frac{v_T}{\sigma_k} \right) \frac{\partial k}{\partial x_j} \right] \quad (3.49)$$

The one-equation model offers some theoretical improvement in the computation of the eddy viscosity due to incorporating nonlocal and historical effects, such as the decay of freestream turbulence, through a transport equation. However, it still suffers from the requirement that the mixing length be specified, and thus offers no practical advantage over the algebraic models, especially in three-dimensional flows, where a plausible length scale can be quite difficult to define.

Baldwin and Barth (1990) and Spalart and Allmaras (1992) have developed more complex versions of the one-equation model where the dependent variable is the eddy viscosity itself. Both models show some improvement in computing separate flows compared to algebraic models, but their performance on other flows remains mediocre.

3.2.3 Two-Equation Models

Since the k-transport equation is well established as providing a reasonable turbulent velocity scale, the two-equation models offer the next step up in theoretical sophistication by supplying a transport equation for a second variable to establish the turbulent length scale. By an overwhelming margin, the most popular choice for the second variable is ε, which appears directly in the k-transport equation; $\omega = \varepsilon/k$ and $\tau = k/\varepsilon$ are the best-known alternatives.

Jones and Launder (1972) developed what is generally considered to be the standard k-ε model, although Chou (1945) and Harlow and Nakayama (1968) made earlier attempts at modeling the ε-transport equation. An exact equation can be derived by differentiating the fluctuating velocity equation with respect to x_j, multiplying the result by $v\partial u_i/\partial x_j$, and finally Reynolds-averaging the result (Kays and Crawford, 1993, pp. 59–60; Wilcox, 1993a, p. 88).

The exact equation contains many new correlation terms impossible to measure experimentally; however, some researchers (Mansour, Kim, and Moin, 1989; Rodi and Mansour, 1993) have attempted to directly model the terms in the ε-transport equation by using DNS results. The standard approach is to ignore the exact equation and to substitute a rational model equation, similar to the k-transport equation, that attempts to balance production, dissipation, convection, and diffusion to produce a reasonable profile of ε that then produces reasonable profiles for k and v_T.

The standard model (Jones and Launder, 1972) consists of the following equations:

$$v_T = \frac{c_\mu k^2}{\varepsilon} \tag{3.50}$$

$$\frac{\partial k}{\partial t} + \bar{u}_j \frac{\partial k}{\partial x_j} = v_T \left(\frac{\partial \bar{u}_i}{\partial x_j}\right)^2 - \varepsilon + \frac{\partial}{\partial x_j}\left[\left(v + \frac{v_T}{\sigma_k}\right)\frac{\partial k}{\partial x_j}\right] \tag{3.51}$$

$$\frac{\partial \varepsilon}{\partial t} + \bar{u}_j \frac{\partial \varepsilon}{\partial x_j} = c_{\varepsilon 1} \frac{\varepsilon}{k} v_T \left(\frac{\partial \bar{u}_i}{\partial x_j}\right)^2 - c_{\varepsilon 2} \frac{\varepsilon^2}{k} + \frac{\partial}{\partial x_j}\left[\left(v + \frac{v_T}{\sigma_\varepsilon}\right)\frac{\partial \varepsilon}{\partial x_j}\right] \tag{3.52}$$

where $c_\mu = 0.09$, $c_{\varepsilon 1} = 1.55$, $c_{\varepsilon 2} = 2.0$, $\sigma_k = 1.0$, $\sigma_\varepsilon = 1.3$. Launder and Sharma (1974) revised two of the constants to their more generally accepted values: $c_{\varepsilon 1} = 1.44$ and $c_{\varepsilon 2} = 1.92$.

Wilcox (1993a, 1993b) has long been a proponent of the k-ω model, where a transport equation for $\omega = \varepsilon/k$ is used as an alternative to the ε-transport equation. This model appears to offer improved performance in boundary layers under pressure gradients.

$$v_T = \frac{k}{\omega} \tag{3.53}$$

$$\frac{\partial k}{\partial t} + \bar{u}_j \frac{\partial k}{\partial x_j} = v_T \left(\frac{\partial \bar{u}_i}{\partial x_j}\right)^2 - \beta^* k\omega + \frac{\partial}{\partial x_j}\left[\left(v + \frac{v_T}{\sigma_k}\right)\frac{\partial k}{\partial x_j}\right] \tag{3.54}$$

$$\frac{\partial \omega}{\partial t} + \bar{u}_j \frac{\partial \omega}{\partial x_j} = \alpha v_T \frac{\omega}{k}\left(\frac{\partial \bar{u}_i}{\partial x_j}\right)^2 - \beta\omega^2 + \frac{\partial}{\partial x_j}\left[\left(v + \frac{v_T}{\sigma_\omega}\right)\frac{\partial \omega}{\partial x_j}\right] \tag{3.55}$$

where $\alpha = 0.556$, $\beta = 0.075$, $\beta^* = 0.09$, $\sigma_k = 2.0$, $\sigma_\omega = 2.0$.

Even though both of the above models include viscous diffusion terms, their constants were evaluated for high-Reynolds-number flows, typically decaying homogeneous turbulence and the logarithmic region of a boundary layer. Neither of the above models gives accurate solutions when integrated through the sublayer to the wall with no-slip boundary conditions.

Any one of three following schemes is typically used to modify the models for the near-wall region.

The first method uses the classic law of the wall to derive algebraic wall functions that specify values for k, ε, and ω in terms of the friction velocity and the distance from the wall for the first grid point off the wall, presumably located in the logarithmic region. Obviously, this method is problematical for separated flows, and the location of the first grid point is largely empirical, even for attached boundary layers.

Once the friction velocity is determined via iterative solution of the law of the wall at the first grid point,

$$\bar{u} = u_\tau \left[\frac{1}{\kappa} \ln\left(\frac{u_\tau y}{v}\right) + B \right] \tag{3.56}$$

the wall functions can be easily computed (Wilcox, 1993a, pp. 126–127) as

$$k = \frac{u_\tau^2}{\sqrt{c_\mu}}, \qquad \varepsilon = c_\mu^{3/4} \frac{k^{3/2}}{\kappa y}, \qquad \omega = \frac{\sqrt{k}}{c_\mu^{1/4} \kappa y} \tag{3.57}$$

The second method uses a robust mixing-length model for the near-wall region. The two-equation model is matched to the mixing-length model at a point generally specified in terms of a fixed y^+ within the logarithmic region or a fixed value of v_T/v large enough to ensure negligible viscous effects. The eddy viscosity computed from the mixing-length model is used to compute boundary values for k and ε at the match point:

$$k = \left(\frac{c_D v_T}{c_\mu \kappa y}\right)^2, \qquad \varepsilon = \frac{c_D k^{3/2}}{\kappa y} \tag{3.58}$$

This method is more general than using wall functions, since it contains no explicit dependence on the law of the wall, which is only valid for attached two-dimensional shear layers. However, most mixing-length models use Van Driest damping through the sublayer and therefore become theoretically invalid for separated flows. The exact location of the match point also remains largely empirical.

The most theoretically satisfactory scheme to include near-wall effects with two-equation models is to modify the standard model transport equations and allow integration right down to the wall with appropriate boundary conditions. This requires more grid points in the near-wall region than the other two near-wall schemes described above, but does permit a continuous solution without the problem of determining an artificial match point.

Low-Reynolds-number (LRN) two-equation models are currently almost as ubiquitous as the algebraic models. Patel, Rodi, and Scheuerer

(1985) published a useful review of eight LRN two-equation models. Lang and Shih (1991), Thangam and Speziale (1992), Michelassi and Shih (1993), Steffen (1993), and Wilcox (1993a, pp. 138–146; 1993b) also published similar reviews.

The general approach for the LRN models is to use damping functions to modify the major constants in the two-equation models to force the desired asymptotic behavior at the wall. These damping functions are typically designed as exponential functions of either $y^+ = u_\tau y/v$, $\mathrm{Re}_k = \sqrt{k}y/v$, or $\mathrm{Re}_T = k^2/v\varepsilon$ to allow a smooth transition from the low-Reynolds-number behavior to the unmodified high-Reynolds-number behavior. Additional terms may also be added to the equations to adjust for a change in the dissipation variable to $\hat{\varepsilon} = \varepsilon - D$, where D is a representation of the asymptotic value subtracted to allow a zero-valued boundary condition.

Three of the most popular and well-tested LRN k-ε models (Launder and Sharma, 1974; Lam and Bremhorst, 1981; Chien, 1982) are summarized in Table 1.

$$v_T = \frac{c_\mu f_\mu k^2}{\hat{\varepsilon}} \tag{3.59}$$

$$\frac{\partial k}{\partial t} + \bar{u}_j \frac{\partial k}{\partial x_j} = v_T \left(\frac{\partial \bar{u}_i}{\partial x_j}\right)^2 - (\hat{\varepsilon} + D) + \frac{\partial}{\partial x_j}\left[\left(v + \frac{v_T}{\sigma_k}\right)\frac{\partial k}{\partial x_j}\right] \tag{3.60}$$

$$\frac{\partial \hat{\varepsilon}}{\partial t} + \bar{u}_j \frac{\partial \hat{\varepsilon}}{\partial x_j} = c_{\varepsilon 1} f_1 \frac{\hat{\varepsilon}}{k} v_T \left(\frac{\partial \bar{u}_i}{\partial x_j}\right)^2 - c_{\varepsilon 2} f_2 \frac{\hat{\varepsilon}^2}{k} + E + \frac{\partial}{\partial x_j}\left[\left(v + \frac{v_T}{\sigma_\varepsilon}\right)\frac{\partial \hat{\varepsilon}}{\partial x_j}\right]$$

$$\tag{3.61}$$

For clarity, the functions are displayed in a form appropriate for a two-dimensional boundary layer, where y is the direction normal to the flow. The Launder-Sharma model requires extra terms involving derivatives, including a second derivative of the velocity, since it exploits the zero-valued boundary condition for ε. The Chien model uses nonderivative formulations for the additional terms. The Lam-Bremhorst uses the true boundary condition for ε and thus requires no additional terms.

While the LRN models above are reasonably robust, the damping functions mimic the desired near-wall behavior solely through viscous effects. The independent kinematic blocking effects of wall proximity on the eddy viscosity are not modeled. Myong and Kasagi (1990) developed a formulation for f_μ that accommodates the wall proximity effect through a rational blending of the dominant dissipation length scales near and away from the wall. Durbin (1991) proposed a novel method involving a third equation for $\overline{v'v'}$ to provide a turbulent velocity scale that naturally incorporates both blocking and viscous effects near the wall.

Table 3.1 LRN k-ε Models

	Launder-Sharma	Lam-Bremhorst	Chien
c_μ	0.09	0.09	0.09
$c_{\varepsilon 1}$	1.44	1.44	1.35
$c_{\varepsilon 2}$	1.92	1.92	1.80
σ_k	1.0	1.0	1.0
σ_ε	1.3	1.3	1.3
f_μ	$\exp(-3.4/[1.0 + \mathrm{Re}_T/50.0]^2)$	$[1.0 - \exp(-0.0165\ \mathrm{Re}_k)]^2$ $\times [1.0 + 20.5/\mathrm{Re}_T]$	$1.0 - \exp(-0.0115 y^+)$
f_1	1.0	$1.0 + (0.05/f_\mu)^3$	1.0
f_2	$1.0 - 0.3\exp(-\mathrm{Re}_T^2)$	$1.0 - \exp(-\mathrm{Re}_T^2)$	$1.0 - 0.22\exp(-\mathrm{Re}_T^2/36.0)$
D	$2\nu(\partial\sqrt{k}/\partial y)^2$	0	$2\nu k/y^2$
E	$2\nu\nu_T(\partial^2\bar{u}/\partial y^2)^2$	0	$[-2\nu\hat{\varepsilon}/y^2]\exp(-0.5 y^+)$
$\hat{\varepsilon}_{\mathbf{wall}}$	0	$\nu(\partial^2 k/\partial y^2)$	0

Speziale and Thangam (1992) used an RNG k-ε model to predict the separated flow over a backward-facing step with excellent results. The RNG model equations were identical to the standard model equations (3.50)–(3.52), but the constants derived through the RNG procedure were $c_\mu = 0.085$, $c_{\varepsilon 1} = 1.42 - [\eta(1 - \eta/\eta_0)]/[1 + \beta\eta^3]$, $c_{\varepsilon 2} = 1.68$, $\sigma_k = 0.7179$, and $\sigma_\varepsilon = 0.7179$, where $\eta = Sk/\varepsilon$, $S = \sqrt{(2\overline{S}_{ij}\overline{S}_{ij})}$, $\overline{S}_{ij} = 0.5(\partial\overline{u}_i/\partial x_j + \partial\overline{u}_j/\partial x_i)$, $\eta_0 = 4.38$, and $\beta = 0.015$. No damping functions were required to integrate the model equations directly to the wall.

3.2.4 Compressibility Effects

The algebraic models require no special modifications to adequately compute boundary layer flows without heat transfer up to Mach 5 and mixing layer flows up to Mach 1. Since the length scale used in one-equation models is essentially similar to the mixing length used in the algebraic models, one-equation models also require no modifications for the same types of compressible flows. Both classes of models incorporate compressibility effects solely through the variable mean density and ignore the effects of the fluctuating density, thus following Morkovin's hypothesis (Morkovin, 1962) that the turbulence structure is unaffected by compressibility for boundary layers below Mach 5 or mixing layers below Mach 1, where the ratio of the density fluctuations to the mean density is small.

For boundary layers above Mach 5 or mixing layers above Mach 1, Morkovin's hypothesis of negligible density fluctuations becomes invalid. The presence of large pressure and temperature gradients due to shocks or significant heat transfer further complicates the turbulence modeling. Higher-order turbulence models are required for such complex compressible flows.

Sarkar et al. (1989), Zeman (1990), and Wilcox (1992) proposed a useful modification to the k-ε model that successfully accounts for compressibility effects. An additional dissipation modeling term is added to the k-equation by splitting the dissipation into two parts $\varepsilon = \varepsilon_s + \varepsilon_d$, where ε_s is the solenoidal component unaffected by compressibility, and ε_d is the dilatational component directly dependent on the divergence of the fluctuating velocity, which vanishes for incompressible flows. The solenoidal components is computed from the usual ε-transport equation, while the dilatational component is modeled as a function of the turbulence Mach number $M_T = \sqrt{2k}/a$ multiplying the solenoidal component.

Sarkar et al. (1989) postulated the simple formulation

$$\varepsilon_d = M_T^2 \varepsilon_s \tag{3.62}$$

while Zeman (1990) proposed the more complex form

$$\varepsilon_d = 0.75\left\{1 - \exp\left[\frac{-0.5(y + 1)(M_T - M_{T0})^2}{\Lambda^2}\right]\right\} H(M_T - M_{T0})\varepsilon_s \quad (3.63)$$

with $\Lambda = 0.60$ and $M_{T0} = 0.25\sqrt{2/(y + 1)}$ for mixing layers and $\Lambda = 0.66$ and $M_{T0} = 0.25\sqrt{2/(y + 1)}$ for boundary layers.

Wilcox (1992) later proposed

$$\varepsilon_d = 1.5\{M_T^2 - M_{T0}^2\}H(M_T - M_{T0})\varepsilon_s \quad (3.64)$$

with $M_{T0} = 0.25$, which produced improved results for both mixing layers and boundary layers when used in the k-ω model.

3.3   STRESS TRANSPORT MODELS

3.3.1   Differential Stress Models

Lakshminarayana (1986), Launder (1989), and Hanjalic (1994) all make compelling arguments for using some form of Reynolds stress transport models to overcome the inherent limitations of using an isotropic eddy viscosity when computing complex flows. Only Reynolds stress transport models can directly capture the effects of additional strain rates, curvature, rotation, and stress anisotropy on the individual turbulent stress terms. However, the potential improvement in fidelity is offset by the increased complexity and computational cost of Reynolds stress transport models compared to algebraic or two-equation eddy viscosity models, since they require six equations (one for each turbulent stress) and a dissipation rate equation, which all incorporate complex modeling terms.

The area of Reynolds stress transport modeling research is extremely active today, and most researchers remain optimistic that these models will eventually supplant the ubiquitous algebraic and two-equation eddy viscosity models as computational power continues to increase. Lumley (1983, 1992) has long argued for including formal mathematical constraints, such as realizability and material frame indifference. Speziale (1991, 1995) recently produced useful reviews concentrating on improved analytical development methods. So et al. (1991) reviewed near-wall Reynolds stress transport modeling. Demuren and Sarkar (1993a, 1993b) systematically evaluated various approaches to modeling the troublesome pressure-strain term.

The exact form of the transport equation for $R_{ij} = \overline{\rho u_i' u_j'}$ can be derived by Reynolds-averaging the product of the instantaneous momentum equation and the fluctuating velocity (Wilcox, 1993a, pp. 17–19):

$$\frac{\partial R_{ij}}{\partial t} + \bar{u}_k \frac{\partial R_{ij}}{\partial x_k} = -\left(R_{ik}\frac{\partial \bar{u}_j}{\partial x_k} + R_{jk}\frac{\partial \bar{u}_i}{\partial x_k}\right) - 2\mu \overline{\frac{\partial u_i'}{\partial x_k}\frac{\partial u_j'}{\partial x_k}} + \overline{p'\left(\frac{\partial u_i'}{\partial x_j} + \frac{\partial u_j'}{\partial x_i}\right)}$$

$$+ \frac{\partial}{\partial x_k}\left[\nu \frac{\partial R_{ij}}{\partial x_k} - (\overline{\rho u_i' u_j' u_k'} + \overline{p' u_i'}\delta_{jk} + \overline{p' u_j'}\delta_{ik})\right] \tag{3.65}$$

From left to right, the terms represent unsteady rate, convection, production (two terms), viscous dissipation, pressure-strain redistribution,
viscous diffusion, and turbulent diffusion (three terms). The pressure-strain
and turbulent diffusion terms require modeling, while the viscous dissipation is obtained from its own transport equation similar to that developed for the k-ε model.

Launder, Reece, and Rodi (1975) developed a high-Reynolds-number
form of a differential stress model (DSM) that has been adopted as a basic
form by the majority of researchers in this area:

$$\frac{\partial R_{ij}}{\partial t} + \bar{u}_k \frac{\partial R_{ij}}{\partial x_k} = P_{ij} - \frac{2}{3}\rho\varepsilon\delta_{ij} + \Pi_{ij} + D_{ij}^T \tag{3.66}$$

$$\Pi_{ij} = -c_1 \frac{k}{\varepsilon}\left(R_{ij} - \frac{2}{3}\delta_{ij}k\right) - c_2\left(P_{ij} - \frac{1}{3}P_{kk}\delta_{ij}\right) \tag{3.67}$$

$$D_{ij}^T = c_s \frac{\partial}{\partial x_k}\left\{\frac{k}{\varepsilon}R_{kl}\frac{\partial R_{ij}}{\partial x_l}\right\} \tag{3.68}$$

$$P_{ij} = -\left(R_{ij}\frac{\partial \bar{u}_i}{\partial x_j} + R_{ij}\frac{\partial \bar{u}_j}{\partial x_i}\right) \tag{3.69}$$

$$\rho\frac{\partial \varepsilon}{\partial t} + \rho\bar{u}_k\frac{\partial \varepsilon}{\partial x_k} = -c_{\varepsilon 1}\frac{\varepsilon}{k}R_{ij}\frac{\partial \bar{u}_i}{\partial x_j} - c_{\varepsilon 2}\rho\frac{\varepsilon^2}{k} + c_\varepsilon \frac{\partial}{\partial x_i}\left(\frac{k}{\varepsilon}R_{ij}\frac{\partial \varepsilon}{\partial x_j}\right) \tag{3.70}$$

where $c_1 = 1.5$, $c_2 = 0.6$, $c_s = 0.25$, $c_{\varepsilon 1} = 1.44$, $c_{\varepsilon 2} = 1.92$, $c_\varepsilon = 0.18$.

Many more complex forms for the pressure-strain term (3.67) and turbulent diffusion term (3.68) have been proposed to improve the DSM predictions (Jones and Musonge, 1988; Demuren and Sarkar, 1993a, 1993b;
Speziale and Gatski, 1994; Speziale, 1995). Near-wall modifications to
eliminate wall-function boundary conditions have been proposed by So and
Yoo (1986), Shima, (1988), Launder and Shima (1989), Lai and So (1990b),
and Durbin (1993). Pollard and Martinuzzi (1989) compared the predictions
of five differential stress models for pipe flow. Demuren (1990) computed
flows in various channels with complex cross section. Shima (1993a, 1993b)
applied the Launder and Shima (1989) DSM to boundary layers with

periodic pressure gradient, transpiration, free-stream turbulence, streamwise curvature, and rotation to demonstrate its ability to capture complex physics. Lien and Leschziner (1994a, 1994b) presented an extensive description of their implementation and validation of a DSM. Rubinstein and Barton (1992) developed an RNG version of a DSM. Hanjalic (1994) showed selected cases where the DSM predictions of complex flows were superior to those from two-equation models. It is clear that there are currently many competing approaches to DSM development and that much future testing is required before a definitive assessment can be attempted.

3.3.2 Algebraic Stress Models

Since the DSM approach may remain prohibitively expensive for some time, it seems natural to attempt a simplified formulation that would improve upon the isotropic eddy viscosity models without the full cost of the additional partial differential equations required by the DSM. There are two distinct approaches to this problem. The first involves developing a nonlinear constitutive relation through formal expansions to extend the iso-tropic eddy viscosity produced by a two-equation model into an aniso-tropic form. The second involves simplification or truncation of the DSM equations into a system of coupled algebraic equations dependent on k, ε, and R_{ij}. Both approaches yield algebraic stress models (ASM) similar in performance and cost.

Lumley (1970) and Pope (1975) were among the first to propose nonlinear extensions to the linear Boussinesq eddy viscosity relation. Yoshizawa (1984), Speziale (1987), Rubinstein and Barton (1990), Taulbee (1992), and Gatski and Speziale (1993) have further developed this theoretical expansion approach in various directions. Shih, Zhu, and Lumley (1994) have shown a variety of successful computations with their ASM model based on invariant theory and realizability constraints. This particular ASM is described below as an example; it is similar to those of Yoshizawa (1984) and Rubinstein and Barton (1990) in its quadratic tensorial form.

$$
\overline{u_i' u_j'} = \frac{2}{3} k\delta_{ij} + v_T \left[\frac{\partial \bar{u}_i}{\partial x_j} + \frac{\partial \bar{u}_j}{\partial x_i} - \frac{2}{3} \frac{\partial \bar{u}_i}{\partial x_i} \delta_{ij} \right]
$$
$$
+ 2a_4 \frac{k^3}{\varepsilon^2} \left[\left(\frac{\partial \bar{u}_i}{\partial x_j} \right)^2 + \left(\frac{\partial \bar{u}_j}{\partial x_i} \right)^2 - \frac{2}{3} \frac{\partial \bar{u}_i}{\partial x_k} \frac{\partial \bar{u}_k}{\partial x_i} \delta_{ij} \right]
$$
$$
+ 2a_6 \frac{k^3}{\varepsilon^2} \left[\frac{\partial \bar{u}_i}{\partial x_k} \frac{\partial \bar{u}_j}{\partial x_k} - \frac{1}{3} \frac{\partial \bar{u}_i}{\partial x_k} \frac{\partial \bar{u}_i}{\partial x_k} \delta_{ij} \right]
$$
$$
+ 2a_7 \frac{k^3}{\varepsilon^2} \left[\frac{\partial \bar{u}_k}{\partial x_i} \frac{\partial \bar{u}_k}{\partial x_j} - \frac{1}{3} \frac{\partial \bar{u}_i}{\partial x_k} \frac{\partial \bar{u}_i}{\partial x_k} \delta_{ij} \right] \tag{3.71}
$$

where

$$2a_4 = \frac{c_{\tau 1}}{A_2 + \eta^3 + \zeta^3}, \quad 2a_6 = \frac{c_{\tau 2}}{A_2 + \eta^3 + \zeta^3}, \quad 2a_7 = \frac{c_{\tau 3}}{A_2 + \eta^3 + \zeta^3}$$

$$\eta = S\frac{k}{\varepsilon}, \quad S = \sqrt{2S_{ij}S_{ij}}, \quad S_{ij} = \frac{1}{2}\left(\frac{\partial \bar{u}_i}{\partial x_j} + \frac{\partial \bar{u}_i}{\partial x_j}\right)$$

$$\xi = \frac{\Omega k}{\varepsilon}; \quad \Omega = \sqrt{\Omega_{ij}^* \Omega_{ij}^*}, \quad \Omega_{ij}^* = \frac{1}{2}\left(\frac{\partial \bar{u}_i}{\partial x_j} - \frac{\partial \bar{u}_i}{\partial x_j}\right) + 4\varepsilon_{mji}\omega_m$$

$$\frac{\partial k}{\partial t} + \bar{u}_j\frac{\partial k}{\partial x_j} = -\overline{u_i'u_j'}\left(\frac{\partial \bar{u}_i}{\partial x_j}\right) - \varepsilon + \frac{\partial}{\partial x_j}\left[\left(\nu + \frac{\nu_T}{\sigma_k}\right)\frac{\partial k}{\partial x_j}\right] \tag{3.72}$$

$$\frac{\partial \varepsilon}{\partial t} + \bar{u}_j\frac{\partial \varepsilon}{\partial x_j} = -c_{\varepsilon 1}\frac{\varepsilon}{k}\overline{u_i'u_j'}\left(\frac{\partial \bar{u}_i}{\partial x_j}\right) - c_{\varepsilon 2}\frac{\varepsilon^2}{k} + \frac{\partial}{\partial x_j}\left[\left(\nu + \frac{\nu_T}{\sigma_\varepsilon}\right)\frac{\partial \varepsilon}{\partial x_j}\right] \tag{3.73}$$

$$\nu_T = c_\mu\frac{k^2}{\varepsilon}, \quad c_\mu = \frac{2/3}{A_1 + \eta}$$

$$c_{\varepsilon 1} = 1.44, \quad c_{\varepsilon 2} = 1.92, \quad \sigma_\varepsilon = 1.3, \quad \sigma_k = 1.0$$

$$c_{\tau 1} = -4.0, \quad c_{\tau 2} = 13.0, \quad c_{\tau 3} = -2.0, \quad A_1 = 5.5, \quad A_2 = 1000.0$$

The first two terms in (3.71) are identical to the standard isotropic eddy viscosity, except that c_μ is now a function of the time scale ratio η. The constants in the k-transport and ε-transport equations take on standard values, while the new constants are calibrated against experimental data. While this type of ASM formulation has an interesting theoretical basis, it will require more testing for a definitive assessment of practical utility.

Rodi (1976) developed a popular simplification of the DSM by assuming that the net convective and diffusive transport of $\overline{u_i'u_j'}$ is locally proportional to the net convective and diffusive transport of k, thus reducing the set of coupled partial differential equations to a set of coupled algebraic equations. Rodi proposed that the local coefficient of proportionality is the ratio $\overline{u_i'u_j'}/k$, leading to the following set of equations for $\overline{u_i'u_j'}$:

$$\frac{\overline{u_i'u_j'}}{k}\left[\overline{u_m'u_n'}\frac{\partial \bar{u}_m}{\partial x_n} - \varepsilon\right] = -\left[\overline{u_i'u_k'}\frac{\partial \bar{u}_j}{\partial x_k} + \overline{u_j'u_k'}\frac{\partial \bar{u}_i}{\partial x_k}\right] - \varepsilon_{ij} + \Pi_{ij} \tag{3.74}$$

Appropriate models for ε_{ij} and Π_{ij} can now be inserted, along with local values of k and ε computed from their transport equations. Launder (1982) proposed a generalized version of this type of ASM to allow preferential transport of the shear stresses compared to the normal stresses.

Lakshminarayana (1986) and Launder (1989) present extensive reviews of this type of ASM, which can also be formulated to include curvature and rotation effects. The ASM formulation is appropriate only for high-

Reynolds-number flows; thus, it must be matched to a low-Reynolds-number model for the near-wall region. It should also be noted that the ASM is notoriously stiff, often requiring extreme underrelaxation.

3.3.3 Compressibility Effects

Few researchers have looked into compressibility effects for DSM. Zhang et al. (1993) showed good results for supersonic boundary layers with adiabatic and cooled walls using a near-wall DSM with no compressibility modifications other than including the variable mean density. The general approach would be to obviously include the direct effects of variable mean density along with the extra dilatational dissipation (Sarkar et al., 1989; Zeman, 1990), as done for the two-equation models. It seems reasonable to presume that new forms of the pressure-strain model will also be required, since most current formulations depend upon the classic Chou (1945) technique involving an incompressible Poisson equation for the pressure.

3.4 THERMAL MODELS

The recent literature shows relatively few descriptions of thermal turbulence modeling, compared to the enormous number of works on dynamic turbulence modeling. Benocci (1991) published a review that concentrated on dynamic turbulence modeling for flows with heat transfer. Launder (1993) argued that second-moment models for the Reynolds stresses should remain the most important part of computing turbulent heat transfer, since the turbulent velocity field appears to have much more influence than the turbulent temperature field. So, Yuan, and Sommer (1992) developed a similar conclusion in a presentation of their hierarchy of thermal turbulence models, although they pointed to the separate computation of thermal turbulence as their goal for ultimate fidelity.

Since the computation of the turbulent temperature field requires a suitable computation of the turbulent velocity field, it would be illogical to use a higher-order model for the temperature field with a lower-order model for the velocity field. However, it seems extremely optimistic to rely on lower-level models for the temperature field with higher-order models for the velocity field in complex flows, where the velocity and temperature fields may involve entirely different physical mechanisms. Although it remains relatively inferior to dynamic turbulence modeling, the continued development of thermal turbulence modeling can offer conceptually parallel modeling of the turbulent velocity and temperature fields, thus permitting more generalized and robust computation of complex flows with heat transfer.

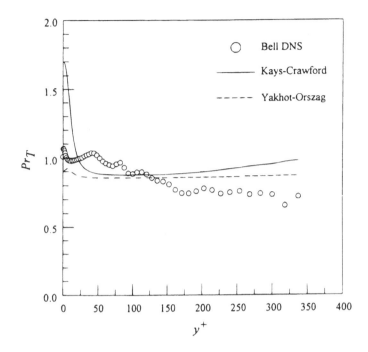

Figure 3.1 Turbulent Prandtl number for a boundary layer at $Re_\theta = 670$.

This section will examine a hierarchy of approaches to thermal turbulence modeling, with particular emphasis on their potential to capture physical effects. It will be limited to fully turbulent, non reacting, single-phase, nonbuoyant flows. The effects of molecular Prandtl number will not be explicitly considered, since they are relatively unimportant, except for liquid metals, where molecular conduction can dominate turbulent diffusion throughout the entire flow field. Although they are extremely interesting and important, transitional flows will not be considered in this Chapter, since their complexity requires a separate detailed treatment beyond the scope of the present work.

For the current work, a fundamental distinction will be utilized to classify thermal turbulence models as either specification models or transport models. Specification models involve the computation of the turbulent thermal diffusivity through the specification of a turbulent Prandtl number, which may be an empirical constant or a complex anisotropic function. Transport models involve the actual computation of thermal turbulence variables through partial differential equations (PDE) governing their transport.

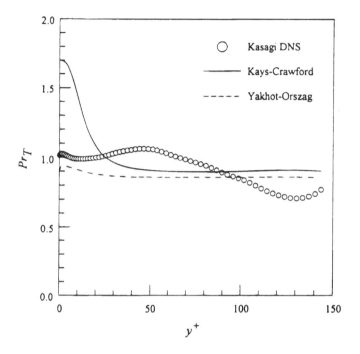

Figure 3.2   Turbulent Prandtl number for a channel flow at $Re_\tau = 150$.

3.4.1   Specification Models

Constant Pr_T

The concept of a constant turbulent Prandtl number is too valuable to abandon for many engineering computations, especially those involving simple boundary layers or fully developed duct flows. Kays and Crawford (1993) and Zukauskas, Slanciauskas, and Karni (1987) show a variety of such computations exhibiting good agreement with experimental data. Launder (1978) cites ubiquitous values of 0.9 for wall-bounded flows and 0.7 for free shear flows.

In a variety of experimental studies summarized by Kays (1994), Pr_T appears to remain fairly constant through the log region of zero-pressure-gradient (ZPG) boundary layers. In fact, Pr_T can be directly computed from the slopes of the nondimensionalized velocity and temperature profiles if one makes the customary assumption of constant turbulent shear stress and heat flux in this region. The evidence supporting a constant Pr_T begins to diminish near the wall or in flows with pressure gradients.

We find conflicting results as we examine Pr_T near the wall. Although many experiments, such as those of Antonia and Kim (1991) and Bagheri,

Strataridakis, and White (1992), show a definite trend of increased Pr_T within the sublayer, we must remember that this is an extremely difficult measurement with relatively large experimental uncertainties. Recent direct numerical simulation (DNS) results presented in Figs. 3.1 and 3.2, for a ZPG boundary layer by Bell and Ferziger (1993) and for a channel flow by Kasagi, Tomita, and Kuroda (1992), show that Pr_T tends toward a constant value near unity through the sublayer. We note that the DNS results were computed at relatively low Reynolds numbers, where the sublayer is relatively thick, which may influence Pr_T in this region.

Figures 3.3 and 3.4 show a clear trend of reduced Pr_T across an adverse-pressure-gradient (APG) boundary layer (Blackwell, Kays, and Moffat, 1972) and enhanced Pr_T across a favorable-pressure-gradient (FPG) boundary layer (Roganov et al., 1984), compared to their respective ZPG cases. Despite this evidence of Pr_T variation with applied pressure gradient, Huang and Bradshaw (1995) show good agreement with experimental temperature profiles using a constant Pr_T. The crucial aspect appears to be the

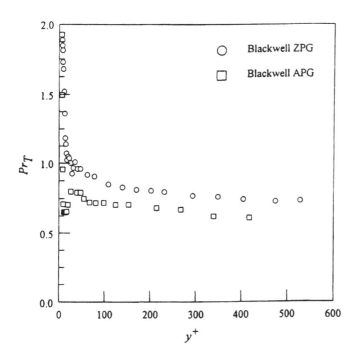

Figure 3.3 Turbulent Prandtl number for an adverse-pressure-gradient boundary layer.

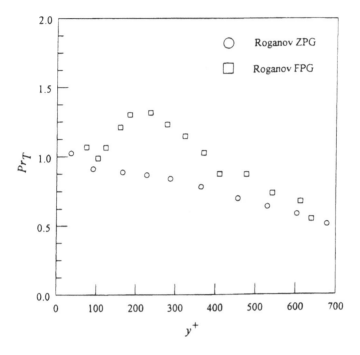

Figure 3.4 Turbulent Prandtl number for a favorable-pressure-gradient boundary layer.

ability of the dynamic turbulence model to model the increased sublayer thickness under FPG and the decreased sublayer thickness under APG.

Variable Pr_T

Since the assumption of a constant Pr_T permits generally accurate predictions of the temperature profiles in simple shear layers, but fails to match the observed variation in Pr_T through the shear layer, we next examine the approach of modeling Pr_T as a variable. Reynolds (1975) published the classic review of over 30 variable Pr_T models over 20 years ago. Two of the most popular modern approaches are discussed below.

The explicit empirical formula of Kays and Crawford (1993, pp. 266–268) depends upon the turbulent Peclet number, an empirical freestream turbulent Prandtl number, and an empirical constant.

$$\mathrm{Pr}_T = \left\{ \frac{0.5}{\mathrm{Pr}_{T_\infty}} + C\,\mathrm{Pe}_T\sqrt{\frac{1}{\mathrm{Pr}_{T_\infty}}} - (C\,\mathrm{Pe}_T)^2 \right.$$

$$\left. \times \left[1 - \exp\left(-\frac{1}{C\,\mathrm{Pe}_T\sqrt{\mathrm{Pr}_{T_\infty}}} \right) \right] \right\}^{-1} \tag{3.75}$$

$$\mathrm{Pe}_T = \frac{\nu_T}{\nu}\,\mathrm{Pr}, \qquad \mathrm{Pr}_{T_\infty} = \left.\frac{\nu_T}{\alpha_T}\right|_\infty = 0.86 \quad \text{(air)}, \qquad C = 0.2 \quad \text{(air)}$$

For Pr near unity, it predicts Pr_T to be nearly constant through the log region, but rapidly increasing toward the wall through the sublayer, thus matching the general trend of the available experimental data, but not the DNS results shown in Figs. 3.1 and 3.2.

The implicit model developed by Yakhot and Orszag (1986) is based on renormalization group theory (RNG) and depends on the turbulent Peclet number and molecular Prandtl number; the constants that appear in the model are derived from fundamental analysis, not from empiricism.

$$\left\{\frac{1/\mathrm{Pr}_{\mathrm{eff}} - 1.18}{1/\mathrm{Pr} - 1.18}\right\}^{0.65} \left\{\frac{1/\mathrm{Pr}_{\mathrm{eff}} + 2.18}{1/\mathrm{Pr} + 2.18}\right\}^{0.35} = \frac{1}{1 + \mathrm{Pe}_T/\mathrm{Pr}} \tag{3.76}$$

$$\mathrm{Pr}_{\mathrm{eff}} = \frac{\mathrm{Pr} + \mathrm{Pe}_T}{1 + \mathrm{Pe}_T/\mathrm{Pr}_T}, \qquad \mathrm{Pe}_T = \frac{\nu_T}{\nu}\,\mathrm{Pr}$$

For Pr near unity, it predicts Pr_T to remain nearly constant throughout the entire shear layer, as shown in Figs. 3.1 and 3.2

Since both the Kays-Crawford and Yakhot-Orszag models predict equally good results for the temperature profile in simple shear layers for air using entirely different approaches, we conclude that specification of a variable Pr_T through the shear layer is of limited utility when Pr is near unity. The variable Pr_T models should be recognized primarily for their ability to predict the higher values of Pr_T associated with low Pr liquid metals.

Generalized Gradient Diffusion Hypothesis (GGDH)

In order to address the obvious shortcomings of specifying an isotropic Pr_T when an anisotropic dynamic turbulence model is used, we now examine a model first described by Daly and Harlow (1970) and later extensively used by Launder (1988) and others. The generalized form of the gradient diffusion hypothesis yields

$$\overline{u_i'\theta'} = -c_\theta \frac{k}{\varepsilon}\,\overline{u_i'u_j'}\,\frac{\partial\bar\theta}{\partial x_j} \tag{3.77}$$

which can be rearranged to form

$$\overline{u_i' \theta'} = -c_\theta \frac{k^2}{\varepsilon} \frac{\overline{u_i' u_j'}}{k} \frac{\partial \overline{\theta}}{\partial x_j} = -c_\theta \nu_T \frac{\overline{u_i' u_j'}}{k} \frac{\partial \overline{\theta}}{\partial x_j} \qquad (3.78)$$

where the ratio $k/\overline{u_i' u_j'}$ can be regarded as an anisotropic turbulent Prandtl number relating an anisotropic thermal diffusivity to an isotropic momentum diffusivity. If the anisotropic turbulent stresses are accurately predicted, this form offers some potential for capturing anisotropy in the turbulent fluxes. Since the turbulent fluxes remain directly dependent on the turbulent stresses, we can view the GGDH as a special form of a specification model. Launder (1993, 1988) has used this model extensively with a differential stress transport model to reasonably capture the anisotropic effects of rotation and curvature as they are manifested in the turbulent stresses.

3.4.2 Eddy Diffusivity Models

Turbulent Diffusivity Concept without Pr_T

We now turn our attention to models involving the computation of actual thermal turbulence variables. Both the one-equation and two-equation thermal turbulence models involve computing a turbulent thermal diffusivity α_T without requiring specification of Pr_T. We assume that α_T, analogous to ν_T, is proportional to some fundamental turbulent scales:

$$\alpha_T \propto [\text{velocity}] \times [\text{length}] \propto [\text{velocity}]^2 \times [\text{time}] \qquad (3.79)$$

The obvious choice for the square of the velocity scale is the turbulent kinetic energy:

$$[\text{velocity}]^2 \propto k \qquad (3.80)$$

It then seems reasonable to propose that the time scale reflects some combination of the time scales for turbulent momentum transport and turbulent thermal transport:

$$[\text{time}] \propto \tau^M \tau_\theta^N \propto \left(\frac{k}{\varepsilon}\right)^M \left(\frac{k_\theta}{\varepsilon_\theta}\right)^N, \qquad M + N = 1 \qquad (3.81)$$

We can now formulate the thermal diffusivity in terms of the momentum diffusivity and the time scale ratio $R = \tau_\theta / \tau$:

$$\alpha_T \propto \frac{k^2}{\varepsilon} (2R)^N \propto \nu_T (2R)^N \qquad (3.82)$$

Commonly proposed values for N include 0.5 (Nagano and Kim, 1988; Sommer, So, and Zhang, 1993; Hattori, Nagano, and Tagawa, 1993), 2.0

(Youssef, Nagano, and Tagawa, 1992; Yoshizawa, 1988), and 1.0 (Schwab and Lakshminarayana, 1994; Schwab and Lakshminarayana, 1995). As R departs from its equilibrium value of 0.5 (Beguier, Dekeyser, and Launder, 1978), we can see that the choice of N will produce clearly different behavior for α_T.

One-Equation Models

The simplest thermal transport model involves only a single partial differential equation governing the transport of the fluctuating temperature variance $k_\theta = \overline{\theta'\theta'}/2$:

$$\frac{\partial k_\theta}{\partial t} + \bar{u}_i \frac{\partial k_\theta}{\partial x_i} = \alpha_T \left(\frac{\partial \bar{\theta}}{\partial x_i}\right)^2 - \varepsilon_\theta + \frac{\partial}{\partial x_i}\left[\left(\alpha + \frac{\alpha_T}{\sigma_{k\theta}}\right)\frac{\partial k_\theta}{\partial x_i}\right] \quad (3.83)$$

We can substitute the following relation:

$$\varepsilon_0 = \frac{\varepsilon k_\theta}{kR} \quad (3.84)$$

but we still have to estimate a value for the time scale ratio, R. Although this model is attractive in its simplicity, and allows separate evolution of the thermal turbulence, the required specification of R remains a severe limiting factor.

Two-Equation Models

The obvious next level of fidelity in modeling involves the computation of ε_θ using a separate transport PDE. While an exact PDE for ε_θ can be derived, it governs the small-scale dissipation process, rather than the large-scale energy transfer process that is more congruent with other modeling assumptions. Therefore, a model PDE is generally used to attempt to empirically balance the convection, production, dissipation, and diffusion terms; the original formulation by Newman, Launder, and Lumley (1981) was later modified by Nagano and Kim (1988) to include the second production term:

$$\frac{\partial \varepsilon_\theta}{\partial t} + \bar{u}_i \frac{\partial \varepsilon_\theta}{\partial x_i} = c_1 f_1 \frac{\varepsilon_\theta}{k_\theta} \alpha_T \left(\frac{\partial \bar{\theta}}{\partial x_i}\right)^2 + c_2 f_2 \frac{\varepsilon_\theta}{k} \nu_T \left(\frac{\partial \bar{u}_j}{\partial x_i}\right)^2$$
$$- c_3 f_3 \frac{\varepsilon_\theta^2}{k_\theta} - c_4 f_4 \frac{\varepsilon \varepsilon_\theta}{k} + \frac{\partial}{\partial x_i}\left[\left(\alpha + \frac{\alpha_T}{\sigma_{\varepsilon\theta}}\right)\frac{\partial \varepsilon_\theta}{\partial x_i}\right] \quad (3.85)$$

We note that the PDE contains multiple production and dissipation terms corresponding to both dynamic and thermal processes.

Schwab and Lakshminarayana (1994, 1995) developed an alternative two-equation model by constructing a PDE for $\tau_\theta = k_\theta/\varepsilon_\theta$ as the second equation in order to exploit the zero-valued wall boundary condition and simple monotonic behavior.

$$
\begin{aligned}
\frac{\partial \tau_\theta}{\partial t} + \bar{u}_i \frac{\partial \tau_\theta}{\partial x_i} = {}& c_1 \frac{\tau_\theta}{k_\theta} \alpha_T \left(\frac{\partial \bar{\theta}}{\partial x_i} \right)^2 + c_2 \frac{\tau_\theta}{k} \nu_T \left(\frac{\partial \bar{u}_j}{\partial x_i} \right)^2 + (c_3 - 1) + c_4 \frac{\tau_\theta}{\tau} \\
& + \frac{\partial}{\partial x_i} \left[\left(\alpha + \frac{\alpha_T}{\sigma_{\tau_\theta}} \right) \frac{\partial \tau_\theta}{\partial x_i} \right] \\
& + \frac{2}{k_\theta} \left(\alpha + \frac{\alpha_T}{\sigma_{\tau_\theta}} \right) \frac{\partial k_\theta}{\partial x_i} \frac{\partial \tau_\theta}{\partial x_i} - \frac{2}{\tau_\theta} \left(\alpha + \frac{\alpha_T}{\sigma_{\tau_\theta}} \right) \frac{\partial \tau_\theta}{\partial x_i} \frac{\partial \tau_\theta}{\partial x_i}
\end{aligned}
\tag{3.86}
$$

The direct gradient transport model used for the first diffusion term requires the two additional cross-diffusion terms for consistency.

Schwab and Lakshminarayana (1995) compared the predictions of four low-Reynolds-number two-equation thermal models for two-dimensional channel flows with heat transfer. The results predicted by the two-equation thermal models show generally improved predictions compared to the constant Pr_T model. Further predictions of more complex flows with heat transfer will be required to show the improved fidelity possible compared to the constant Pr_T model.

Since the near-wall flow structure is extremely important for the computation of surface heat transfer, the near-wall turbulence modeling can be crucial. Although thermal turbulence wall functions can be developed in a manner similar to those developed for the k-ε model, this approach seems disjointed. If a separate thermal turbulence transport model is used to provide improved thermal predictions, it should incorporate low-Reynolds-number effects rather than relying upon wall functions.

3.4.3 Flux Transport Models

Flux transport models, with an individual PDE governing the transport of each turbulent flux, represent the ultimate fidelity in thermal turbulence models. They require the accurate prediction of $\overline{u_i' u_j'}$ from a concurrent turbulent stress transport model, so they have the highest potential for capturing anisotropic interactions between the turbulent fluxes and stresses that we might expect in complex flows subjected to rotation and curvature. Unfortunately, the development of flux transport models greatly lags the development of stress transport models, especially for predictions in the near-wall region. The models of Shih and Lumley (1986), Jones and Musonge (1988), and Lai and So (1990a) have achieved reasonable success, but much more development and validation will be required to bring them to a level suitable for practical applications.

The exact transport equation can be written as

$$\frac{\partial \overline{u_i' \theta'}}{\partial t} + \bar{u}_j \frac{\partial \overline{u_i' \theta'}}{\partial x_j} = - \overline{u_i' u_j'} \frac{\partial \bar{\theta}}{\partial x_j} - \overline{u_j' \theta'} \frac{\partial \bar{u}_i}{\partial x_j} - (\alpha + \nu) \overline{\frac{\partial \theta'}{\partial x_j} \frac{\partial u_i'}{\partial x_j}} - \frac{\overline{\theta'}}{\rho} \frac{\partial p'}{\partial x_i}$$

$$+ \frac{\partial}{\partial x_j} \left[\overline{\nu \theta' \frac{\partial u_i'}{\partial x_j}} + \overline{\alpha u_i' \frac{\partial \theta'}{\partial x_j}} \right] - \frac{\partial}{\partial x_j} \overline{u_i' u_j' \theta'} \qquad (3.87)$$

or more compactly as

$$\frac{D \overline{u_i' \theta'}}{Dt} = P_{i\theta\,1} + P_{i\theta\,2} - \varepsilon_{i\theta} + \Phi_{i\theta} + D_{i\theta}^M + D_{i\theta}^T \qquad (3.88)$$

where the right-hand-side terms represent production by mean temperature gradients, production by mean velocity gradients, molecular dissipation, pressure scrambling, molecular diffusion, and turbulent diffusion. The modeling of the pressure scrambling and molecular dissipation terms is considered to be especially critical in the near-wall region (Lai and So, 1990a).

So, Yuan, and Sommer (1992) found that the prediction of their combined stress transport and flux transport model was inferior to that of their combined two-equation models when compared to DNS results for a channel flow. They attribute this situation to the more extensive validation of the two-equation models compared to the stress and flux transport models, which have only recently been adapted for low-Reynolds-number near-wall flow regions.

If we assume that we can ignore the viscous diffusion and molecular dissipation terms for high-Reynolds-number flow, we can obtain an algebraic form of the transport equation by postulating that the combined convection and turbulent diffusion terms for $\overline{u_i' \theta'}$ are proportional to those for the product $\sqrt{kk_\theta}$, and that the proportionality factor is $\overline{u_i' \theta'}/\sqrt{kk_\theta}$ (Gibson, 1978; Launder, 1988):

$$\frac{D \overline{u_i' \theta'}}{Dt} - D_{i\theta}^T = \frac{\overline{u_i' \theta'}}{\sqrt{kk_\theta}} \left[\frac{D\{\sqrt{kk_\theta}\}}{Dt} - D^T\{\sqrt{kk_\theta}\} \right]$$

$$= \frac{\overline{u_i' \theta'}}{k_\theta} [P_{k\theta} - \varepsilon_\theta] + \frac{\overline{u_i' \theta'}}{k} [P_k - \varepsilon] \qquad (3.89)$$

After substituting back into the $\overline{u_i' \theta'}$ PDE, we obtain an entirely algebraic model for $\overline{u_i' \theta'}$:

$$\frac{\overline{u_i' \theta'}}{k_\theta} [P_{k\theta} - \varepsilon_\theta] + \frac{\overline{u_i' \theta'}}{k} [P_k - \varepsilon] = P_{i\theta\,1} + P_{i\theta\,2} + \Phi_{i\theta} \qquad (3.90)$$

This model requires the computation of k, ε, k_θ, and ε_θ, and is limited to high-Reynolds-number flows by the underlying assumption. It would seem reasonable to combine this algebraic flux model with an algebraic stress model and appropriate low-Reynolds-number two-equation dynamic and thermal models. However, algebraic stress and flux models are notoriously stiff, often requiring extreme under relaxation that may increase the computational effort almost as much as the additional effort that differential flux transport models would require. The current research trend appears to be either to step down to a GGDH model or step up to a differential flux transport model, rather than continuing to develop an algebraic flux model.

3.4.4 Compressibility Effects

Since few researchers have even considered thermal turbulence transport models, it is to be expected that even fewer have extended them to compressible flows. At the current state of development and validation for thermal turbulence models, the incorporation of gross compressibility effects through the variable mean density appears to be a reasonable approach. Sommer, So, and Zhang (1993) have shown some improved results for high-Mach-number boundary layer flows over adiabatic and cooled walls using their compressible two-equation thermal turbulence model compared to a constant turbulent Prandtl number.

3.5 CONCLUSIONS AND RECOMMENDATIONS

The algebraic eddy viscosity models are computationally robust and economical. They are generally adequate for two-dimensional shear flows with mild pressure gradients. For more complex flows involving multiple strain rates, the mixing-length concept derived for a simple shear layer becomes inadequate.

The one-equation models offer some theoretical improvement in the computation of the eddy viscosity due to incorporating nonlocal and historical effects, such as the decay of freestream turbulence, through a transport equation. However, they still suffer from the requirement that the mixing length be specified, and thus offer no practical advantage over the algebraic models, especially in three-dimensional flows, where a plausible length scale can be quite difficult to define.

The two-equation models are nearly as ubiquitous as the algebraic models for practical engineering computations. Extensive testing has been reported, especially for the low-Reynolds-number versions that eliminate wall functions. They are adequate for two-dimensional flows with pressure gradients and two-dimensional recirculating flows. While offering some distinct advantages over the algebraic and one-equation models, the two-equation models become inadequate whenever the underlying assumption of an isotropic eddy viscosity fails in highly three-dimensional flows with multiple strain rates. Coupling a two-equation model to an algebraic stress model offers the potential to capture some anisotropic effects, especially rotation and curvature.

The differential stress models are the only modeling approach capable of directly capturing the anisotropic effects of multiple strain rates upon each individual turbulent stress. Very complex, highly three-dimensional flows may require this level of modeling for accurate predictions, but it remains computationally expensive and will not soon replace lower levels of modeling for most practical engineering computations. Many researchers are currently investigating the difficult pressure-strain term and near-wall effects; thus, extensive testing will be required for any definitive assessment.

The algebraic stress models are an economical alternative to the differential stress models. Either the nonlinear constitutive relation approach or the proportional transport approach can be formulated to include rotation and curvature effects, thus extending the isotropic eddy viscosity limitations of the two-equation models to allow capturing some anisotropic effects. This level of modeling can provide good engineering predictions when isotropic two-equation models are obviously inadequate, and is recommended until the differential stress models become practical for routine computations.

The specification of a constant Pr_T remains extremely useful with isotropic eddy viscosity models, and is quite adequate for boundary layers and fully developed duct flows without anisotropic curvature or rotation effects. Pressure gradient effects are not explicitly captured with a constant Pr_T, but correct temperature profiles can be predicted if the effective momentum sublayer thickness is made a function of the pressure gradient.

The specification of a variable Pr_T as a function of the turbulent Peclet number appears to have little discernible effect, since the proposed functions tend toward a constant value away from the wall for Prandtl numbers near unity. Such models should be primarily recognized for their ability to predict the higher values of Pr_T associated with low-Prandtl-number liquid metal flows.

The generalized gradient diffusion hypothesis offers a rational relation of turbulent fluxes to turbulent stresses and can incorporate anisotropic interactions with turbulent stresses affected by rotation and curvature. It is recommended for use with stress transport models until flux transport models reach sufficient maturity.

The one-equation k_θ transport model eliminates the specification of Pr_T

and allows separate evolution of the thermal turbulence. However, its utility is severely limited by the required specification of the dynamic/thermal time scale ratio in order to relate the thermal dissipation to the momentum dissipation. Since we know that the dissipation plays a crucial role in turbulent transport, and we are attempting to model the thermal transport separately from the momentum transport, we should therefore model the thermal dissipation separately from the momentum dissipation.

The two-equation transport models eliminate the specification of Pr_T and allow separate evolution of the thermal turbulence in a complete manner conceptually parallel to the well-tested two-equation dynamic models. They can be used through the near-wall region and offer the potential to incorporate anisotropic effects when used with algebraic stress and flux transport models. The two-equation level of thermal turbulence modeling is recommended for further exploration and application in complex flows where the physical processes affecting momentum transport and thermal transport may be quite different.

The flux transport models represent the ultimate fidelity in modeling thermal turbulence. They require accurate prediction of the turbulent stresses through a concurrent stress transport model, and have the potential to directly capture the anisotropic interaction effects caused by rotation and curvature. However, the development and validation of flux transport models lag far behind that of stress transport models, especially for incorporating low-Reynolds-number effects in the near-wall region. The algebraic flux model offers a simpler formulation for use with a two-equation thermal model. Until the flux transport PDE models reach sufficient maturity, the generalized gradient diffusion hypothesis model is recommended for use with stress transport models.

NOMENCLATURE

a	local speed of sound
A^+	van Driest constant
k	turbulent kinetic energy; thermal conductivity
k_θ	turbulent temperature variance
l	length scale
M_T	turbulent Mach number: $M_T = \sqrt{2k}/a$
p	pressure
Pe_T	turbulent Peclet number: $Pe_T = Pr\, v_T/v$
Pr	Prandtl number: $Pr = v/\alpha$
Pr_T	turbulent Prandtl number: $Pr_T = v_T = \alpha_T$
q	instantaneous value of flow variable

$\bar{q}$	mean component of flow variable (Reynolds averaging)
$\tilde{q}$	mean component of flow variable (Favre averaging)
q'	fluctuating component of flow variable (Reynolds averaging)
q''	fluctuating component of flow variable (Favre averaging)
q_i	heat flux vector
R	turbulent time scale ratio: $R = \tau_\theta/\tau$
R_{ij}	Reynolds stress tensor
Re_k	turbulent Reynolds number: $\text{Re}_k = \sqrt{k}y/v$
Re_T	turbulent Reynolds number: $\text{Re}_T = k^2/v\varepsilon$
$\text{Re}_{\delta*}$	displacement thickness Reynolds number: $\text{Re}_{\delta*} = u_\infty \delta*/v$
t, T	time
u, v, w	Cartesian velocity components
u_τ	friction velocity: $u_\tau = \sqrt{\tau_0/\rho}$
v_0	transpiration velocity
x, y, z	Cartesian space coordinates
y^+	sublayer-scaled wall-normal distance: $y^+ = u_\tau y/v$

Greek

α	thermal diffusivity
α_T	turbulent thermal diffusivity
γ	specific heat ratio
δ	boundary layer thickness
δ_{ij}	Kronecker delta function
$\delta*$	boundary layer displacement thickness
δ_v^*	boundary layer velocity thickness
ε	dissipation rate of turbulent kinetic energy
ε_{ijk}	permutation tensor
ε_θ	dissipation rate of turbulent temperature variance
θ	temperature
κ	von Karman constant
λ	Escudier constant
μ	dynamic viscosity
v	kinematic viscosity: $v = \mu/\rho$
v_T	turbulent viscosity
ρ	density
τ	turbulent time scale: $\tau = k/\varepsilon$
τ_θ	turbulent thermal time scale: $\tau_\theta = k_\theta/\varepsilon_\theta$
τ_{ij}	viscous stress tensor
τ_0	wall shear stress
Φ	dissipation function
ω	vorticity; specific dissipation rate: $\omega = \varepsilon/k$

REFERENCES

Anderson DA, Tannehill JC, Pletcher RH. Computational Fluid Mechanics and Heat Transfer. Hemisphere, 1984.

Antonia RA, Kim J. Turbulent Prandtl number in the near-wall region of a turbulent channel flow. Int J Heat Mass Transfer 34: 1905–1908, 1991.

Bagheri N, Strataridakis CJ, White BR. Measurements of turbulent boundary layer Prandtl numbers and space-time temperature correlations. AIAA J 30: 35–42, 1992.

Baldwin BS, Barth TJ. A one-equation turbulence transport model for high Reynolds number wall-bounded flows. NASA TM-102847, 1990.

Baldwin BS, Lomax H. Thin layer approximation and algebraic model for separated turbulent flows. AIAA 78-257, 1978.

Beguier C, Dekeyser I, Launder BE. Ratio of scalar and velocity dissipation time scales in shear flow turbulence. Phys Fluids 21: 307–310, 1978.

Bell DM, Ferziger JH. Turbulent boundary layer DNS with passive scalars. In: So RMC, Speziale CG, Launder BE, eds. Near-Wall Turbulent Flows. Elsevier, 1993, pp 327–336.

Benocci C, Modeling of turbulent heat transport. VKI Technical Memorandum 47, 1991.

Blackwell BF, Kays WM, Moffat RJ. The turbulent boundary layer on a porous plate: an experimental study of the heat transfer behavior with adverse pressure gradients. NASA CR-130291, 1972.

Cebeci T, Smith AMO. Analysis of Turbulent Boundary Layers, Applied Mathematics and Mechanics. Vol. 15. Academic Press, 1974.

Chien K-Y. Predictions of channel and boundary-layer flows with a low Reynolds-number turbulence model. AIAA J 20: 33–38, 1982.

Chou PY. On the velocity correlations and the solution of the equations of turbulent fluctuation. Quart J Appl Math 3: 38–54, 1945.

Daly BJ, Harlow FH. Transport equations in turbulence. Phys Fluids 13: 2634–2649, 1970.

Demuren AO. Calculation of turbulent flow in complex geometries with a second-moment closure model. ASME FED-Vol. 94. In: Bower WM, Morris MJ, Saming M, eds. Forum on Turbulent Flows—1990. 1990, pp 163–167.

Demuren AO, Sarkar S. Perspective: systematic study of Reynolds stress closure models in the computations of plane channel flows. ASME J Fluids Eng 115: 5–12 1993a.

Demuren AO, Sarkar S. Study of second-moment closure models in computations of turbulent shear flows. In: Rodi W, Martelli F, eds. Engineering Turbulence Modeling and Experiments 2. Elsevier, pp 1993b, 53–62.

Durbin PA. Near-wall turbulence closure modeling without damping functions. Theoret Comput Fluid Dynamics 3: 1–13, 1991.

Durbin PA. A Reynolds stress model for near-wall turbulence. J Fluid Mech 249: 465–498, 1993.

Gatski TB, Speziale CG. On explicit algebraic stress models for complex turbulent flows. J Fluid Mech 254: 59–78, 1993.

Germano M, Piomelli U, Moin P, Cabot WH. A dynamic subgrid-scale eddy viscosity model. Phys Fluids A 3: 1760–1765, 1991.

Gibson MM. An algebraic stress and heat-flux model for turbulent shear flow with streamline curvature. Int J Heat Mass Transfer 21: 1609–1617, 1978.

Granville PS. Baldwin-Lomax factors for turbulent boundary layers in pressure gradients. AIAA J 25: 1624–1627, 1987.

Hanjalic K. Advanced turbulence closure models: a view of current status and future prospects. Int J Heat Fluid Flow 15: 178–203, 1994.

Harlow FH, Nakayama PI. Transport of turbulence energy decays rate. University of California Report LA-3854, 1968.

Hattori H, Nagano Y, Tagawa M. Analysis of turbulent heat transfer under various thermal conditions with two-equation models. In: Rodi W, Martelli F, eds. Engineering Turbulence Modeling and Experiments 2. Elsevier, 1993, pp 43–52.

Huang PG, Bradshaw P. Law of the wall for turbulent flows in pressure gradients. AIAA J 33: 624–632, 1995.

Jones WP, Launder BE. The prediction of laminarization with a two-equation model of turbulence. Int J Heat Mass Transfer 15: 301–314, 1972.

Jones WP, Musonge P. Closure of the Reynolds stress and scalar flux equations. Phys Fluids 31: 3589–3604, 1988.

Kasagi N, Tomita Y, Kuroda A. Direct numerical simulation of passive scalar field in a turbulent channel flow. ASME J Heat Transfer 114: 598–606, 1992.

Kays WM. Turbulent Prandtl number—where are we?, ASME J Heat Transfer 116: 284–295, 1994.

Kays WM, Crawford ME. Convective Heat and Mass Transfer. 3rd ed. McGraw-Hill, 1993.

Kays WM, Moffat RJ. The behavior of transpired turbulent boundary layers. In: Studies in Convection: Theory, Measurement, and Applications, Vol. 1. Academic Press, 1975, pp 213–319.

Kirtley KR. An algebraic RNG-based turbulence model for three-dimensional turbomachinery flows. AIAA J 30: 1500–1506, 1992.

Lai YG, So RMC. Near-wall modeling of turbulent heat fluxes. Int J Heat Mass Transfer 33: 1429–1440, 1990a.

Lai YG, So RMC. On near-wall turbulent flow modeling, J Fluid Mech 221: 641–673, 1990b.

Lakshminarayana B. Turbulence modeling for complex shear flows. AIAA J 24: 1900–1917, 1986.

Lam CHG, Bremhorst KA. Modified form of the k-ε model for predicting wall turbulence. ASME J Fluids Eng 103: 456–460, 1981.

Lang NJ, Shih T-H. A critical comparison of two-equation turbulence models. NASA TM-105237, 1991.

Launder BE. Heat and mass transport. In: Bradshaw P, ed. Turbulence. 2nd ed.

Topics in Applied Physics. Vol. 12. Springer-Verlag, 1978, Chap. 6, pp 231–287.

Launder BE. A generalized algebraic stress transport hypothesis. AIAA J 20: 436–437, 1982.

Launder BE. On the computation of convective heat transfer in complex turbulent flows. ASME J Heat Transfer 110: 1112–1128, 1988.

Launder BE. Second-moment closure and its use in modeling turbulent industrial flows. Int J Numer Meth Fluids. 9: 963–985, 1989.

Launder BE. Modeling convective heat transfer in complex turbulent flows. In: Rodi W, Martelli F, eds. Engineering Turbulence Modeling and Experiments 2. Elsevier, 1993, pp 3–22.

Launder BE, Sharma BI. Application of the energy dissipation model of turbulence to the calculation of flow near a spinning disk. Lett Heat Mass Transfer 1: 131–138, 1974.

Launder BE, Shima N. Second-moment closure for the near-wall sublayer: development and application. AIAA J 27: 1319–1325, 1989.

Launder BE, Reece GJ, Rodi W. Progress in the development of a Reynolds stress turbulence model. J Fluid Mech 68: 537–566, 1975.

Lien FS, Leschziner MA. A general non-orthogonal collocated finite volume algorithm for turbulent flow at all speeds incorporating second-moment turbulence-transport closure. Part 1: Computational implementation. Comput Meth Appl Mech Eng 114: 123–148, 1994a.

Lien FS, Leschziner MA. A general non-orthogonal collocated finite volume algorithm for turbulent flow at all speeds incorporating second-moment turbulence-transport closure. Part 2: Application. Comput Meth Appl Mech Eng 114: 149–167, 1994b.

Lumley JL. Toward a turbulent constitutive relation. J Fluid Mech 41: 413–434, 1970.

Lumley JL. Turbulence modeling. ASME J Appl Mech 50: 1097–1103, 1983.

Lumley JL. Some comments on turbulence. Phys Fluids A 4: 2: 203–211, 1992.

Mansour NN, Kim J, Moin P. Near-wall k-ε turbulence modeling. AIAA J 27: 1068–1073, 1989.

Martinelli L, Yakhot V. RNG-based turbulence transport approximations with applications to transonic flows. AIAA 89-1950-CP, 1989.

Michelassi V, Shih T-H. Elliptic flow computation by low Reynolds number two-equation turbulence models. NASA TM-105376, 1993.

Morkovin MV. Effects of compressibility on turbulent flow. In: Favre A, ed. The Mechanics of Turbulence. Gordon and Breach, 1962, pp 367–380.

Myong HK, Kasagi N. A new approach to the improvement of k-ε turbulence model for wall-bounded shear flows. AMSE Int J 33: 63–72, 1990.

Nagano Y, Kim C. A two-equation model for heat transport in wall turbulent shear flows. ASME J Heat Transfer 110: 583–589, 1988.

Newman GR, Launder BE, Lumley JL. Modeling the behaviour of homogeneous scalar turbulence. J Fluid Mech 111: 217–232, 1981.

Patel VC, Rodi W, Scheuerer G. Turbulence models for near-wall and low Reynolds number flows: a review. AIAA J 23: 1308–1319, 1985.

Pollard A, Martinuzzi R. Comparative study of turbulence models in predicting turbulent pipe flow. Part II: Reynolds stress and k-ε models. AIAA J 27: 1714–1721, 1989.

Pope SB. A more general effective-viscosity hypothesis. J Fluid Mech 72: 331–340, 1975.

Rai MM, Moin P. Direct simulations of turbulent flow using finite-difference schemes. J Comput Phys 96: 15–53, 1991.

Reynolds AJ. The prediction of turbulent Prandtl and Schmidt numbers. Int J Heat Mass Transfer 18: 1055–1069, 1975.

Rodi W. A new algebraic relation for calculating the Reynolds stresses. ZAMM 56: T219–T221, 1976.

Rodi W, Mansour NN. Low Reynolds number k-ε modeling with the aid of direct simulation data. J Fluid Mech 250: 509–524, 1993.

Rogallo RS, Moin P. Numerical simulation of turbulent flows. Annu Rev Fluid Mech 16: 99–137, 1984.

Roganov PS, Zabolotsky VP, Shishov EV, Leontiev AI. Some aspects of turbulent heat transfer in accelerated flows on permeable surfaces. Int J Heat Mass Transfer 27: 1251–1259, 1984.

Rubinstein R, Barton JM. Nonlinear Reynolds stress models and the renormalization group. Phys Fluids A 2: 1472–1476, 1990.

Rubinstein R, Barton JM. Renormalization group analysis of the Reynolds stress transport equation. NASA TM-105588, 1992.

Sarkar S, Erlebacher G, Hussaini MY, Kreiss HO. The analysis and modeling of dilatational terms in compressible homogeneous turbulence. NASA CR-181959, 1989.

Schwab JR, Lakshminarayana B. Dynamic and thermal turbulent time scale modeling for homogeneous shear flows. (NASA TM-106635), ASME FED-Vol. 184. In: Donovan JF, Dutton JC, eds Boundary Layer and Free Shear Flows. 1994, pp 75–86.

Schwab JR, Lakshminarayana B. Dynamic and thermal turbulent time scale modeling for wall-bounded shear flows. ASME HTD-Vol. 318. In: Anand NK, Amano RS, Armaly BF, eds. Heat Transfer in Turbulent Flows. 1995, pp 111–118.

Shih T-H, Lumley JL. Influence of timescale ratio on scalar flux relaxation: modeling Sirivat & Warhaft's homogeneous passive scalar fluctuations. J Fluid Mech 162: 211–222, 1986.

Shih T-H, Zhu J, Lumley JL. Modeling of wall-bounded complex flows and free shear flows (NASA TM-106513), ASME FED-Vol. 184. In: Donovan JF, Dutton JC, eds. Boundary Layer and Free Shear Flows. 1994, pp 105–112.

Shima N. A Reynolds-stress model for near-wall and low-Reynolds-number regions. ASME J Fluids Eng 110: 38–44, 1988.

Shima N. Prediction of turbulent boundary layers with a second-moment closure. Part 1. Effects of periodic pressure gradient, wall transpiration, and free-stream turbulence. ASME J Fluids Eng 115: pp. 56–63, 1993a.

Shima N. Prediction of turbulent boundary layers with a second-moment closure. Part 2. Effects of streamwise curvature and spanwise rotation. ASME J Fluids Eng 115: 64–69, 1993b.

So RMC, Yoo GJ. On the modeling of low-Reynolds-number regions. NASA CR-3994, 1986.

So RMC, Yuan SP, Sommer TP. A hierarchy of near-wall closures for turbulent heat transfer. Trends Heat, Mass, Momentum Transfer 2: 203–221, 1992.

So RMC, Lai YG, Zhang HS, Hwang BC. Second-order near-wall turbulence closures: a review. AIAA J 29: 1819–1835, 1991.

Sommer TP, So RMC, Zhang HS. Near-wall variable-Prandtl-number turbulence model for compressible flows. AIAA J 31: 27–35, 1993.

Spalart PR, Allmaras SR. A one-equation turbulence model for aerodynamic flows. AIAA 92–439, 1992.

Speziale CG, On nonlinear k-l and k-ε models of turbulence. J Fluid Mech 178: 459–475, 1987.

Speziale CG. Analytical methods for the development of Reynolds-stress closures in turbulence. Ann Rev Fluid Mech 23: 107–157.

Speziale CG. A review of Reynolds stress models for turbulent shear flows. NASA CR-195054, 1995.

Speziale CG, Gatski TB. Assessment of second-order closure models in turbulent shear flows. AIAA J 32: 2113–2115, 1994.

Speziale CG, Thangam S. Analysis of an RNG-based turbulence model for separated flows. NASA CR-189600, 1992.

Steffen CJ. A critical comparison of several low Reynolds number k-ε turbulence models for flow over a backward-facing step. NASA TM-106173, 1993.

Taulbee DB. An improved algebraic Reynolds stress model and corresponding nonlinear stress model. Phys Fluids A 4: 2555–2561, 1992.

Thangam S, Speziale CG. Turbulent flow past a backward facing step: a critical evaluation of two-equation models. AIAA J 30: 1314–1320, 1992.

Van Driest ER. On turbulent flow near a wall. J Aeronaut Sci 23: 1007–1011, 1956.

Wilcox DW. Dilatation-dissipation corrections for advanced turbulence models. AIAA J 30: 2639–2646, 1992.

Wilcox DW. Turbulence Modeling for CFD. DCW Industries, Inc, 1993a.

Wilcox DW. Application of low Reynolds number two-equation turbulence models to high Reynolds number flows. In: So RMC, Speziale CG, Launder BE, eds. Near-Wall Turbulent Flows. Elsevier, 1993b, pp 155–164.

Yakhot V, Orszag SA. Renormalization group analysis of turbulence. I: Basic theory. J Sci Comput 1: 3–51, 1986.

Yoshizawa A. Statistical analysis of the deviation of the Reynolds stress from its eddy-viscosity representation. Phys Fluids 27: 1377–1387, 1984.

Yoshizawa A. Statistical modeling of passive-scalar diffusion in turbulent shear flows. J Fluid Mech 195: 541–555, 1988.

Youssef MS, Nagano Y, Tagawa M. A two-equation heat transfer model for predicting turbulent thermal fields under arbitrary wall thermal conditions. Int J Heat Mass Transfer 35: 3095–3104, 1992.

Zeman O. Dilatational dissipation: the concept and application in modeling compressible mixing layers. Phys Fluids A 2: 178–188, 1990.

Zhang HS, So RMC, Gatski TB, Speziale CG. A near-wall second-order closure for compressible turbulent flows. In: So RMC, Speziale CG, Launder BE, eds. Near-Wall Turbulent Flows. Elsevier, 1993, pp 209–218.

Zukauskas A, Slanciauskas A, Karni J. Heat Transfer in Turbulent Fluid Flows. Hemisphere, 1987.

4

Grid Generation

Vijay K. Garg

AYT Corporation/NASA Lewis Research Center, Cleveland, Ohio

Philip C. E. Jorgenson

NASA Lewis Research Center, Cleveland, Ohio

4.1 INTRODUCTION

The numerical solution of partial differential equations requires some discretization of the field into a collection of points or elemental volumes (cells). The differential equations are approximated as a set of finite difference or finite element or finite volume equations on this collection, and the resulting set of algebraic equations is then solved for the discrete values of the variables on this grid. The algebraic equations can be obtained on an organized or an unorganized distribution of points or cells. The former leads to a *structured* grid, while the latter leads to an *unstructured* grid. Though unstructured grids are increasingly in use these days, they cannot resolve the gradients near the walls, and are thus unable to provide heat transfer results. We will devote most of the chapter to the generation of structured grids, leaving the discussion on unstructured grid generation to the last section. The reader is referred to Thompson et al. (1982, 1985) for a comprehensive review of structured grid generation techniques, and to Choo (1995) and Carey (1997) for recent updates including unstructured grid generation.

The organized discretization of the field has been handicapped by two problems. The first stems from the fact that most fields of interest are arbitrarily shaped regions, and thus accurate application of the boundary conditions requires that the discretization conform to the boundaries. This has led to the use of boundary-fitted coordinate systems, as we will see. The second problem in the numerical solution of partial differential equations is the resolution of grid points in regions of the field where very high changes in the solution occur. These high gradient regions are not known a priori, and therefore presetting of a fine mesh in these regions is out of question. Lack of prior information about these high gradients can render the grid useless since it is not able to resolve these high gradient regions satisfactorily. This makes the numerical simulation itself a waste since important physical phenomena do occur in the high gradient regions. Due to nonlinear phenomena associated with various physical processes such as boundary layer, turbulence, shock formation, and others, there is a tendency for the most important physical processes to occur in high gradient regions. These regions may or may not be associated with solid boundaries, and can also move in time. Thus, the problem of accurate resolution of high gradient regions is important not only from truncation error considerations but also from the physical point of view. The need to have an accurate physical simulation of these high gradient regions, and the lack of a priori information about these regions have led to the development of *adaptive grid* techniques. This technique automatically controls the grid size depending upon the driving function—usually the gradient of the solution—making the grid very fine in high gradient regions and relatively coarse in low gradient regions.

The generation of grids for one-dimensional (1-D) problems is straightforward. There are many functions or other methods that can be used to generate a suitable 1-D grid. Moreover, the problem of complex boundaries does not arise in 1-D problems. That is why most work in structured grid generation has been done in two dimensions. Grid generation in three dimensions is very complicated and is generally achieved by stacking several two-dimensional (2-D) grids in the third direction. The general problem of grid generation is that of determining the mapping that transforms the grid points from the complex physical domain into the regular (generally rectangular in 2-D) computational domain. Such a mapping needs to satisfy several criteria, some of which are:

1. The mapping must be one-to-one.
2. The grid lines (surfaces in three dimensions) should be smooth to provide continuous transformation derivatives.
3. Grid points should be closely spaced in the physical domain where large numerical errors are expected.
4. Excessive grid skewness should be avoided. It has been shown (Raithby, 1976) that grid skewness exaggerates truncation errors.

For an arbitrary three-dimensional domain, it is not possible to generate an orthogonal grid. However, a grid that is almost orthogonal near the boundaries facilitates the application of boundary conditions. Although strict orthogonality is not necessary, the accuracy deteriorates if the departure from orthogonality is too large (Thompson et al., 1985). The implementation of turbulence models is more reliable with near-orthogonality at the boundary, since information on local boundary normals is usually required in such models. Algorithms based on the parabolized Navier-Stokes equations require that coordinate lines approximate the flow streamlines and the lines normal thereto, especially near solid boundaries. It is thus better in general if grid lines can be nearly orthogonal to the boundaries.

Structured grid generation techniques can be classified into two categories.

1. Partial differential equation methods
2. Algebraic methods

In the former, the partial differential system may be elliptic, parabolic, or hyperbolic. Included in the elliptic systems are both the conformal and the quasi-conformal mappings, the former being orthogonal. Orthogonal systems, not necessarily conformal, may be generated from hyperbolic or elliptic systems. Some procedures are designed to produce nearly orthogonal coordinates. The algebraic procedures include simple normalization of boundaries, transfinite interpolation from boundary surfaces, the use of intermediate interpolating surfaces, and other related techniques.

4.1.1 Metric Tensor and Physical Features of the Coordinate Transformation

The generalized coordinates were introduced in section 1.11 when the governing equations were expressed in that coordinate system. In order to understand the following discussion better, let us link the generalized coordinates to the orthogonal and conformal coordinates. To do so, we introduce the metric tensor g_{ij}, which is related to the Jacobian matrix J in Eq. (1.95). We represent the physical domain by Cartesian coordinates x^i ($\equiv x, y, z$), $i = 1, 3$, and the computational domain by generalized coordinates $\xi^i (\equiv \xi, \eta, \zeta)$, $i = 1,3$.

The small distance Δs between two points in physical space can be written in terms of the coordinate displacements as

$$\Delta s^2 = \sum_{k=1}^{3} \Delta x^k \, \Delta x^k \tag{4.1}$$

The physical coordinate displacements Δx^k can be related to displacements in the generalized coordinates $\Delta \xi^i$ by

$$\Delta x^k = \frac{\partial x^k}{\partial \xi^i} \, \Delta \xi^i \qquad \text{(summation over } i \text{ implied)} \tag{4.2}$$

Thus the small distance Δs related to generalized coordinates becomes

$$\Delta s^2 = \sum_{k=1}^{3} \left(\frac{\partial x^k}{\partial \xi^i} \Delta \xi^i \right) \left(\frac{\partial x^k}{\partial \xi^j} \Delta \xi^j \right)$$

$$= g_{ij} \Delta \xi^i \Delta \xi^j \quad \text{(summation over } i \text{ and } j \text{ implied)} \quad (4.3)$$

where

$$g_{ij} = \sum_{k=1}^{3} \frac{\partial x^k}{\partial \xi^i} \frac{\partial x^k}{\partial \xi^j} \quad (4.4)$$

The metric tensor g_{ij}, discussed at length by Aris (1962, p. 142), relates the distance Δs to small changes in the generalized coordinates $\Delta \xi^i$. The metric tensor expressed in matrix form is related to the inverse Jacobian by

$$g = (J^{-1})^T J^{-1}$$

Taking determinants yields

$$|g|^{1/2} = |J^{-1}| \quad (4.5)$$

The metric tensor g_{ij} can be interpreted in terms of physical features of the computational grid. We will demonstrate this in two dimensions here. The three-dimensional forms are given by Kerlick and Klopfer (1982). According to Fig. 4.1, the grid cell area is given by

$$\text{Area} = |g|^{1/2} \Delta \xi \Delta \eta \quad (4.6)$$

which, from (4.5), gives a physical interpretation of the inverse Jacobian determinant. The physical orientation of the computational grid (tangent to a ξ-coordinate line) relative to the x-axis is given by the direction cosine

$$\cos \alpha = \frac{x_\xi}{(g_{11})^{1/2}} \quad (4.7)$$

where $x_\xi \equiv \partial x / \partial \xi$. The grid aspect ratio AR is given by the ratio of the magnitude of the tangent vectors (with $\Delta \xi = \Delta \eta$)

$$\text{AR} = \frac{\Delta s_\eta}{\Delta s_\xi} = \left(\frac{g_{22}}{g_{11}} \right)^{1/2} \quad (4.8)$$

The local distortion of the grid is determined by the angle θ between the ξ and η-coordinate lines, given by

$$\cos \theta = \frac{g_{12}}{(g_{11} g_{22})^{1/2}} \quad (4.9)$$

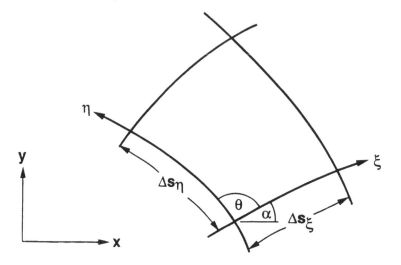

Figure 4.1 Physical features of the 2-D computational grid.

In two dimensions it is also convenient to write (4.4) in matrix form as

$$
g = \begin{bmatrix} (x_\xi^2 + y_\xi^2) & (x_\xi x_\eta + y_\xi y_\eta) \\ (x_\xi x_\eta + y_\xi y_\eta) & (x_\eta^2 + y_\eta^2) \end{bmatrix}
$$
$$
= \frac{1}{|J|^2} \begin{bmatrix} (\eta_x^2 + \eta_y^2) & -(\xi_x \eta_x + \xi_y \eta_y) \\ -(\xi_x \eta_x + \xi_y \eta_y) & (\xi_x^2 + \xi_y^2) \end{bmatrix} \tag{4.10}
$$

where $|J|$ is the determinant of the Jacobian matrix in (1.95).

4.1.2 Orthogonal and Conformal Coordinate Systems

While the use of generalized coordinates permits quite arbitrary domains to be analyzed, it is well known that the accuracy of the solution deteriorates as the grid gets distorted. For high accuracy the grid should be orthogonal or near orthogonal. For orthogonal coordinate systems, some of the transformation terms disappear and the governing equations get simplified. If the coordinate system is conformal, the governing equations simplify further. Unfortunately, the use of orthogonal or conformal coordinate systems is restricted to relatively simple physical domains, together with some limitations on the disposition of the grid points. For a two-dimensional *orthogonal* grid, we must have $\theta = 90°$ (Fig. 4.1), or, from (4.9),

$$
g_{12} = x_\xi x_\eta + y_\xi y_\eta = 0 \tag{4.11}
$$

In three dimensions the orthogonality condition becomes

$$g_{ij} = 0, \qquad i \neq j \tag{4.12}$$

Thus the metric tensor contains only diagonal terms g_{ii} for an orthogonal coordinate system. In such a case, it is the convention to define

$$h_i = (g_{ii})^{1/2}, \qquad i = 1, 3 \quad \text{(no summation)} \tag{4.13}$$

The terms h_i can be interpreted as scale factors since a small change in the ξ^i coordinate, on an orthogonal grid, produces a scaled overall movement given by

$$\Delta s = h_i \, \Delta \xi^i \qquad \text{(no summation)} \tag{4.14}$$

In two dimensions the condition of orthogonality implies, from (4.10), that

$$x_\eta = -y_\xi \text{AR} \qquad \text{and} \qquad y_\eta = x_\xi \, \text{AR} \tag{4.15}$$

where AR is the grid aspect ratio (4.8).

If AR = 1, Eq. (4.15) reduces to the Cauchy-Riemann conditions, and the grid is *conformal*. If AR is constant, but not unity, a simple scaling of ξ or η will produce a related conformal coordinate system.

The level of complexity in the governing equations in the various coordinate systems has been demonstrated by Fletcher (1991, Chap. 12). If the physical domain is simple enough and a grid can be generated that is able to concentrate grid points in regions of large gradients, conformal grids are to be preferred since they imply fewer terms in the governing equations, and thus, a more economical algorithm. In three dimensions, however, completely conformal or even orthogonal grids are not usually possible, leading to the use of generalized coordinates. The *strongly conservative* form of the governing equations is given by (1.98). Leaving the task of proper numerical evaluation of metric terms to the reader (see, for example, Anderson et al., 1984, and Fletcher, 1991), we turn to the problem of grid generation.

4.2 GRID GENERATION VIA PARTIAL DIFFERENTIAL EQUATION SOLUTION

For a conformal, orthogonal or generalized grid, the transformation between the physical and computational domains can be obtained by solving a partial differential equation as described in this section.

4.2.1 Conformal Mapping: General Considerations

Conformal mapping has traditionally been used to obtain potential flow solutions about relatively complicated shapes (Milne-Thomson, 1968) given the flow behavior about a simple shape, such as a unit circle. Here, of course, conformal mapping is used as a grid generation technique, with no restriction on the type of flow. If the generated grid lines can be chosen to coincide with the streamlines of an equivalent potential flow problem, the stability of the computational method used to solve the more general flow problem is improved. When a conformal grid or an orthogonal grid using a conformal mapping followed by independent algebraic stretching is used, a by-product is the solution for the two-dimensional incompressible flow. This incompressible solution may often be used with a compressibility transformation as an accurate first guess for an iterative solution of the compressible problem, reducing the computer time needed for convergence of the iterative process.

For a conformal mapping, the relation between the physical (x, y) and the computational (ξ, η) domains, in two dimensions, is given by

$$\begin{bmatrix} dx \\ dy \end{bmatrix} = \begin{bmatrix} h \cos \alpha & -h \sin \alpha \\ h \sin \alpha & h \cos \alpha \end{bmatrix} \begin{bmatrix} d\xi \\ d\eta \end{bmatrix} \tag{4.16}$$

The scale factor h is related to the components of the metric tensor by $h = g_{11}^{1/2} = g_{22}^{1/2}$. The angle α is the angle between the tangent to the ξ-coordinate line and the x-axis [Fig. 4.1 and Eq. (4.7)]. Once h and α are known, $x_\xi (= h \cos \alpha)$, etc., follow from (4.16). For a conformal transformation, the computational grid (ξ, η) is linked to the physical grid (x, y) by the Laplace equations

$$\xi_{xx} + \xi_{yy} = 0, \qquad \eta_{xx} + \eta_{yy} = 0 \tag{4.17}$$

and the Cauchy-Riemann conditions $\xi_x = \eta_y$ and $\xi_y = -\eta_x$. Since the Laplace equations are linear, it is possible to construct the solutions $\xi(x, y)$ and $\eta(x, y)$ by superposition of simple solutions and by complex transformations (Milne-Thomson, 1968). Using complex variables, $z = x + iy$ and $\zeta = \xi + i\eta$, a conformal transformation can be expressed symbolically as $Z = F(\zeta)$, or as

$$dZ = H \, d\zeta \qquad \text{or} \qquad Z = \int H \, d\zeta \tag{4.18}$$

where

$$H = h \, \exp(i\alpha) = h(\cos \alpha + i \sin \alpha) \tag{4.19}$$

Thus, from Eq. (4.16), H contains the transformation parameters x_ξ, etc.

Since a conformal mapping is basically a surface-to-surface correspondence, the complete grid generation technique can be considered to consist of two stages:

1. The construction of a single mapping or a sequence of mappings to obtain the correspondence between *boundary* points in the physical and computational domains.
2. The generation of interior points given the boundary correspondence from the first stage. Elliptic techniques can be used for generating the interior points.

Two approaches will be described. The first approach (section 4.2.2) is appropriate to streamlined shapes like airfoils or turbine blades which can be transformed via a sequence of mappings into a rectangle. The second approach (section 4.2.3) uses a one-step mapping based on a Schwarz-Christoffel transformation of a polygon with N straight sides into a straight line. Various modifications of the Schwarz-Christoffel transformation are now available for quite general shapes; some of these are described in section 4.2.3. Ives (1982) discusses some general techniques that can be used to create new conformal mappings and some restrictions to keep in mind while doing so. While conformal mapping is basically a surface technique, it can be used as one component of a three-dimensional grid generation system. Some examples of three-dimensional grids are given in Jameson (1974), Jameson and Caughey (1977), Grossman and Siclari (1980), and in Ives and Menor (1981).

4.2.2   Sequential Conformal Mapping

For a complicated mapping of the form $Z = F(\zeta)$, it is easier to break up the process into an equivalent sequence of simpler mappings. For instance, the von Karman-Trefftz transformation (von Karman and Trefftz, 1918)

$$\frac{\zeta' - a}{\zeta' - b} = \left(\frac{Z - A}{Z - B}\right)^k \tag{4.20}$$

can be restated as the sequence

$$\omega = \frac{Z - A}{Z - B} \tag{4.21a}$$

$$\psi = \omega^k \tag{4.21b}$$

$$\zeta' = \frac{a - b\psi}{1 - \psi} \tag{4.21c}$$

where a, b in the ζ'-plane and A and B in the Z-plane are chosen to suit the geometry considered. This transformation maps an airfoil in the Z-plane into a near circle in the ζ'-plane. For the airfoil shown in Fig. 4.2, the

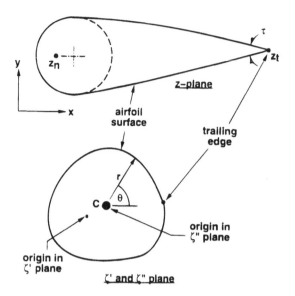

Figure 4.2 Sequential mapping for an airfoil.

parameters in (4.20) or (4.21) are chosen as

$$A = Z_t, \quad B = Z_n, \quad a = -b = \frac{k(Z_n - Z_t)}{2}, \quad \frac{1}{k} = 2 - \frac{\tau}{\pi} \tag{4.22}$$

where Z_t corresponds to the location of the trailing edge of the airfoil, Z_n is a point midway between the nose of the airfoil and its center of curvature, and τ is the trailing edge included angle. The transformation is singular at Z_t and Z_n. The choice of parameters given by (4.22) maps the airfoil in the Z-plane to a near circle in the ζ'-plane (Fig. 4.2), approximately centered at C. It is obvious that the angle τ has expanded to 180° in the ζ'-plane.

In the form (4.21), the bilinear transformations, Eqs. (4.21a) and (4.21c), require no root selection, while Eq. (4.21b) introduces the complication that multiple values of ψ exist for each value of ω. Strategies for selecting the correct value of ψ are discussed by Ives (1982). Broadly the transformation is tracked to the point of interest from a "safe" point such as upstream infinity in the physical Z-plane.

The near circle in the ζ'-plane is transformed to a near circle in the ζ''-plane centered at the origin by

$$\zeta'' = \zeta' - C \tag{4.23}$$

This mapping improves the convergence of the Theodorsen-Garrick transformation, described next, from the near circle in the ζ''-plane to a unit circle in the ζ-plane.

Near Circle–to–Circle Mapping

If an orthogonal grid is desired, the mapping of a near circle in the ζ''-plane to a unit circle in the ζ-plane is often chosen as one step of the mapping process. Although there are a number of ways to accomplish this mapping (including panel techniques, Schwarz-Christoffel techniques, and elliptic techniques with orthogonal boundary control), the classical technique is that of using the Theodorsen-Garrick (1933) transformation, given by

$$\frac{d\zeta''}{d\zeta} = \exp\left[\sum_{j=0}^{N}(A_j + iB_j)\zeta^j\right] \tag{4.24}$$

The coefficients A_j and B_j in (4.24) are chosen by mapping $2N$ equally spaced points around the unit circle in the ζ-plane to equivalent points in the ζ''-plane. An efficient technique, based on the discrete fast Fourier transform (Cooley and Tukey, 1965), is described by Ives (1976).

Mapping to a Rectangular Computational Domain

The region outside of the unit circle in the ζ-plane is mapped to the inside of a rectangle ($1 \le R \le R_{\max}, 0 \le \beta \le 2\pi$) by letting

$$\zeta = r \exp(i\phi)$$

and setting

$$R = \exp(\ln r) \quad \text{and} \quad \beta = \phi \tag{4.25}$$

Thus the sequence of transformations (4.21), (4.23) to (4.25) maps the region exterior to an isolated streamlined body such as airfoil in the (x, y) plane to the interior of a rectangle in the (R, β) plane.

In principle a uniform grid can be laid out in the (R, β) plane and the inverse mapping used to provide the corresponding grid in the physical domain. However, while the above sequence of transformations to establish the boundary correspondence between the physical and computational domains is relatively efficient, the inverse transformation often is not (Ives, 1982). Thus, for the general case, a fast elliptic solver (Temperton, 1979) is recommended by Ives to generate the interior grid by solving

$$x_{\xi\xi} + x_{\eta\eta} = 0 \quad \text{and} \quad y_{\xi\xi} + y_{\eta\eta} = 0 \tag{4.26}$$

with the boundary values already determined via stage 1.

The sequential mapping procedure described above can be extended to multiple isolated bodies, for example, an airfoil and flap (Ives, 1976).

4.2.3 One-Step Conformal Mapping

A one-step mapping maps a contour onto a canonical shape, such as a circle or the real axis, in a single step. The classic Schwarz-Christoffel transformation allows a region bounded by a simple closed polygon in the physical plane Z to be mapped into the upper half of the transform plane ζ. The polygon coincides with the real axis in the ζ-plane. By introducing a branch cut, it is possible to consider the region between the polygon and infinity in the Z-plane as the bounded region, thus allowing the Schwarz-Christoffel transformation to be used for exterior grids. However, this transformation can also be used for an internal flow geometry (Fletcher, 1991, Chap. 13).

In the conventional form, the Schwarz-Christoffel transformation for a polygon with N straight sides takes the form (Milne-Thomson, 1968, p. 277)

$$\frac{dZ}{d\zeta} = M \prod_{j=1}^{N} (\zeta - b_j)^{-\alpha_j/\pi} \tag{4.27}$$

where the α_j's are the angles turned through each corner (counterclockwise positive). The b_j's are unknown locations on the real axis in the transform plane ζ; three of the b_j's may be chosen arbitrarily. M is a complex constant, typically related to the geometry of the physical domain. The body in the physical domain need not be closed. Thus the region inside an arbitrary duct can be mapped to the upper half ζ-plane.

For a polygon with curved sides, Davis (1979) used a differential form of the product term in (4.27) so that

$$\frac{dZ}{d\zeta} = \exp\left[\frac{1}{\pi} \int \ln(\zeta - b) \, d\beta\right] \tag{4.28}$$

where β is an angle-related variable. Davis employed a composite integration formula to resolve the curvature effects. The Davis technique has been successfully used by Grossman (1979) and by Sridhar and Davis (1981) to map a wide range of complicated configurations.

An alternative is to use the form

$$\frac{dZ}{d\zeta} = g(\zeta)\exp\left[\sum_{j=0}^{N} (a_j + ib_j)\zeta^j\right] \tag{4.29}$$

where $g(\zeta)$ is chosen to resolve angles or general behavior, while the exponential term accounts for the curvature. The $g(\zeta)$ function is easily constructed using the Schwarz-Christoffel technique, as illustrated by Nehari

(1952), Kober (1957), and Milne-Thomson (1968). This form can take advantage of fast Fourier techniques to evaluate the coefficients a_j and b_j. An example of such a mapping is given by

$$\frac{dZ}{d\zeta} = \left(1 - \frac{1}{\zeta}\right)^k \exp\left[\sum_{j=0}^{N}(a_j + ib_j)\zeta^{-j}\right] \tag{4.30}$$

which was used by Bauer et al. (1975) to map an airfoil in the Z-plane to a circle in the ζ-plane.

Another mapping of this type is given by

$$\frac{dZ}{d\zeta} = -\frac{\zeta - iR_0}{(1 - iR_0)(\zeta - 1)(\zeta + 1)^2}\exp\left[\sum_{j=0}^{N}(a_j + ib_j)\zeta^j\right] \tag{4.31}$$

where

$$R_0 = 1 + 2\sqrt{\pi\varepsilon} \quad \text{and} \quad \varepsilon = \frac{\text{nose radius}}{\text{inlet interior radius}}$$

This mapping was developed by Ives (1982) to map the region exterior to a semi-infinite inlet and above the centerline to the interior of a circle as illustrated in Fig. 4.3.

The one-step mapping of Eq. (4.29) is far simpler to program than the classic sequential mapping technique described in section 4.2.2. It converges stably and rapidly and is easy to modify for a new class of geometries. For the use of fast Fourier techniques, Eq. (4.29) is recommended as *the* method of choice (Ives, 1982) rather than using a sequence resulting in a near circle followed by a derivative transform like Eq. (4.24). The computer program in Bauer et al. (1975) can be easily modified to implement the mapping in Eq. (4.29) instead of Eq. (4.30) for the airfoil. Ives (1982) states that the use of Fourier series for Eqs. (4.24) and (4.29) is unable to resolve a slope discontinuity larger than about 5 degrees even if a large number of Fourier terms

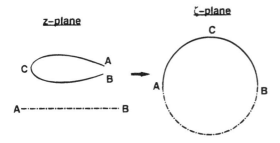

Figure 4.3   Inlet-to-circle mapping.

are used. The accuracy and convergence of the transformation suffer for larger slope discontinuities. However, it is easy to handle such slope discontinuities using the hinge point transformation described next.

Hinge Point Transformation

The hinge point transformations due to Moretti (1976) and Hall (1980) break up the problem into a large number of sequential applications of a single mapping. For problems with a plane of symmetry, Moretti applied a von Karman-Trefftz transformation to each pair of slope discontinuities in turn, to produce a smooth near circle. Hall (1980) repeatedly applied the transformation $Z = (\zeta - \zeta_0)^k$, where ζ_0 and k were chosen to remove the leftmost remaining slope discontinuity at each stage, to produce a smooth near half-plane. Once a near circle or near half-plane is obtained, a conformal mapping can be used to produce a canonical domain.

4.2.4 Orthogonal Grid Generation

For orthogonal coordinate systems the metric parameters satisfy $g_{ij} = 0$, $i \neq j$. Thus, in two dimensions, the physical (x, y) and the computational (ξ, η) domains are related via

$$\left(\xi_x \frac{h_1}{h_2}\right)_x + \left(\xi_y \frac{h_1}{h_2}\right)_y = 0 \quad \text{and} \quad \left(\eta_x \frac{h_2}{h_1}\right)_x + \left(\eta_y \frac{h_2}{h_1}\right)_y = 0 \quad (4.32)$$

in the interior, and

$$h_1 \xi_x = h_2 \eta_y \quad \text{and} \quad h_1 \xi_y = -h_2 \eta_x \quad (4.33)$$

on the boundary. The ratio h_2/h_1 is the grid aspect ratio (4.8) where

$$h_1 = (x_\xi^2 + y_\xi^2)^{1/2} \quad \text{and} \quad h_2 = (x_\eta^2 + y_\eta^2)^{1/2} \quad (4.34)$$

There are basically two types of orthogonal grid generation systems; one is based on the construction of an orthogonal system from a nonorthogonal system, and another involves field solutions of partial differential equations in a manner similar to that described in section 4.2.5. The first approach involves the construction of orthogonal trajectories on a given nonorthogonal system, and is described in some detail here following Fletcher (1991). For details of the second approach, the reader is referred to Thompson et al. (1985, p. 332).

The orthogonal trajectory method permits points to be specified on three surfaces, say AB, BCD, and DE in Fig. 4.4, with points on the fourth surface (EA in Fig. 4.4) determined by the orthogonal trajectory as follows. First, a family of curves is laid out as in Fig. 4.4. The location along a

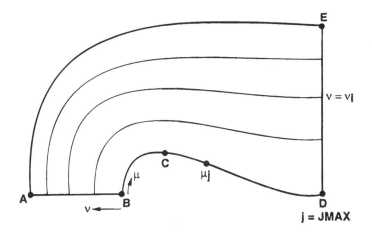

Figure 4.4   Preliminary configuration for orthogonal trajectory method.

particular member $(v = v_i)$ of the family may be determined by a simple shearing transformation

$$x = (1 - \mu')AB^x(v_i) + \mu'DE^x(v_i)$$
$$y = (1 - \mu')AB^y(v_i) + \mu'DE^y(v_i) \qquad (4.35)$$
$$\mu' = \frac{\mu - \mu_1}{\mu_{JMAX} - \mu_1}$$

The prescribed functions $AB^x(v)$, etc., determine the grid point distribution on AB and DE (Fig. 4.4) and the clustering of grid lines adjacent to BCD. For the $v = v_i$ lines to be orthogonal to AB and DE, (4.35) can be replaced by a cubic representation in μ' (Eiseman 1982a, p. 209). The distribution of points on BCD $(\mu = \mu_j)$ is prescribed.

For the construction of an orthogonal grid (ξ, η), we need to define trajectories starting at $\mu = \mu_j$, finishing at the target boundary EA and intersecting each intervening coordinate line $(v = v_i)$ at right angles. This yields both the locations of the interior grid points as well as the grid point locations on EA. Letting $\eta_i = v_i$, we seek $\xi = \xi_j$ (constant) lines that are orthogonal to the $\eta = \eta_i$ lines as follows.

Given that $x = x(\mu, v)$ and $y = y(\mu, v)$, the slope of a $v = v_i$ line can be expressed as

$$\left.\frac{dy}{dx}\right|_{v = v_i} = \frac{\partial y / \partial \mu}{\partial x / \partial \mu} \qquad (4.36)$$

Since the constant ξ-line is orthogonal to the η-line, its slope, from (4.36), must be

$$\frac{dy}{dx}\bigg|_{\xi=\xi_j} = -\frac{\partial x/\partial\mu}{\partial y/\partial\mu} \tag{4.37}$$

Since $\xi = \xi(\mu, v)$ and $\eta = v$, we can write

$$\frac{dy}{dx}\bigg|_{\xi=\xi_j} = \frac{[(\partial y/\partial\mu)(d\mu/dv)_{\xi=\xi_j} + \partial y/\partial v]}{[(\partial x/\partial\mu)(d\mu/dv)_{\xi=\xi_j} + \partial x/\partial v]} \tag{4.38}$$

Equating (4.37) and (4.38) yields the following ordinary differential equation defining the $\xi = \xi_j$ trajectory in the (μ, v) plane

$$\frac{d\mu}{dv}\bigg|_{\xi=\xi_j} = -\left(\frac{\partial x}{\partial v}\frac{\partial x}{\partial\mu} + \frac{\partial y}{\partial\mu}\frac{\partial y}{\partial v}\right)\bigg/\left[\left(\frac{\partial x}{\partial\mu}\right)^2 + \left(\frac{\partial y}{\partial\mu}\right)^2\right] \tag{4.39}$$

Upon comparison with (4.10), Eq. (4.39) can be written in the general form

$$\frac{d\mu}{dv}\bigg|_{\xi=\xi_j} = -\frac{g_{12}}{g_{11}} \tag{4.40}$$

where g_{11} and g_{12} are metric components associated with mapping from the physical (x, y) plane to the nonorthogonal (μ, v) plane. Since the (μ, v) grid is known (Fig. 4.4), the right side of Eq. (4.39) can be evaluated. Initial values for (4.39) are given by $\mu = \mu_j$ on BCD. Generally (4.39) is integrated numerically. At the intersection with each $v = v_i$ line, the physical coordinates (x, y) follow from the equation defining the (μ, v) grid, e.g., (4.35).

Equation (4.40) is a characteristic equation for the hyperbolic equation

$$\frac{\partial\xi}{\partial v} - \frac{g_{12}}{g_{11}}\frac{\partial\xi}{\partial\mu} = 0 \tag{4.41}$$

so that the orthogonal trajectory method is essentially a method of characteristics. For more details, the reader is referred to Eiseman (1982a). In numerical solutions, the concept of numerical orthogonality, i.e., that the off-diagonal metric coefficients vanish when evaluated numerically, is usually more important than strict analytical orthogonality. This is specially true when the governing equations to be solved for the system are in the conservative law form (Thompson et al., 1985). For details on numerical orthogonality, see Eiseman (1982a, p. 202). Strictly orthogonal grids are generally restricted to two dimensions, since it is not possible to construct completely orthogonal grids in three dimensions and still retain sufficient control over boundary grid point locations (Eiseman, 1982a). It is, however, possible to construct grids that are orthogonal to specific surfaces and to construct three-dimensional grids from a stack of two-dimensional orthogonal grids. In such a case the grid distribution in the third direction may not be very smooth (Thompson, 1984a). When a strictly orthogonal grid cannot be generated, a near orthogonal grid may be constructed easily (Fletcher, 1991).

4.2.5 Elliptic Grid Generation

A more general technique for grid generation is presented by the solution of Poisson equations

$$\frac{\partial^2 \xi}{\partial x^2} + \frac{\partial^2 \xi}{\partial y^2} = P(\xi, \eta), \qquad \frac{\partial^2 \eta}{\partial x^2} + \frac{\partial^2 \eta}{\partial y^2} = Q(\xi, \eta) \tag{4.42}$$

where P and Q are known as control functions used to control clustering of interior grid points and angles at the boundaries. Specification of desired grid points (x, y) on the boundary of the physical domain provide the boundary conditions for the solution of (4.42). The technique generates grids that are not necessarily conformal or orthogonal but does permit arbitrary specification of spacing normal to the boundaries. The use of an elliptic partial differential equation to generate the interior grid points has some advantages. First, the grid is smoothly varying even if the domain boundary has a slope discontinuity. If, for example, a hyperbolic partial differential equation were used to generate the interior grid, any slope discontinuity at the boundary would propagate into the interior. Second, elliptic equations like (4.42) satisfy the maximum principle for reasonable values of P and Q, implying thereby that the maximum and minimum values of ξ and η must occur on the boundary. According to Thompson (1982a), this normally guarantees a one-to-one mapping. Extreme choices for P and Q, however, may cause a local grid overlap.

The Laplace equations (with $P = 0 = Q$) can also be used in place of the Poisson equations. However, it implies loss of control over grid clustering and angle at the boundaries. The choice of Laplace or Poisson equations can be physically understood by considering the solution of a steady heat conduction problem in two dimensions with Dirichlet boundary conditions. The solution of this problem produces isotherms that are smooth (C^2 continuous) and are nonintersecting. The number of isotherms in a given region can be increased by adding a source term. Thus, if the isotherms were used as grid lines, they would be smooth, nonintersecting, and could be densely packed in any region by control of the source terms.

The actual solution of (4.42) is carried out in the computational (ξ, η) domain. In this domain (4.42) transform to

$$\begin{aligned} \alpha x_{\xi\xi} - 2\beta x_{\xi\eta} + \gamma x_{\eta\eta} + \delta(Px_\xi + Qx_\eta) &= 0 \\ \alpha y_{\xi\xi} - 2\beta y_{\xi\eta} + \gamma y_{\eta\eta} + \delta(Py_\xi + Qy_\eta) &= 0 \end{aligned} \tag{4.43}$$

where $\alpha = g_{22}$, $\beta = g_{12}$, $\gamma = g_{11}$, and $\delta = g$, the determinant of the metric tensor (4.10). Solution of (4.43) for a particular choice of control or forcing functions P and Q and for a particular set of boundary conditions leads to the generation of the boundary-fitted grid.

A great latitude exists in the grids so generated due to the ability to choose the functions P and Q. If $P = Q = 0$, a basic grid results, as the Poisson equations degenerate to the Laplace equations. The challenge is to choose P and Q so that a desirable grid results with a reasonable amount of effort. Thomas and Middlecoff (1980) developed a method for approximate control of interior grid points by evaluating P and Q according to a desired point distribution on the boundary. Sorenson (1980) defines P and Q in terms of four new variables, and sets up four geometrical constraints that translate into four new equations for the four variables. Including the Poisson equations, six equations are solved iteratively for the six unknowns. The four geometrical constraints pertain to the desired intersection angle of the grid line at the boundary, and the spacing of the first grid line off the boundary specified at each boundary point. This provides a highly automated technique for control of grid spacing near boundaries and for near-orthogonality at the boundaries.

Thompson et al. (1985) suggest the following form for P in order to attract grid lines to other grid lines and/or points:

$$P(\xi, \eta) = -\sum_{i=1}^{N} a_i \, \text{sign}(\xi - \xi_i)\exp(-c_i|\xi - \xi_i|)$$

$$- \sum_{i=1}^{M} b_i \, \text{sign}(\xi - \xi_i)\exp\{-d_i[(\xi - \xi_i)^2 + (\eta - \eta_i)^2]^{1/2}\} \quad (4.44)$$

and an analogous form for $Q(\xi, \eta)$ with ξ and η interchanged. In this form, the control functions are functions of the curvilinear coordinates only. In (4.44), the effect of amplitude a_i is to attract ξ-line toward the ξ_i-line, while the effect of amplitude b_i is to attract ξ-lines toward the single point (ξ_i, η_i), as shown in Fig. 4.5. Note that the attraction to a point is really attraction of ξ-lines to a point on another ξ-line, and thus acts normal to the ξ-line through the point. There is no attraction of ξ-lines to this point via the P function. In each case the effect of attraction decays with distance in ξ-η space from the attraction site via the decay factors c_i and d_i. This decay depends only on the ξ-distance from the ξ_i-line in the first term of (4.44), so that entire ξ-lines are attracted to the entire ξ_i-line. In the second term of (4.44), however, the decay depends on both the ξ and η distances from the attraction point (ξ_i, η_i), so that the effect is limited to portions of the ξ-lines. With the inclusion of the sign-changing function, the attraction occurs on both sides of the ξ-line, or the (ξ_i, η_i) point, as the case may be. Without

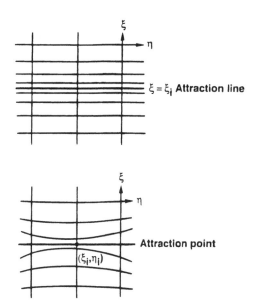

Figure 4.5   Attraction to grid lines and points.

this function, attraction occurs only on the side toward increasing ξ, with repulsion occuring on the other side. A negative amplitude reverses all these effects. The effect of function Q on η-lines follows analogously. This technique does provide the means to move the grid lines around but does not provide any direction as to the proper amplitudes and decay factors necessary to achieve desired spacing distribution. For attraction to lines or points in space, rather than on a grid line, see Thompson et al. (1985).

This grid generation technique has been incorporated in several codes. Sorenson (1980) describes the code GRAPE, an acronym derived from GRids about Airfoils using Poisson Equations. It can generate curvilinear grids of the O-type or C-type about airfoils or about any other user-specified shape. This code is available from NASA through its COSMIC library. While GRAPE is a 2-D code, its three-dimensional (3-D) version is also available as 3DGRAPE (Sorenson, 1989), and more recently, as 3DGRAPE/AL (Sorenson and Alter, 1995). The latter is a significantly improved version of 3DGRAPE, with many of the improvements taken from the grid generator program 3DMAGGS (Alter and Weilmuenster, 1993).

The WESCOR code of Thompson (1982b) is applicable to two-dimensional regions with any number of interior obstacles and/or boundary intrusions, which are transformed into slits and/or slabs using an elliptic generation system. The attraction scheme of Thompson et al. (1985) described above [Eq. (4.44)] has been used in the TOMCAT code (Thompson et al., 1977) in which second-order central differences are used to approximate all derivatives in (4.43). The resulting set of nonlinear difference equations, two for each point, are solved in TOMCAT by accelerated Gauss-Seidel iteration.

While the above codes can generate grids for both inviscid and viscous flows, the computational effort increases considerably for viscous grids,* due to poorer convergence, as the Reynolds number increases. This problem is overcome by the codes JERRY and TOM of Arnone (1992). While JERRY generates the inviscid grid over a 2-D body, such as a section of an airfoil, by solving the discretized Poisson equations using a point relaxation scheme, TOM creates the viscous grid by embedding grid lines near the wall within the inviscid grid generated by JERRY. These codes have been designed specifically for turbomachinery applications. Control functions like those proposed by Sorenson (1980) are used to control the grid spacing and orientation at the boundary. Two versions of each code exist; JERRYC and TOMC for generating a C-grid, and JERRYH and TOMH for generating an H-grid. For a 3-D body such as a turbine blade, 2-D grids can be generated at several sections of the airfoil using JERRY and TOM. A code STACK3D then stacks these 2-D grids to yield a 3-D grid for flow over the airfoil. The author has used this combination of JERRYC, TOMC, and STACK3D codes to generate grids for analyzing heat transfer on film-cooled turbine blades. As an example, Fig. 4.6 shows the grid on the film-cooled ACE rotor surface (Garg, 1997) with suction surface in the front, and a C-grid on the hub. The entire grid consists of over 2.25 million grid points; only alternate grid lines are shown in Fig. 4.6 for the sake of clarity. Also, extension of the C-grid all the way to the shroud is not shown to avoid confusion.

4.2.6 Multiblock Grids

For complicated geometries, various choices for the grid structure include Chimera grid system (Steger et al., 1983), unstructured grids discussed in section 4.5, and multiblock grids. A multiblock grid system is essentially an unstructured assembly of structured hexahedra blocks. Thus the multiblock

---

* A viscous grid is dense near a boundary in order to resolve large gradients normal to the boundary due to the presence of the boundary layer.

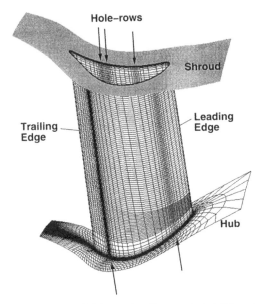

Figure 4.6   Grid on the film-cooled ACE rotor surface and C-grid on the hub. Suction surface is in the front. Alternate grid lines shown.

grid can be thought of as globally unstructured but locally structured. It offers several advantages over the other two grid structures. Compared to Chimera grids, communication of data between blocks is straight forward and without loss of accuracy at block interfaces in the multiblock grid system. This allows easy implementation of higher-order schemes and guarantees conservation at grid interfaces. Compared to unstructured grid systems, multiblock grids require less memory, and, due to regularity of the grid structure, various efficient schemes based on dimensional splitting can be applied to structured grids (Steinthorsson et al., 1997). Moreover, structured body-fitted grids are well suited for simulations of viscous flows due to good resolution of the boundary layer regions.

The main disadvantage of using multiblock grids is that they have traditionally been the hardest to generate. However, recent advances in grid generation are changing the way grids are generated (Choo, 1995). For instance, the introduction of the commercial code GridPro/az3000 (Eiseman, 1995; Program Development Corporation, 1997) has ma easier to generate good-quality multiblock grids. It is a general purpose. 3-D, multiblock, structured grid generator using an advanced smoothing technique. The process of grid generation using GridPro/az3000 is accomplished via an iterative updating scheme, thus requiring multiple sweeps. In this sense, it is more CPU intensive than algebraic grid generation methods. However, it reduces dramatically the amount of required user input and generates grids with excellent quality. The approach of generating a viscous grid via embedding grid lines within an inviscid grid has been implemented in this code.

Another useful system for generation of 3-D multiblock, structured grids is Gridgen (Steinbrenner and Chawner, 1995; Chawner and Steinbrenner, 1995). It consists of two codes: Gridgen, an interactive program that provides capabilities ranging from geometry model import through volume grid initialization and analysis software preprocessing; and Gridgen3D, a batch program for volume grid refinement. Transfinite interpolation, elliptic and hyperbolic partial differential equation methods are available for controlling the distribution of grid points within the blocks.

The generation of a multiblock grid system generally leads to a large number of small blocks. Such a situation is counterproductive for computational efficiency and storage. If two layers of ghost cells are used to store data that is needed from neighboring blocks, which is typical for second- and third-order discretizations, average dimensions of blocks should be kept larger than about 25 in order to keep storage overhead within reasonable limits (Steinthorsson et al., 1997). It is therefore important to be able to merge small blocks to form larger blocks that contain several thousand cells each. It is equally important to be able to merge the blocks automatically, since manual merging is too tedious and error prone except for a handful of small blocks.

At least two block-merging algorithms are currently available. One is based on the simulated annealing optimization method (Dannenhoffer, 1995) that reorganizes the blocks into an optimum configuration, that is, one which minimizes a user-defined objective function such as the number of clusters or the differential in the sizes of all the clusters. A computationally efficient algorithm, however, was proposed by Rigby (1996) for merging small blocks down to a minimum number of larger blocks. It is a heuristic algorithm, called the method of weakest descent, that uses special rules to determine which blocks to merge, combined with some random choices to break ties if they occur. The algorithm has been shown to produce close to the minimum number of blocks in various applications starting from grids containing several hundreds of small blocks. The algorithm has been implemented by Rigby et al. (1997), together with schemes for analyzing connectivity in a multiblock grid system after performing the actual merging of the small blocks. This package has proven very useful for merging of multiblock grid systems. For example, Fig. 4.7 shows a 283-block branch duct topology. The original block boundaries are shown as thin lines, while the merged 27-block result is shown as heavy dashed lines.

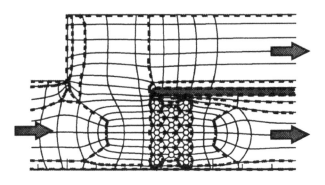

Figure 4.7 Branch duct topology (283 blocks merged to 27).

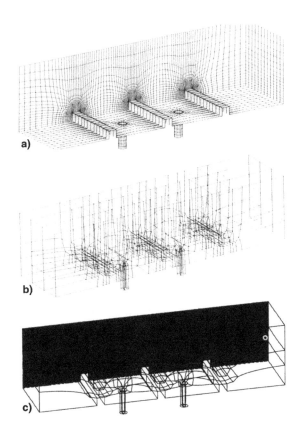

Figure 4.8 Internal passage with ribs and bleed topology (282 blocks merged to 22).

While the absolute minimum number of blocks for the branch duct is not
known, it is believed to be 27 or very close to it (Rigby et al., 1997). Another
example is shown in Fig. 4.8 for a complex internal coolant passage with
ribs and bleed holes (Rigby et al., 1997). Figure 4.8a shows the surface grid
generated using the GridPro software (Program Development Corporation,
1997). Note that high-quality grids can be formed around the ribs and holes
by taking full advantage of the multiblock capability. Figure 4.8b shows the
block boundaries for the initial 282-block grid from GridPro, while Fig.
4.8c shows the block boundaries of the merged grid system, which has only
22 blocks.

4.3   ALGEBRAIC GRID GENERATION

Algebraic mapping techniques can be used to interpolate the boundary data
in one or more dimensions in order to generate the interior grid. Clearly,
the grid so generated should be smoothly varying, near-orthogonal at least
close to the boundaries, and with local grid aspect ratios close to unity. The
distribution of grid points in the interior is governed mainly by the normalized one-dimensional stretching functions along boundary segments in the
computational domain. For details on algebraic grid generation, see Smith
(1982, 1983), Thompson et al. (1982, 1985), Eiseman (1979, 1982b, 1982c,
1982d, 1985), Eiseman and Smith (1980), and Fletcher (1991, Chap. 13).
Much of the development in this section follows Fletcher (1991).

4.3.1   One-Dimensional Stretching Functions

One-dimensional stretching functions are used for distributing points along
a boundary in order to resolve accurately certain regions of the domain.
Both the dependent and independent variables in these functions are generally expressed in the normalized form. For a one-dimensional stretching
function applied to AB in Fig. 4.9 an appropriate normalized independent
variable would be

$$\xi^* = \frac{\xi - \xi_1}{\xi_2 - \xi_1} \tag{4.45}$$

so that $0 \leq \xi^* \leq 1$ as $\xi_1 \leq \xi \leq \xi_2$. An effective stretching function suggested by Eiseman (1979) is

$$s = P\xi^* + (1 - P)\left(1 - \frac{\tanh[Q(1 - \xi^*)]}{\tanh Q}\right) \tag{4.46}$$

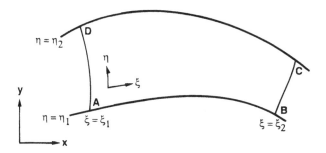

Figure 4.9   Curved two-dimensional channel.

where P and Q are parameters to provide grid point control. Basically P provides the slope of the distribution, $s \approx P\xi^*$, close to $\xi^* = 0$. Q is a damping factor that controls the departure from the linear s versus ξ^* behavior. Small values of Q cause small departures from linearity. However, if P is close to unity, the departure from linearity will be small and will occur only for ξ^* close to unity.

Once s is obtained it is used to specify the distribution of x and y. For example, defining

$$\frac{x - x_A}{x_B - x_A} = f(s), \qquad \frac{y - y_A}{y_B - y_A} = g(s) \tag{4.47}$$

generates $x(s)$ and $y(s)$ directly. A simple choice is $f(s) = g(s) = s$, so that (4.47) gives

$$x = x_A + s(x_B - x_A), \qquad y = y_A + s(y_B - y_A) \tag{4.48}$$

Vinokur (1983) suggested a two-parameter stretching function wherein the two parameters are the slopes $ds/d\xi^*$ at each end of the interval $\xi^* = 0$ and $\xi^* = 1$. Since multiple contiguous stretching functions can be used to cover a particular boundary, these yield continuity of s and $ds/d\xi^*$ at the interfaces. However, Vinokur's stretching function cannot be expressed in a single equation like (4.47); thus coding is a bit more complicated.

4.3.2   Two-Boundary Technique

Consider a curved two-dimensional channel shown in Fig. 4.9. Assume that stretching functions, $s_{AD}(\eta^*)$ and $s_{BC}(\eta^*)$, have been defined to control the distribution of points on the inlet and outlet boundaries AD and BC, respectively, where η^* is the normalized parameter given by $\eta^* = (\eta$

$-\eta_1)/(\eta_2 - \eta_1)$. Equations equivalent to (4.47) could be used to generate $s_{AD}(\eta^*)$ and $s_{BC}(\eta^*)$. A simple interpolation

$$s = s_{AD} + \xi^*(s_{BC} - s_{AD}) \tag{4.49}$$

can be used to obtain the value s between surfaces AD and BC, where $\xi^* = (\xi - \xi_1)/(\xi_2 - \xi_1)$. Similarly assume that the distribution of grid points along AB and CD are controlled by one-dimensional stretching functions $r_{AB}(\xi^*)$ and $r_{DC}(\xi^*)$. Interpreting r_{AB} and r_{DC} as normalized coordinates measured along the surface, $x_{AB}(r_{AB})$, $y_{AB}(r_{AB})$ follow directly, and similarly for $x_{DC}(r_{DC})$, $y_{DC}(r_{DC})$.

The two-boundary technique provides a means of interpolating the interior between the two boundaries, AB and DC. This specifies the interior grid completely. A simple interpolation is

$$\begin{aligned}
x(\xi, \eta) &= (1 - s)x_{AB}(r_{AB}) + sx_{DC}(r_{DC}) \\
y(\xi, \eta) &= (1 - s)y_{AB}(r_{AB}) + sy_{DC}(r_{DC})
\end{aligned} \tag{4.50}$$

where s is given by (4.49). Considerable control over the clustering of the grid points in the interior can be obtained through the boundary stretching functions, s_{AD}, s_{BC}, r_{AB}, and r_{DC}.

Equation (4.50) may distort the grid close to the surface if the boundary points (x_{AB}, y_{AB}) and (x_{CD}, y_{CD}) are out of alignment. In order to generate a grid that is locally orthogonal to the surfaces AB and CD (Fig. 4.9), Eq. (4.50) should be replaced by

$$\begin{aligned}
x(\xi, \eta) &= \beta_1(s)x_{AB}(r_{AB}) + \beta_2(s)x_{DC}(r_{DC}) + H_1\beta_3(s)\left(\frac{dy_{AB}}{dr_{AB}}(r_{AB})\right) \\
&\quad + H_2\beta_4(s)\left(\frac{dy_{DC}}{dr_{DC}}(r_{DC})\right)
\end{aligned} \tag{4.51}$$

$$\begin{aligned}
y(\xi, \eta) &= \beta_1(s)y_{AB}(r_{AB}) + \beta_2(s)y_{DC}(r_{DC}) - H_1\beta_3(s)\left(\frac{dx_{AB}}{dr_{AB}}(r_{AB})\right) \\
&\quad - H_2\beta_4(s)\left(\frac{dx_{DC}}{dr_{DC}}(r_{DC})\right)
\end{aligned} \tag{4.52}$$

where

$$\begin{aligned}
\beta_1(s) &= 2s^3 - 3s^2 + 1, \quad \beta_2(s) = -2s^3 + 3s^2, \\
\beta_3(s) &= s^3 - 2s^2 + s, \quad \beta_4(s) = s^3 - s^2
\end{aligned}$$

and where H_1 and H_2 are used to control grid orthogonality into the interior. Choosing too large a value for H_1 and H_2 may cause a double-valued mapping in the interior (Smith, 1982). For extension of the method to three-dimensional grid generation, see Smith (1982).

4.3.3   Multisurface Method

Eiseman (1979) proposed the multisurface method in order to provide more control over the interior grid distribution by introducing intermediate surfaces between the boundary surfaces AB and CD in Fig. 4.9. By adjusting the grid point correspondence between a boundary surface, say AB, and its neighboring intermediate surface, it is possible to make the grid locally orthogonal at the boundary. Also, the grid distribution in s is obtained by integrating the interpolation of the sequence of directions specified by connecting corresponding points on adjacent surfaces. This provides a very smooth s distribution. While there is no limit to the number of intermediate surfaces, good control over the interior grid is obtained with just two intermediate surfaces.

In two dimensions let us denote the distribution of points on the ith surface by a vector function $F_i(r)$ with components $x_i(r)$ and $y_i(r)$. A sequence of surfaces is shown in Fig. 4.10. For a total of N surfaces (counting the boundary surfaces) there will be $N - 2$ intermediate surfaces. Let the parameter r define the location in all surfaces simultaneously. However, different choices for the functions $F_i(r)$ will allow adjustment of the relative orientation of (x_i, y_i) on each surface for a given r. Assume that corresponding points (x_i, y_i) on each surface, associated with a particular value of $r = r_j$, are joined by straight lines between surfaces (Fig. 4.10).

Let the tangents to these straight lines between the surfaces be denoted by a family of vector functions $V_i(r)$ for $i = 1, \ldots, N - 1$. Let us define a

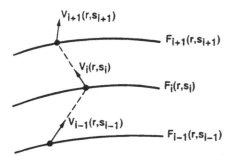

Figure 4.10   Intermediate surfaces $\boldsymbol{F}_i$ and tangent vectors $\boldsymbol{V}_i$.

relation between the tangent vector functions V_i and surface vector functions F_i via

$$V_i(r) = B_i[F_{i+1}(r) - F_i(r)], \qquad i = 1, \ldots, N-1 \tag{4.53}$$

The parameters B_i are determined later so as to fit the final grid interpolation properly into the interval $0 \leq s \leq 1$. Let us also denote the interpolation through the family of semidiscrete tangent vector functions $V_i(r)$ by a tangent vector function $V(r, s)$ that is continuous in both r and s. Then

$$V(r, s) = \sum_{i=1}^{N-1} \phi_i(s) V_i(r) \tag{4.54}$$

where $\phi_i(s)$ are interpolating functions, to be determined, such that $\phi_i(s_k) = 1$ if $i = k$ and is zero if $i \neq k$. Clearly, we have

$$\frac{\partial F}{\partial s}(r, s) = V(r, s) = \sum_{i=1}^{N-1} \phi_i(s) V_i(r) \tag{4.55}$$

where $F(r, s)$ is the continuous function required to generate the grid in the physical domain for given values of r and s, and hence ξ and η. It is obtained by integrating (4.55) over the interval $0 \leq s \leq 1$. This interval corresponds to $\eta_1 \leq \eta \leq \eta_2$ in Fig. 4.9. Using (4.53), we get

$$F(r, s) = F_1(r) + \sum_{i=1}^{N-1} B_i G_i(s)[F_{i+1}(r) - F_i(r)] \tag{4.56}$$

where

$$G_i(s) = \int_0^s \phi_i(\psi) \, d\psi \tag{4.57}$$

The parameters B_i are chosen so that $B_i G_i(1) = 1$, since then (4.56) yields $F(r, s) = F_N(r)$ when $s = 1$, as desired. Thus (4.56) can be written as

$$F(r, s) = F_1(r) + \sum_{i=1}^{N-1} \frac{G_i(s)}{G_i(1)} [F_{i+1}(r) - F_i(r)] \tag{4.58}$$

which is the general multisurface transformation.

The interpolating function $\phi_i(s)$ must be continuously differentiable up to order one less than the level of smoothness required in the grid. An appropriate family for ϕ_i is

$$\phi_i(s) = \prod_{\substack{j=1 \\ j \neq 1}}^{N-1} (s - s_j) \tag{4.59}$$

The simplest case is $N = 2$. For this case there are no intermediate surfaces, and (4.58) reduces to

$$F(r, s) = F_1(r) + s[F_2(r) - F_1(r)] \tag{4.60}$$

which is equivalent to the linear two-boundary relation (4.50). For $N = 3$, one intermediate surface is introduced, and using (4.59) in (4.58) yields

$$F(r, s) = (1 - s)^2 F_1(r) + 2s(1 - s)F_2(r) + s^2 F_3(r) \tag{4.61}$$

With $N = 4$ (two interior surfaces), it is possible to generate a grid orthogonal to both bounding surfaces AB and CD in Fig. 4.9. The use of N surfaces provides N degrees of freedom. Two degrees of freedom are used in requiring the grid to match the boundary surface specifications, while the rest can be used to control the characteristics of the interior grid.

For extension of the multisurface method to three dimensions, see Eiseman (1982a, 1982b).

4.3.4   Transfinite Interpolation

Both the two-boundary and the multisurface methods interpolate in only one direction (s or η) and assume that a continuous mapping is prescribed on the bounding surfaces $\eta = \eta_1$ and $\eta = \eta_2$ in the other (r or ξ) direction. Transfinite interpolation (Gordon and Hall, 1973), however, allows one to specify continuous mappings $F_{AB}(\xi, \eta_1)$ on AB, $F_{DC}(\xi, \eta_2)$ on DC as well as $F_{AD}(\xi_1, \eta)$ on AD and $F_{BC}(\xi_2, \eta)$ on BC (Fig. 4.9). An interpolation in both ξ and η, or equivalently r and s, is performed in the interior. It is assumed that r and s are normalized coordinates, i.e.,

$$\begin{aligned} 0 \leq r \leq 1 \quad &\text{as} \quad \xi_1 \leq \xi \leq \xi_2 \\ 0 \leq s \leq 1 \quad &\text{as} \quad \eta_1 \leq \eta \leq \eta_2 \end{aligned} \tag{4.62}$$

The following blending functions are defined:

$$\psi_i(r) = \delta_{ir}, \quad i = 0, 1 \quad \text{and} \quad \phi_j(s) = \delta_{js}, \quad j = 0, 1 \tag{4.63}$$

where

$$\begin{aligned} \delta_{ir} &= 1 \quad \text{if } i = r \\ &= 0 \quad \text{if } i \neq r \end{aligned} \quad \text{and} \quad \begin{aligned} \delta_{js} &= 1 \quad \text{if } j = s \\ &= 0 \quad \text{if } j \neq s \end{aligned}$$

Thus $\psi_0 = 1, \psi_1 = 0$ on AD; $\psi_0 = 0, \psi_1 = 1$ on BC; $\phi_0 = 1, \phi_1 = 0$ on AB; and $\phi_0 = 0, \phi_1 = 1$ on CD.

If $F_r(r, s)$ denotes the continuous mapping produced by interpolating between F_{AD} and F_{BC} for intermediate values of r, we can write

$$F_r(r, s) = \psi_0(r)F_{AD}(0, s) + \psi_1(r)F_{BC}(1, s) \tag{4.64}$$

where F_{AD}, F_{BC} are the continuous mappings between the (ξ, η) and (x, y) planes on the two boundaries $\xi = \xi_1$ and $\xi = \xi_2$. Similarly, in the s-direction, we have

$$F_s(r, s) = \phi_0(s)F_{AB}(r, 0) + \phi_1(s)F_{CD}(r, 1) \tag{4.65}$$

Note that F_r and F_s are mappings equivalent to those used in the two-boundary and multisurface ($N = 2$) methods. For the two-dimensional interpolation, a product interpolation may be used as

$$F_{rs}(r, s) = F_r \cdot F_s \tag{4.66}$$

However, the product interpolation agrees with the boundary functions, F_{AB}, etc., only at the four corners $(0, 0)$, $(0, 1)$, $(1, 0)$, and $(1, 1)$, and *not along* the boundaries AB, etc. In order to achieve exact matching with the mapping functions everywhere on the boundaries, it is necessary to define a Boolean sum interpolation (Gordon and Hall, 1973)

$$F(r, s) = F_r(r, s) + F_s(r, s) - F_{rs}(r, s) \tag{4.67}$$

In practice (4.67) is implemented in two stages. In the first stage

$$F_r(r, s) = \sum_{i=0}^{1} \psi_i(r)F_b(i, s) \tag{4.68}$$

where b denotes the appropriate boundary, AD or BC. In the second stage

$$F(r, s) = F_r(r, s) + \sum_{j=0}^{1} \phi_j(s)[F_b(r, j) - F_r(r, j)] \tag{4.69}$$

The blending functions ψ_i and ϕ_j can be chosen just as for the two-boundary and multisurface methods. For example, the choice

$$\psi_0(r) = 1 - r, \quad \psi_1(r) = r, \quad \phi_0(s) = 1 - s, \quad \phi_1(s) = s \tag{4.70}$$

produces a transfinite bilinear interpolation (4.69) that suffers from the same problem as (4.50). Thus (4.69) and (4.70) generate an undesirable nonorthogonal grid adjacent to boundary surfaces. Gordon and Thiel (1982) proposed introducing interior surfaces to obtain better control of the interior grid. Eriksson (1982) specified parametric derivatives, e.g., $\partial^n F/\partial s^n$ up to $n = 3$, to provide near-orthogonal smoothly varying grids. This provides better control over the grid, specially in three dimensions.

Extension of the transfinite interpolation method to three dimensions is straightforward. The algorithm, (4.68) and (4.69), then has a third stage

$$F(r, s, t) = F_2(r, s, t) + \sum_{k=0}^{1} \omega_k(t)[F_b(r, s, k) - F_2(r, s, k)] \tag{4.71}$$

where $F_2(r, s, t)$ is equivalent to $F(r, s)$ in (4.69), and $\omega_k(t)$ are blending functions similar to $\psi_i(r)$ and $\phi_j(s)$. For more details, see Rizzi and Eriksson (1981).

4.4 ADAPTIVE GRID GENERATION

An adaptive grid is one that adapts itself to the solution of the physical problem by automatically clustering grid points in regions of high-flow-field gradients. It uses the properties of the flow-field solution to relocate the grid points in the physical domain. Adaptive grid generation must provide a means of communication between points so that a smooth distribution is maintained as points are repositioned. If the points must concentrate in some regions, they must not do so excessively as to make other regions devoid of points. The grid must not get highly skewed, which not only results in increased truncation error (Thompson et al., 1985), but excessive skewness may lead to the loss of linear independence of the coordinates. This implies that points cannot move independently; instead, each point is somehow coupled at least to its neighbors. Moreover, the grid points must not move too far or too fast or oscillations may occur. Also, the solution error or some other driving measure must be sensed and translated into the motion of the grid points via some mechanism. It should be noted that the use of an adaptive grid may not necessarily increase the computer time required for solution of the physical problem. Even though computations are necessary to determine the adapted grid, convergence properties of the solution may be improved and fewer grid points may be needed for a pre-specified accuracy.

Based on the above considerations, several methods for grid adaptation have been proposed. Most of them can be categorized under two approaches:

1. Moving the grid points under some sort of attraction and repulsion
2. Solving some elliptic system via the method of equidistribution

A brief discussion of the methods suggested under both these categories follows. The reader is referred to Thompson (1984b) for a survey.

4.4.1 Method of Attraction and Repulsion

The main contributors to this method have been Anderson and Rai (1982). In this method, the movement of points is accomplished by assigning to each point an attraction proportional to the difference between the magnitude of some measure of the error (or solution variation) and the average magnitude of this measure over all points. The points with values of this measure exceeding the average attract other points, thus reducing the spacing, while points with values of the measure less than the average repel other points, thereby increasing the spacing. The collective attraction of all points is then made to induce a velocity for each point. Since each point is influenced by all other points, it is effectively an elliptic grid generation system as well. In this approach, the variation measure used has been either the first or second or higher derivative of the physical solution. The use of a higher derivative is erratic both from the lack of accuracy in computing higher derivatives and because the error measure may be of degree higher than the physical solution itself. Moreover, this procedure does not take care of the smoothness or orthogonality of the grid. Thus, considerable distortion is possible.

4.4.2   The Principle of Equidistribution

Grid adaptation by solving some elliptic system basically involves the *equidistribution principle*. The grid points must be distributed so as to equidistribute some positive weight function, say $w(x)$ for a one-dimensional problem, over the field, i.e.,

$$\int_{x_i}^{x_{i+1}} w(x)\, dx = \text{constant} \qquad (4.72)$$

or, in discrete form,

$$w_i\, \Delta x_i = \text{constant} \qquad (4.73)$$

where Δx_i is the grid interval $(x_{i+1} - x_i)$, and w_i is the weight function assumed to be constant over the interval.

Ideally $w(x)$ should be the magnitude of the truncation error. Pereyra and Sewell (1975) and Davis and Flaherty (1982) have used truncation error as the weight function. Denny and Landis (1972) adapt the grid such that the truncation error vanishes at the grid points. This, however, concentrates the grid points where the solution is smooth, rather than in the high-gradient region. In general, the accurate calculation of the truncation error is difficult since the discrete representation of higher-order derivatives is progressively less accurate and subject to computational noise. Therefore, the most common approach has been to equidistribute some appropriate derivatives of the solution, as noted by Russell and Christiansen (1978). Since the largest numerical errors are found near high-gradient regions, the gradient of the solution is often the most suitable driving mechanism for grid adaptation.

The equidistribution of (4.73) is generally achieved via a transformation is which the physical domain (x) is transformed into the computational domain (ξ) of uniform grid through $x(\xi)$. Then (4.73) takes the form

$$x_\xi\, w(\xi) = \text{constant} \tag{4.74}$$

which is basically the Euler-Lagrange equation for the minimization of the integral

$$I_w = \int_0^1 w(\xi) x_\xi^2 \, d\xi \tag{4.75}$$

This can be taken to represent the energy of a spring system with spring constant $w(\xi)$ spanning each grid interval, assuming all points to have been expanded from a common point. The grid point distribution resulting from the equidistribution, thus, represents the equilibrium state of such a spring system. An alternative view point results from consideration of (4.74) in the form

$$x_\xi\, w(x) = \text{constant} \tag{4.76}$$

This is the Euler-Lagrange equation for minimization of the integral

$$I_s = \int_0^1 \frac{\xi_x^2}{w(x)} \, dx \tag{4.77}$$

which is the measure of smoothness of the point distribution; ξ_x may be considered as point density. This yields the smoothest grid point distribution attainable.

In practice, the weight function is expressed as an explicit function of the solution variable and its derivatives, so that both (4.74) and (4.76) simply become

$$x_\xi\, w = \text{constant} \tag{4.78}$$

and either approach can be used. White (1979) and Ablow (1982) have used the first approach with various forms of the weight function, but for one-dimensional adaptation only.

Dwyer (1984) and Dwyer et al. (1980, 1982) have applied the smoothness approach to a two-dimensional curvilinear coordinate system with the grid points constrained to move along one family of fixed coordinate lines. The weight function is taken in the form

$$w = 1 + \alpha \left| \frac{\partial f}{\partial s} \right| + \beta \left| \frac{\partial^2 f}{\partial s^2} \right| \tag{4.79}$$

where s is the arc length along the fixed coordinate curve, f is the field solution, and α and β are parameters adjusted to improve grid quality. The method, however, has some handicaps. It requires the generation and choice of a fixed set of grid lines in multidimensional problems. Not all problems have such preferred directions. Moreover, the second derivative term, due to large truncation errors, many a time leads to oscillations when the solution is too rough for accurate discrete representation of this derivative. Above all, the method exercises no control over the grid orthogonality and may lead to highly skewed and unstable grid systems. More choices for weight functions are available in Thompson et al. (1985).

4.4.3 Variational Approach

Brackbill (1982) and Saltzman and Brackbill (1982) developed a variational approach for adaptive grid generation. In their scheme, a function containing a measure of grid smoothness, orthogonality, and error distribution is minimized using variational principles. The smoothness of a two-dimensional grid is represented by the integral

$$I_s = \int_D [(\nabla \xi)^2 + (\nabla \eta)^2] \, dV \tag{4.80}$$

while a measure of orthogonality is provided by

$$I_0 = \int_D (\nabla \xi \cdot \nabla \eta)^2 J^3 \, dV \tag{4.81}$$

and the error distribution measure is given by

$$I_w = \int_D wJ \, dV \tag{4.82}$$

where w is a given weight function, J is the Jacobian of the transformation, and dV is an infinitesimal volume in the physical domain D. The transformation relating D and the computational domain is determined by minimizing a linear combination of the above three integrals, i.e.,

$$I = I_w + \lambda_0 I_0 + \lambda_s I_s \tag{4.83}$$

where λ_0 and λ_s are weighting coefficients. In order to minimize I, the Euler-Lagrange equations must be formed. Consider, for example, the minimization of I_s alone. When the variables are interchanged and integration is performed over the computational domain, the smoothness measure in (4.80) can be expressed as

$$I_s = \int \int \left[\frac{x_\xi^2 + x_\eta^2 + y_\xi^2 + y_\eta^2}{J} \right] d\xi \, d\eta \tag{4.84}$$

The Euler-Lagrange equations corresponding to minimization of I_s are

$$\left(\frac{\partial}{\partial x} - \frac{\partial}{\partial \xi}\frac{\partial}{\partial x_\xi} - \frac{\partial}{\partial \eta}\frac{\partial}{\partial x_\eta}\right)\left(\frac{x_\xi^2 + x_\eta^2 + y_\xi^2 + y_\eta^2}{J}\right) = 0$$
$$\left(\frac{\partial}{\partial y} - \frac{\partial}{\partial \xi}\frac{\partial}{\partial y_\xi} - \frac{\partial}{\partial \eta}\frac{\partial}{\partial y_\eta}\right)\left(\frac{x_\xi^2 + x_\eta^2 + y_\xi^2 + y_\eta^2}{J}\right) = 0$$

(4.85)

Performing the differentiation leads to

$$A(\alpha x_{\xi\xi} - 2\beta x_{\xi\eta} + \gamma x_{\eta\eta}) - B(\alpha y_{\xi\xi} - 2\beta y_{\xi\eta} + \gamma y_{\eta\eta}) = 0$$
$$- B(\alpha x_{\xi\xi} - 2\beta x_{\xi\eta} + \gamma x_{\eta\eta}) + C(\alpha y_{\xi\xi} - 2\beta y_{\xi\eta} + \gamma y_{\eta\eta}) = 0$$

(4.86)

where the coefficients A, B, C, α, β, and γ are functions of the metrics. If $B^2 \neq AC$, these equations may be written as

$$\alpha x_{\xi\xi} - 2\beta x_{\xi\eta} + \gamma x_{\eta\eta} = 0$$
$$\alpha y_{\xi\xi} - 2\beta y_{\xi\eta} + \gamma y_{\eta\eta} = 0$$

(4.87)

This is the basic form of the elliptic grid generation discussed in section 4.2.5. If I as defined in (4.83) is minimized, each integral I_o and I_w also contributes terms to yield a significantly more complicated set of Euler-Lagrange equations given by

$$b_1 x_{\xi\xi} + b_2 x_{\xi\eta} + b_3 x_{\eta\eta} + a_1 y_{\xi\xi} + a_2 y_{\xi\eta} + a_3 y_{\eta\eta} = \frac{-J^2}{2w} w_x$$
$$a_1 x_{\xi\xi} + a_2 x_{\xi\eta} + a_3 x_{\eta\eta} + c_1 y_{\xi\xi} + c_2 y_{\xi\eta} + c_3 y_{\eta\eta} = \frac{-J^2}{2w} w_y$$

(4.88)

where the a_i, b_i, c_i are functions of the metric coefficients and λ_o and λ_s. These equations are elliptic, and to obtain their solution is as formidable a task as obtaining the solution to the flow-field equations themselves. Thus, while the variational approach provides a sound mathematical basis for the grid, it involves additional effort in solving the complex partial differential equations just for generating the grid.

4.4.4 Tension and Torsion Spring Approach

Nakahashi and Deiwert (1984, 1986, 1987) developed a method that avoids solving the complex elliptic system (4.88) and still controls the orthogonality and smoothness along with error distribution of the grid. They suggest a concept of tension and torsion springs to model their technique; the spring constants being functions of the solution itself. Herein, we first describe its development for 1-D problems and follow its extension to 2-D and 3-D problems.

One-Dimensional Adaptation

For the 1-D adaptation, we follow the equidistribution approach discussed in section 4.4.2. The constant on the right side of Eq. (4.74) is evaluated by integrating (4.74) as follows:

$$\int_0^L x_\xi \, d\xi = \int_0^1 \frac{\text{const}}{w(\xi)} \, d\xi$$

or

$$\int_0^L dx = \int_0^1 \frac{\text{const}}{w(\xi)} \, d\xi$$

assuming the range of x to be from 0 to L and that of ξ to be normalized from 0 to 1. The constant is then evaluated to be

$$\text{const} = \frac{L}{\int_0^1 d\xi/w(\xi)} \tag{4.89}$$

Equation (4.74) now becomes

$$x_\xi = \frac{L}{w(\xi) \int_0^1 d\xi/w(\xi)} \tag{4.90}$$

With a uniform grid in the transformed space ξ, so that $\Delta\xi = 1/N$, where N is the number of grid intervals, Eq. (4.90) can be rewritten as

$$\Delta x_i = \frac{L}{w(\xi_i)N \int_0^1 d\xi/w(\xi)} \tag{4.91}$$

If x is also normalized from 0 to 1 and $w(\xi_i)$ is held fixed over the interval $(i, i + 1)$, Eq. (4.91), using (4.73), leads to

$$\Delta x_i = \frac{L}{w_i \sum_{j=1}^N (1/w_j)} \tag{4.92}$$

Nakahashi and Deiwert (1987) suggest the following form for the weight function:

$$w_i = 1 + A\bar{f}^B, \qquad \bar{f} = \frac{f_i - f_{\min}}{f_{\max} - f_{\min}} \tag{4.93}$$

which is normalized so that the first constant is unity, and when $A = 0$, the grid points are equally spaced. A and B are positive constants, f_i is the nonnegative flow solution or its gradient as discussed in section 4.4.2, and $f_{\min}$ and $f_{\max}$ are the minimum and maximum values of f_i. The grid spacing is determined by (4.92).

From (4.73), it follows that the ratio of maximum to minimum grid spacing is equal to the ratio of maximum to minimum weight function. Thus the constant A in (4.93) is given by

$$A = \frac{\Delta x_{max}}{\Delta x_{min}} - 1 \qquad (4.94)$$

The constant B in (4.93) is chosen such that the minimum grid spacing $\text{Min}(\Delta x_i)$ is equal to the specified minimum Δx_{min}. This can be achieved by finding solution to the equation

$$F(B) = \text{Min}(\Delta x_i) - \Delta x_{min} = 0 \qquad (4.95)$$

The Newton-Raphson method can be used to solve this equation iteratively using the relations

$$B^{k+1} = B^k - \Delta B^k \qquad (4.96)$$

where

$$\Delta B^k = \frac{F(B^k)}{\partial F(B^k)/\partial B} = \frac{\text{Min}(\Delta x_i) - \Delta x_{min}}{\partial \, \text{Min}(\Delta x_i)/\partial B} \qquad (4.97)$$

The quantity $\partial F/\partial B$ or $\partial \, \text{Min}(\Delta x_i)/\partial B$ can be evaluated using (4.92) and (4.93). Leaving the details to the reader (Das, 1986), we give the final result

$$\frac{\partial F}{\partial B} = \frac{\partial \, \text{Min}(\Delta x_i)}{\partial B} = A \sum_{j=1}^{N} \frac{\Delta x_i^2 w_i}{L w_j^2} \bar{f}_j^B \ln \bar{f}_j \qquad (4.98)$$

where w_i and Δx_i correspond to $\text{Min}(\Delta x_i)$.

The solution for B starts with a guess for B and Eqs. (4.92) through (4.98) are solved to find a new value of B. The physical problem solution is interpolated on the new grid resulting from (4.92) with the new value of B, and the whole process is repeated until an acceptable low value for $|1 - \text{Min}(\Delta x_i)/\Delta x_{min}|$ is achieved.

Two-Dimensional Adaptation

The self-adaptive concept developed above can be extended to two or three dimensions via a tension-torsion spring analogy, as discussed below. Basically, 2-D or 3-D adaptations are achieved as a sequence of 1-D adaptation in each direction.

Consider the adaptation in ξ-direction along a fixed η_j-curve (Fig. 4.11). The grid point A is connected to its neighboring points B and C along the η_j-curve through tension springs of constant $w_{i-1, j}$ and $w_{i, j}$ which control the movement of A along the η_j-curve, as in the case of 1-D adaptation via

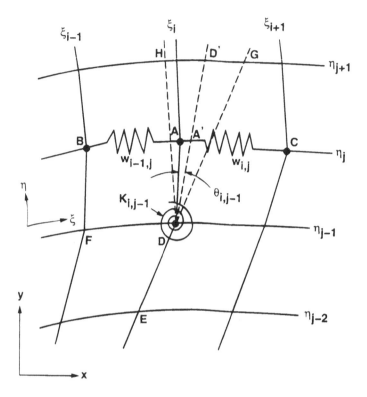

Figure 4.11 Spring analogy for 2-D grid adaptation.

the spring constant w_i. Equation (4.93) now changes to

$$w_{i,j} = 1 + A\bar{f}_{i,j}^{B}, \qquad \bar{f}_{i,j} = \frac{f_{i,j} - f_{min}}{f_{max} - f_{min}} \tag{4.99}$$

where $f_{i,j}$ is the field solution (or gradient of the solution) used to drive the grid adaptation, and f_{min} and f_{max} are the minimum and maximum values, respectively, of $f_{i,j}$ along the η_j-curve.

The constants A and B in (4.99) are calculated in a way similar to that in 1-D adaptation along the $j = 1$ boundary (η_1-curve). The same values of A and B are used for other η_j-curves in order to avoid additional computational effort. Equations (4.92) through (4.98) are used except that (i) Δx_i are replaced by $\Delta s_{i,j}$—the arc length between the points (i, j) and $(i + 1, j)$—and are approximated by arc secants, and (ii) x_i are replaced by $s_{i,j}$, where $s_{i,j}$ is the arc length along η_j-curve from the point $(1, j)$ to the point (i, j), so that

(4.92) takes the form

$$\Delta s_{i,j} = \frac{L}{w_{i,j} \sum_{k=1}^{N} (1/w_{k,j})} \tag{4.100}$$

In order to control smoothness *and* orthogonality, the movement of point A is also influenced by a torsional spring of constant $K_{i,j-1}$ attached at point D (Fig. 4.11). The torsional spring constant $K_{i,j-1}$ is taken to be proportional to the average of the tension spring constants $w_{i,j}$ along the line of adaptation (η_j) and is determined automatically by

$$K_{i,j-1} = \frac{\lambda}{N} \sum_{k=1}^{N} w_{k,j} \tag{4.101}$$

where λ is a user-specified parametric constant and is determined empirically. A value of $\lambda = 0.005$ was used by Nakahashi and Deiwert (1987).

The equilibrium of point A is maintained if the total force on A is zero; i.e., if

$$w_{i,j}(s_{i+1,j} - s_{i,j}) - w_{i-1,j}(s_{i,j} - s_{i-1,j}) + K_{i,j-1}\theta_{i,j-1} = 0 \tag{4.102}$$

where $\theta_{i,j-1}$ is the angle between the grid line DA and the reference line DD' (cf. Fig. 4.11). The reference line is taken to be the average of (i) the extension of the grid line ED (i.e., DG) and (ii) the normal to the grid line DF (i.e., DH), so as to enforce smoothness and orthogonality, respectively. Other reference lines, such as streamlines or shocks, could be used as well.

The solution of (4.102) is simplified if the last term on the left side of (4.102) is rewritten as

$$K_{i,j-1}\theta_{i,j-1} = T_{i,j-1}(s'_{i,j} - s_{i,j}) \tag{4.103}$$

where $s'_{i,j}$ is the arc length from (i, j) to the point A', which is the intersection of the reference line DD' with the η_j-line (cf. Fig. 4.11). The constant $T_{i,j-1}$ can be evaluated from (4.103), realizing that the term $(s'_{i,j} - s_{i,j})/\theta_{i,j-1}$ is the length of reference line DA'. Equations (4.102) and (4.103) finally yield

$$w_{i,j}s_{i+1,j} - (w_{i-1,j} + w_{i,j} + T_{i,j-1})s_{i,j} + w_{i-1,j}s_{i-1,j} = -T_{i,j-1}s'_{i,j} \tag{4.104}$$

Equation (4.104) represents a tridiagonal set of equations for $s_{i,j}$ along the η_j-curve and can be easily solved. We can then locate the new points in the physical space (x, y) via proper geometric relations (Das, 1986). Since the torsional force constraint is considered along with the equidistribution constraint, it is not possible to enforce equidistribution precisely. Hence the grid spacings are checked along each η_j-curve, and extreme clustering or coarseness is adjusted by modifying the weighting function $w_{i,j}$ locally. In this readjustment of $w_{i,j}$, the equidistribution concept of (4.100) can be applied locally.

We can now observe the advantage of using only one-sided influence of the torsional spring $T_{i,j-1}$, which led to the above tridiagonal system so that adaptation along all η_j-lines can be done in a marching fashion starting at the $j = 1$ line. Had we used the influence of torsional springs $T_{i,j-1}$ and $T_{i,j+1}$ from both sides, system (4.104) would no longer be tridiagonal due to the introduction of ellipticity, and the computational effort would have increased considerably without any additional benefit.

After adaptation along each η_j-line the solution is interpolated at the new points so as to be used for further adaptations along other η_j- and ξ_i-lines. After one complete adaptation in ξ- and η-directions, we can either re-solve the governing differential equations for the physical problem on the new grid or march ahead with the already interpolated values in a marching fashion.

Three-Dimensional Adaptation

The concept and analysis of 2-D adaptation can be easily extended to 3-D adaptation. Figure 4.12 shows the adaptation along $\zeta_{i,j,k}$-line whose configuration is fixed. The movement of point A is controlled by tension springs $w_{i,j,k}$ and $w_{i,j,k-1}$, as well as by torsional springs $K_{i-1,j,k}$ and $K_{i,j-1,k}$, instead of only one as in 2-D adaptation. Once again, for the equilibrium of point A, we have

$$w_{i,j,k}(s_{i,j,k+1} - s_{i,j,k}) - w_{i,j,k-1}(s_{i,j,k} - s_{i,j,k-1})$$
$$+ K_{i-1,j,k}\theta_{i-1,j,k} + K_{i,j-1,k}\theta_{i,j-1,k} = 0 \qquad (4.105)$$

where $\theta_{i-1,j,k}$ and $\theta_{i,j-1,k}$ are angles from the two reference lines DA' and EA'' as shown in Fig. 4.12. The last two terms of (4.105) can be rewritten as

$$K_{i-1,j,k}\theta_{i-1,j,k} = T_{i-1,j,k}(s'_{i,j,k} - s_{i,j,k})$$
$$K_{i,j-1,k}\theta_{i,j-1,k} = T_{i,j-1,k}(s''_{i,j,k} - s_{i,j,k}) \qquad (4.106)$$

where $s'_{i,j,k}$ and $s''_{i,j,k}$ are arc lengths from $(i, j, 1)$ to the points A' and A''—the points of intersection of DA' and EA'' with the $\zeta_{i,j,k}$-line, and $T_{i-1,j,k}$ and $T_{i,j-1,k}$ are set equal to $K_{i-1,j,k}$ and $K_{i,j-1,k}$ divided by DA' and EA'', respectively. Equations (4.105) and (4.106) then yield

$$w_{i,j,k-1}s_{i,j,k-1} - (w_{i,j,k-1} + w_{i,j,k} + T_{i,j-1,k} + T_{i-1,j,k})s_{i,j,k}$$
$$+ w_{i,j,k}s_{i,j,k+1}$$
$$= - T_{i-1,j,k}s'_{i,j,k} - T_{i,j-1,k}s''_{i,j,k} \qquad (4.107)$$

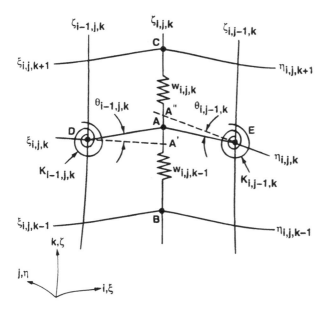

Figure 4.12   Spring analogy for 3-D grid adaptation.

This again is a tridiagonal set of equations for $s_{i,j,k}$ along $\zeta_{i,j,k}$ and can be easily solved. Further considerations are similar to those for 2-D adaptation. Some examples are given in Nakahashi and Deiwert (1986).

4.5   UNSTRUCTURED GRID GENERATION: AN INTRODUCTION

4.5.1   Introduction

One recent and currently maturing topic of research in computational fluid dynamics is the generation of unstructured grids over complex domains. The geometric information from the grid generator is then used to solve a system of partial differential equations governing a given flow regime in discretized form. Unstructured grid technology allows a general computer code to be written that can handle complex geometries. In the subsections that follow the unstructured grid will be defined and reasons for its usefulness stated. An example of unstructured grid generation in two dimensions is developed, and a data structure for unstructured triangular grids is discussed. Finally, some special computational techniques for use with unstructured grids are observed.

4.5.2   Unstructured Grids

What is an unstructured grid? Perhaps it can be best understood by comparing its features with those of a structured grid. As shown earlier there is a certain visual order in a structured mesh. In a given structured grid block a nodal point can reference its neighbors or any grid point in the entire domain by simply adding or subtracting integer values from its index number. This provides the necessary geometric information to compute distances between nodes and between cells. An unstructured grid has no such connection. An unstructured node can reference its neighborhood only by looking at a connectivity matrix where such information is stored.

The principal reason for using an unstructured grid generator in computational fluid dynamic simulations is its ability to generate grids about complex geometry. Another benefit is the ability to adaptively refine the mesh in areas that contain complex flow structures. For a strictly structured grid, an entire line of grid points must be added thus refining regions of the domain where resolution is not required. Grid embedding has been used successfully in structured grids by Davis and Dannenhoffer (1994) where a structured grid is subdivided selectively cell by cell to get the desired local grid resolution. This type of grid generation has been addressed earlier in this chapter.

4.5.3   Triangular Grid Generation

There are several techniques for generating an unstructured grid. One scheme involves covering a domain with triangular shapes. The next several subsections will review one method of generating a two-dimensional triangular unstructured mesh. The method presented below is similar to that of Holmes (1989).

Dirichlet Tessellation

There are several ways to generate an unstructured triangular mesh. One method is to simply place points in the domain of interest and connect the points by hand. This is easy for a small number of points in two dimensions but becomes very inefficient as the number of points increases. It becomes quite complex in three dimensions.

Another method is that of advancing front. Here a background grid is used to control the placement of points through some sort of interpolation. A grid is then generated by advancing triangles, in two dimensions, away from the boundaries generating a new front as the triangulation proceeds. The front continues to advance into the domain until the fronts meet and the entire region is triangulated.

The method described here is Delaunay triangulation. One can imagine a two-dimensional region of space surrounded by an outer boundary and possibly including an arbitrary number of interior boundaries. The domain can then be covered with points with some density distribution. Take any two points A and B (Fig. 4.13). A perpendicular bisection of the line connecting points A and B can be constructed. Any point in the plane to the logical right of intersection point C is closer to point B and any point in the plane to the logical left of the intersecting point C is closer to point A. A similar geometric construction can be made with a set of points. However, for a set of points the line is terminated when it intersects another perpendicular bisection of a different set of points since it would violate the same geometric construction requirements of that pair of points. This construction can be seen in Fig. 4.14. These regions are called Dirichlet polygons, and the entire polygon-covered region is called a Dirichlet tessellation. An interesting property of these polygons is that the vertices of the polygons are the centers of circles that pass through the three points whose edges are bisected by the sides of the polygon. Point A in Fig. 4.14 is an example of this property. The circle that has its center at the vertex of a Dirichlet polygon contains no other points of the domain other than the three points that can then be connected to form a triangle. The resulting triangle is said to satisfy the Delaunay criterion; for example, in Fig. 4.15, triangle 2 fails to meet the criterion since its circumcircle contains other points in the domain while triangle 1 satisfies the criterion.

Delaunay Triangulation

The procedure described above can be stated in the following way. A region can be triangulated by considering any three points to be the vertices of a

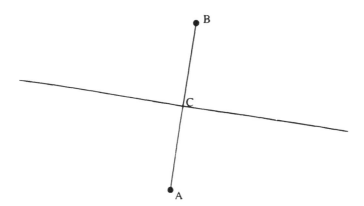

Figure 4.13 Perpendicular bisection of the line connecting points A and B.

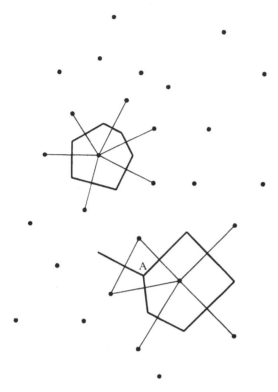

Figure 4.14 Dirichlet polygons.

valid triangle whose circumcircle, the circle that passes through the three points, contains no other points of the domain. A procedure can be developed to cover a domain with points and then triangulate the domain using the above criterion.

The method described in the following sections uses Delaunay criterion, but points will be inserted one at a time until the domain is covered with a given quality of triangulation. This allows the triangulation to be tailored to the type of problem that is being considered. A specific geometry will be used as an example of each step in generating a triangulation by point insertion. Clearly, the technique can be applied to other complex domains.

4.5.4 Triangulation by Point Insertion

In this section the steps to triangulate a region by point insertion will be described in detail. The first step is to define and discretize the boundaries.

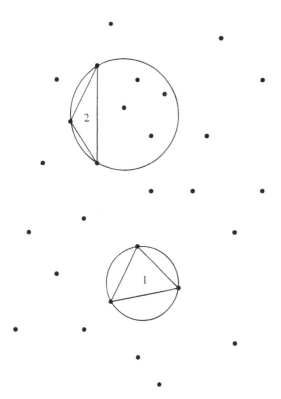

Figure 4.15 Delaunay triangulation.

Next the boundary discretized points are triangulated. Finally points are inserted into the domain using various geometric constraints. The region to be used in detailing the steps is shown in Fig. 4.16.

Boundary Definition/Discretization

As with any grid generation technique one must input the boundary description of the problem. The boundary is usually described as successive sets of points in two dimensions. Typically the distribution and number of points along a given boundary element is not optimum. This requires the capability of adding points to the boundary and redistributing them along the boundary to give a good initial boundary point resolution. This can be done using a cubic spline. Other techniques are also applicable (cf. section 4.3; also Thompson et al., 1985, Chap. 8). The initial point distribution that defines the boundaries for the example is shown in Fig. 4.17.

Figure 4.16 Boundary definition.

Spline Definition

After the initial discretization of the boundary segments, cubic splines can be used to curve-fit these boundary points. It only takes two points to define a straight line, but a cubic spline needs four points to be defined. For a given curve the given points are first parameterized and then spline-fit.

Figure 4.17 Boundary discretization.

For more information on cubic spline curve fitting, see Conte and de Boor (1980). The number and distribution of points along a given boundary segment is not ideal; so each splined segment is resolved to meet the expected flow or evident geometric complexity. A point distribution function can be used as described below for each boundary segment.

Point Redistribution

After the boundaries are curve-fit with the cubic spline, the points can be redistributed along the boundary by a given distribution function. These functions can allow stretching and clustering to be used to perhaps capture flow features known in advance. A simple yet effective distribution function that clusters grid points to the left is

$$s = t^2$$

or a function that clusters grid points to the right is

$$s = t(2 - t)$$

or a function that clusters grid points to the left and right is

$$s = t^2(3 - 2t)$$

In all cases, t can be varied from 0 to 1. Upon combining this function with a linear function, the amount of clustering can be controlled. For example, the equation that clusters grid points to the left and right is modified to give

$$s = \alpha t^2(3 - 2t) + (1 - \alpha)t$$

The parameter α varies from 0 to 1 and translates from the linear function (for $\alpha = 0$) to the nonlinear function (for $\alpha \neq 0$).

Similarly one can cluster at the ends and between the ends. The control of clustering at the ends and the location of the central clustering are given by the function

$$s = \alpha\beta t^2(3 - 2t) + (1 - \alpha)\beta t$$

for $0 \leq t \leq \beta$, and by

$$s = \alpha(1 - \beta)t^2(3 - 2t) + (1 - \alpha)(1 - \beta)t + \beta$$

for $\beta \leq t \leq 1$. The parameter α is defined as before, and β controls the location of the clustering between the ends.

The above allows a desired number of nodes to be distributed along the boundaries. The clustering of these nodes for the domain is shown in Fig. 4.18.

Initial Domain Triangulation

Once the boundaries are discretized, the boundary nodes are triangulated using the Delaunay criterion. Effectively this means that all boundary points are triangulated with other boundary points as long as the Delaunay criterion is not violated. A result of this is to give an initial triangulation of the boundary domain.

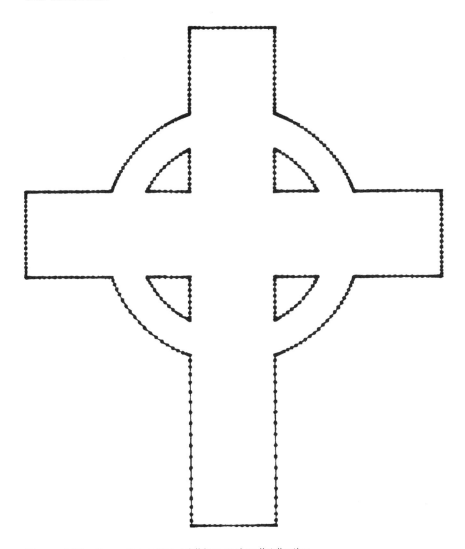

Figure 4.18 Boundary point addition and redistribution.

Point Insertion Algorithms

The next step is to determine how to selectively insert points in the domain to give a reasonable computational grid. For example, a different grid will be generated if the fluid flow of interest is viscous or inviscid. The result can be used as an initial grid for a code that uses adaptive grid refinement scheme to capture detailed flow features. What follows is a list of typical point insertion criteria. The first three criteria insert the point at the center of the circumcircle. The next two place the points at specific locations relative to the boundary data. These geometric constraints can be used in any combination, and the definition of each criterion will be given below. This list is not exhaustive. Another criterion could be to input points from a list and generate a grid using the same triangulation algorithm.

The aspect ratio of a triangle is one criterion that can be used to insert points in a domain. It is defined as

$$\text{Aspect ratio} = \frac{r_i}{2r_c}$$

where r_i is the radius of the inscribed circle of the triangle,

$$r_i = \frac{\sqrt{s(s-a)(s-b)(s-c)}}{s}$$

and r_c is the radius of the circumscribed circle of the triangle,

$$r_c = \frac{abc}{4\sqrt{s(s-a)(s-b)(s-c)}}$$

The variable s is defined as the semiperimeter of a triangle,

$$s = \frac{a+b+c}{2}$$

and the quantities a, b, c refer to the lengths of the sides of the triangle. A new point is inserted at the circumcenter of the triangle with the smallest aspect ratio.

Another refinement criterion is based on the area of the triangle,

$$\text{Area} = \sqrt{s(s-a)(s-b)(s-c)}$$

where s is the semiperimeter. A new point is placed at the circumcenter of the triangle with the largest area.

A third criterion is based on the radius of the circumcircle of the triangle defined above. Again the domain is searched and a new point is inserted at the circumcenter of the triangle with the largest circumcircle radius. Triangles with both bad aspect ratios and large areas are affected. An example of the combined use of the above criteria is shown in Fig. 4.19.

Points can also be inserted based on the boundary data. Specifically, if a flow code needs certain types of cells near a boundary, points can be inserted to generate that type of cell. The two criteria described below generate points that are normal to a boundary edge and points that are normal to boundary points (in an average sense).

Figure 4.19 Triangulation of domain.

A point is inserted at a location normal to an edge to generate triangles along a surface. This algorithm requires no searching of the domain. In general, no stretching normal to the edge is used for these points. This type of point insertion results in a good inviscid grid. An example of using this type of point insertion algorithm is shown in Fig. 4.20.

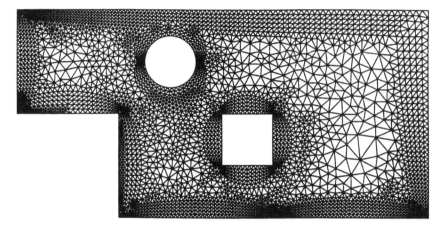

Figure 4.20   Point insertion via edge normal and aspect ratio.

Points can also be inserted normal to the points along a boundary. Here the normal to a point is defined as the average of the normals of edges that meet to form the point. Again no search routine is needed. Points are added to specified boundaries. Here a stretching parameter can be used in the normal direction to generate boundary layer type grid resolution. An example of this point insertion algorithm can be seen in Fig. 4.21. Here the boundary layer type of grid is generated along solid wall boundaries.

It should be noted that many of the above geometric computations use the square root function. This is an expensive operation and can be elimi-

Figure 4.21   Point insertion via point normal and maximum area.

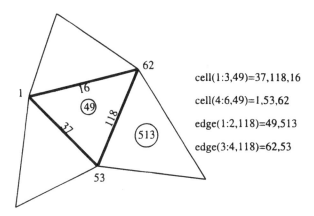

cell(1:3,49)=37,118,16

cell(4:6,49)=1,53,62

edge(1:2,118)=49,513

edge(3:4,118)=62,53

Figure 4.22   Data structure.

nated from the exhaustive search algorithm. Also special attention must be paid to inside corners. Special logic can be implemented to ensure that points are not generated which are in close proximity to one another.

When the initial boundary grid triangulation as well as when a point insertion algorithm is used, it is important to make sure that the resulting triangle and the new point lie within the domain. If the proper checks are not performed, the grid can break through the domain boundaries resulting in a failure. There are several methods, usually referred to as a point-in-polygon check, to make sure this does not happen. Two point-in-polygon checking procedures are outlined by O'Rourke (1994).

4.5.5   Data Structures

Different simulation codes require different geometric data for use in a computation. It is the job of the grid generator to produce complete data files that these codes can read and manipulate into a specific data structure. Grid data files can be composed of floating point data and integer data. The only floating point quantities produced by a grid generator are the x, y node point coordinates. All other grid quantities are of integer type. Typically a cell data structure will contain the three node numbers and the three edge numbers that make up a given triangular element. Another data structure is associated with the edge. An edge is shared by two cells and two nodes. Finally the node is the most basic element. It can be shared by a variable number of edges and a variable number of cells. This information is stored in arrays that have lengths based on the total number of cells, edges, and nodes. With these arrays it is easy to obtain the necessary grid geometric quantities. For example, Fig. 4.22 shows a cell with its node numbers and edge numbers. It is straightforward to find the immediate neighbor of cell i (49) by referencing edge k (118). The two cell integer numbers that border edge k can then be added and the neighboring cell number is obtained by subtracting the integer value of the current cell, i (e.g., $(49 + 513) - 49 = 513$). With this structure it is not necessary to store the neighbors of cell i.

Boundary conditions such as solid wall, inlet, and exit information may be defined explicitly through the edge array. This can be accomplished by setting the second storage element of the edge array to a negative integer number that can be used to set the boundary condition in the flow code.

4.5.6 Special Features

Grid Coloring/Removing Recursion

It is often advantageous to write algorithms to take advantage of special computer hardware. Computational codes often suffer from recursion where the solution in a given cell is dependent on a level of neighboring cells. This property does not allow the code to vectorize. The solution is to color the grid such that the solution can be computed on a given color array which does not touch any other cells of the same color. The four-color theorem has been applied successfully to unstructured grids to remove recursion from immediate neighbor cells. Special techniques must be used to make the number of cells in each color array approximately equal. This has been used successfully by Jorgenson and Pletcher (1996) to vectorize an implicit viscous flow solver using unstructured grids.

Grid Reordering/Bandwidth Reduction

For implicit flow solvers with structured grids the discretized equations produce a banded matrix equation of the form $\mathbf{Ax} = \mathbf{b}$, where $\mathbf{A}$ is a block-banded matrix. For unstructured grids the bandwidth of $\mathbf{A}$ could potentially be the entire domain; for example, cell 1 could neighbor the cell with the highest integer number. For some implicit methods it is advantageous to renumber the cells to reduce this bandwidth. The Cuthill-McKee (1969) algorithm can be used to reduce the bandwidth of the matrix by reordering the cell numbering.

Parallel Considerations/Load Balancing

Reducing the bandwidth of the matrix has the added benefit of simplifying the parallelization and load balancing of an unstructured code. Many structured codes that use parallel architecture to give timely solutions do so by using the natural divisions given by a multiblock grid or by dividing a single grid into several subgrids for computing on multiple processors. This causes some problems and extra coding for doing the load balancing with structured grids which may include the need for interpolation if block interfaces are of different size or misaligned. With unstructured triangular grids the neighbors of a cell are known through the grid connectivity. Also load balancing is straightforward since the entire grid can be simply divided by the number of processors. Structured grids may require reblocking or subdividing. There are interface minimization techniques which reduce the amount of communication between processors for unstructured grids. Though it is not covered in this introduction, a review of this material can be found in Venkatakrishnan et al. (1992).

4.5.7   Future Directions for Unstructured Grid Generation

The techniques in this section are all extendable to three dimensions. Baker (1994) has shown this extension. Recently there has been an effort in mixing the positive aspects of structured grids with those seen in unstructured grids. This is especially important when high-speed viscous flows are considered. Connell and Braaten (1994) have developed a grid generator that produces prismatic cells for the viscous flow near boundaries and tetrahedral cells to fill the volume core of the flow domain. It is impossible to list all the contributors to unstructured grid generation. A list and links to researchers doing work in the general area of grid generation can be obtained from the World Wide Web.*

4.6   CLOSURE

In this chapter various techniques for generation of structured as well as unstructured grids have been described. While conformal grids, if possible to construct, lead to the simplest form of transformed governing equations for the physical problem, they sometimes generate extreme clustering and extreme sparsity of the grid. In such a case, one-dimensional stretching

---

* The current web address for this list is http://137.226.136.7/ ~ roberts/ peoplelist.html.

functions (section 4.3.1) can generate a more evenly spaced grid but at the expense of producing an orthogonal rather than a conformal grid. Strict orthogonality along with adequate control of the grid point distribution is difficult to achieve, specially for complex geometries. However, a grid that is nearly orthogonal to the boundaries is highly desirable in order to be able to enforce the boundary conditions of the physical problem with minimum error. Excessive skewness of the grid should be avoided throughout the domain.

Grid generation via the solution of elliptic partial differential equations has two merits: (1) discontinuities on the boundary are not propagated into the interior, and (2) smoothness of the interior grid allows numerical evaluation of the metric coefficients with smaller truncation errors. It also allows good control over grid point distribution and orientation of the grid with the boundaries.

The algebraic grid generation scheme offers good control over the interior grid besides being computationally efficient. It also enables generation of locally orthogonal grids at the boundaries. Unstructured grids are the easiest to generate but are unable to resolve viscous layers or shocks, i.e., regions of high gradients.

NOMENCLATURE

a, b, c	length of the sides of a triangle (section 4.5.4)
$\mathbf{F}$	surface vector function (section 4.3)
g_{ij}	metric tensor
h	scale factor
J	Jacobian of the coordinate transformation
K	torsional spring constant (section 4.4.4)
P, Q	control functions (section 4.2.5)
r	radial coordinate; also position vector
s	arc length; also semiperimeter of a triangle (section 4.5.4)
$\mathbf{V}$	tangent vector function (section 4.3)
w	weight function
x, y, z	Cartesian coordinates; z also a complex variable $(= x + iy)$

Greek

α	angle
Δ	a small change
θ	angle

ξ, η, ζ curvilinear coordinates

Subscripts

x, y partial derivative with respect to x or y
ξ, η partial derivative with respect to ξ or η

REFERENCES

Ablow CM. Equidistant mesh for gas dynamic calculations. In: Thompson JF, ed. Numerical Grid Generation. Amsterdam: North-Holland, 1982, pp 859–863.

Alter SJ, Weilmuenster KJ. The three-dimensional multi-block advanced grid generation system (3DMAGGS). NASA TM 108985, 1993.

Anderson DA, Rai MM. The use of solution adaptive grids in solving partial differential equations. In: Thompson JF, ed. Numerical Grid Generation. Amsterdam: North-Holland, 1982, pp 317–338.

Anderson DA, Tannehill JC, Pletcher RH. Computational Fluid Mechanics and Heat Transfer. Washington DC: Hemisphere, 1984.

Aris R. Vectors, Tensors and the Basic Equations of Fluid Dynamics. Englewood Cliffs NJ: Prentice Hall, 1962.

Arnone A. Notes on the use of the JERRY and TOM grid generation codes, 1992, unpublished.

Baker T. Point placement and control of triangle quality for inviscid and viscous mesh generation. Proceedings of 4th International Grid Generation Conference, 1994, pp 137–149.

Bauer F, Garabedian P, Korn D, Jameson A. Supercritical wing sections II. Lecture Notes in Economics and Mathematical Systems. Vol. 108. New York: Springer-Verlag, 1975.

Brackbill JU. Coordinate system control: adaptive meshes. In: Thompson JF, ed. Numerical Grid Generation. Amsterdam: North-Holland, 1982, pp 277–294.

Carey GF. Computational Grids, Generation, Adaptation, and Solution Strategies. Bristol, PA: Taylor & Francis, 1997.

Chawner JR, Steinbrenner JP. Automatic structured grid generation using Gridgen (some restrictions apply). In: Choo YK, comp. Surface Modeling, Grid Generation, and Related Issues in Computational Fluid Dynamic (CFD) Solutions. NASA CP-3291, 1995, pp 463–476.

Choo YK (compiler). Surface Modeling, Grid Generation, and Related Issues in Computational Fluid Dynamic (CFD) Solution. Cleveland, OH: NASA CP-3291, 1995.

Connell S, Braaten M. Semi-structured mesh generation for 3D Navier-Stokes calculations. GE Report 94CRD154, 1994.

Conte SE, de Boor C. Elementary Numerical Analysis. New York: McGraw-Hill, 1980, pp 289–293.

Cooley JW, Tukey JW. An algorithm for the machine calculation of complex Fourier series. Math Comput 19: 297–301, 1965.

Cuthill E, McKee J. Reducing the bandwidth of sparse symmetric matrices. Proceedings of 24th National Conference, Association for Computing Machinery, 1969, pp 157–172.

Dannenhoffer JF III. A technique for optimizing grid blocks. In: Choo YK, comp. Surface Modeling, Grid Generation, and Related Issues in Computational Fluid Dynamic (CFD) Solutions. NASA CP-3291, 1995, pp 751–762.

Das AK. Adaptive grid generation and its applications. M Tech thesis, Department of Mechanical Engineering, Indian Institute of Technology, Kanpur, India, 1986.

Davis RL, Dannenhoffer JF. Three-dimensional adaptive grid-embedding Euler technique. AIAA J 21: 1167–1174, 1994.

Davis RT. Numerical methods for coordinate generation based on Schwarz-Christoffel transformation. AIAA paper 79-1463, 1979.

Davis SF, Flaherty JE. An adaptive finite element method for initial-boundary value problems for partial differential equations. SIAM J Sci Stat Comput 3: 6–27, 1982.

Denny VE, Landis RB. A new method for solving two-point boundary value problems using optimal node distribution. J Comput Phys 9: 120–137, 1972.

Dwyer HA. Grid adaptation for problems in fluid dynamics. AIAA J 22: 1705–1712, 1984.

Dwyer HA, Kee RJ, Sanders BR. Adaptive grid method for problems in fluid mechanics and heat transfer. AIAA J 18: 1205–1212, 1980.

Dwyer HA, Smooke HD, Kee RJ. Adaptive gridding for finite difference solution to heat and mass transfer problems. In: Thompson JF, ed. Numerical Grid Generation. Amsterdam: North-Holland, 1982, pp 339–356.

Eiseman PR. Automatic algebraic coordinate generation. In: Thompson JF, ed. Numerical Grid Generation. Amsterdam: North-Holland, 1982d, pp 447–463.

Eiseman PR. Coordinate generation with precise controls over mesh properties. J Comput Phys 47: 331–351, 1982b.

Eiseman PR. Grid generation for fluid mechanics computations. Ann Rev Fluid Mech 17: 487–522, 1985.

Eiseman PR. High level continuity for coordinate generation with precise controls. J Comput Phys 47: 352–374, 1982c.

Eisenman PR. Multiblock grid generation with automatic zoning. In: Choo YK, comp. Surface Modeling, Grid Generation, and Related Issues in Computational Fluid Dynamic (CFD) Solutions, NASA CP-3291, 1995, pp 143–162.

Eiseman PR. A multi-surface method of coordinate generation. J Comput Phys 33: 118–150, 1979.

Eiseman PR. Orthogonal grid generation. In: Thompson JF, ed. Numerical Grid Generation. Amsterdam: North-Holland, 1982a, pp 193–234.

Eiseman PR, Smith RE. Mesh generation using algebraic techniques. In: Smith RE, ed. Numerical Grid Generation Techniques. NASA CP-2166, 1980, pp 73–120.

Eriksson LE. Generation of boundary-conforming grids around wing-body configurations using transfinite interpolation. AIAA J 20: 1313–1320, 1982.

Fletcher CAJ. Computational Techniques for Fluid Dynamics. 2nd ed. Berlin: Springer, 1991.

Garg VK. Comparison of predicted and experimental heat transfer on a film-cooled rotating blade using a two-equation turbulence model. ASME Paper 97-GT-220, 1997.

Gordon WJ, Hall CA. Construction of curvilinear coordinate systems and application to mesh generation. Int J Numer Meth Eng 7: 461–477, 1973.

Gordon WJ, Thiel LC. Transfinite mappings and their application to grid generation. In: Numerical Grid Generation, Thompson JF, ed. Amsterdam: North-Holland, 1982, pp 171–192.

Grossman B. Numerical procedure for the computation of irrotational conical flows. AIAA J 17: 828–837, 1979.

Grossman B, Siclari MJ. The non-linear supersonic potential flow over delta wings. AIAA paper 80-0269, 1980.

Hall DW. A three-dimensional body-fitted coordinate system for flow field calculations on asymmetric nosetips. In: Smith RE, ed. Numerical Grid Generation Techniques. NASA CP-2166, 1980, pp 315–328.

Holmes DG. Inviscid 2D solutions on unstructured, adaptive grids. von Karman Institute for Fluid Dynamics Lecture Series, 1989–06, 1989.

Ives DC. Conformal grid generation. In: Numerical Grid Generation, Thompson JE, ed. Amsterdam: North-Holland, 1982, pp 107–135.

Ives DC. A modern look at conformal mapping including multiply connected regions. AIAA J 14: 1006–1011, 1982.

Ives DC, Menor WA. Grid generation for inlet and inlet-centerbody configurations using conformal mapping and stretching. AIAA paper 81-0997, 1981.

Jameson A. Iterative solution of transonic flows over airfoils and wings, including flows at Mach 1. Commun Pure Appl Math 27: 283–309, 1974.

Jameson A, Caughey DA. A finite volume method for transonic potential flow calculations. AIAA paper 77–635, 1977.

Jorgenson PCE, Pletcher RH. An implicit numerical scheme for the simulation of internal viscous flows on unstructured grids. Computers & Fluids 25: 447–466, 1996.

Kerlick DG, Klopfer GH. Assessing the quality of curvilinear coordinate meshes by decomposing the Jacobian matrix. In: Numerical Grid Generation. Thompson JF, ed. Amsterdam: North-Holland, 1982, pp 787–807.

Kober H. Dictionary of Conformal Representations. New York: Dover, 1957.

Milne-Thomson L. Theoretical Hydrodynamics. London: Macmillan, 1968.

Moretti G. Conformal mappings for computations of steady, three-dimensional, supersonic flows. Numerical/Laboratory Computer Methods in Fluid Mechanics, ASME, 1976.

Nakahashi K, Deiwert GS. A practical adaptive-grid method for complex fluid flow problems. Lecture Notes in Physics. Vol. 218, Springer-Verlag, 1984, pp 422–426; also NASA TM 85989.

Nakahashi K, Deiwert GS. Self-adaptive-grid method with application to airfoil flow. AIAA J 25: 513–520, 1987.

Nakahashi K, Deiwert GS. Three-dimensional adaptive grid method. AIAA J 24: 948–954, 1986.

Nehari Z. Conformal Mapping. New York: Dover, 1952.

O'Rourke J. Computations Geometry in C. Oxford: Cambridge University Press, 1994, pp 233–238.

Pereyra V, Sewell EG. Mesh selection for discrete solution of boundary value problems in ordinary differential equations. Numer Math 23: 261–268, 1975.

Program Development Corporation. GridPro™/az3000—User's Guide and Reference Manual. White Plains NY, 1997.

Raithby GD. Skew upstream differencing schemes for problems involving fluid flow. Comput Methods Appl Mech Eng 9: 153–164, 1976.

Rigby DL. Method of weakest descent for automatic block merging. 15th International Conference on Numerical Methods in Fluid Dynamics. Monterey CA; 1996.

Rigby DL, Steinthorsson E, Coirier W. Automatic block merging using the method of weakest descent. AIAA Paper 97-0197, 1997.

Rizzi A, Eriksson LE. Transfinite mesh generation and damped Euler equation algorithm for transonic flow around wing-body configurations. AIAA paper 81-0999, 1981.

Russel RD, Christiansen J. Adaptive mesh selection strategies for solving boundary value problems. SIAM J Numer Anal 15: 59–80, 1978.

Saltzman J, Brackbill J. Applications and generalizations of variational methods for generating adaptive meshes. In: Thompson JF, ed. Numerical Grid Generation. Amsterdam: North-Holland, 1982, pp 865–884.

Smith RE. Algebraic grid generation. In: Thompson JF, ed. Numerical Grid Generation. Amsterdam: North-Holland, 1982, pp 137–170.

Smith RE. Three-dimensional algebraic grid generation. AIAA paper 83-1904, 1983.

Sorenson RL. A computer program to generate two-dimensional grids about airfoils and other shapes by the use of Poisson's equations. NASA TM 81198, 1980.

Sorenson RL. The 3DGRAPE Book: Theory, Users' Manual, Examples. NASA TM 102224, 1989.

Sorenson RL, Alter SJ. 3DGRAPE/AL: The Ames/Langley technology upgrade. In: Choo YK, comp. Surface Modeling, Grid Generation, and Related Issues in Computational Fluid Dynamic (CFD) Solutions. NASA CP-3291, 1995, pp 447–462.

Sridhar KP, Davis RT. A Schwarz-Christoffel method for generating internal flow grids. Symposium on Computers in Flow Prediction and Fluid Dynamic Experiments. ASME, 1981, pp 35–44.

Steger JL, Dougherty FC, Benek JA. A chimera grid scheme. In: Ghia KN, Ghia U, eds. Advances in Grid Generation. ASME FED. Vol 5, 1983, pp 59–70.

Steinbrenner JP, Chawner JR. The Gridgen User Manual: Version 10. Bedford, TX: Pointwise, Inc., 1995.

Steinthorsson E, Ameri AA, Rigby DL. TRAF3DMB—a multi-block flow solver for turbomachinery flows. AIAA paper 97-0996, 1997.

Temperton C. Direct methods for the solution of the discrete Poisson equation: some comparisons. J Comput Phys 31: 1–20, 1979.

Theodorsen T, Garrick IE. General potential theory of arbitrary wing sections. NACA TR 452, 1933.

Thomas PD, Middlecoff JF. Direct control of the grid point distribution in meshes generated by elliptic equations. AIAA J 18: 652–656, 1980.

Thompson JF. General curvilinear coordinates systems. In: Thompson JF, ed. Numerical Grid Generation. Amsterdam: North-Holland, 1982a, pp 1–30.

Thompson JF. Grid generation techniques in computational fluid dynamics. AIAA J 22: 1505–1523, 1984a.

Thompson JF. A survey of dynamically-adaptive grids in the numerical solution of partial differential equations. AIAA paper 84–1606, 1984b.

Thompson JF. WESCOR—boundary-fitted coordinate code for general 2D regions with obstacles and boundary intrusions. Contractor Rep, Vicksburg, MS: US Army Engineer Waterways Experiment Station, 1982b.

Thompson JF, Thames FC, Mastin CW. TOMCAT—a code for numerical generation of boundary-fitted curvilinear coordinate systems on fields containing any number of arbitrary two-dimensional bodies. J Comput Phys 24: 274–302, 1977.

Thompson JF, Warsi ZUA, Mastin CW. Boundary-fitted coordinate systems for numerical solution of partial differential equations—a review. J Comput Phys 47: 1–108, 1982.

Thompson JF, Warsi ZUA, Mastin CW. Numerical Grid Generation, Foundations and Applications. New York: North-Holland, 1985.

Venkatakrishnan V, Simon HD, Barth TJ. A MIMD implementation of a parallel Euler solver for unstructured grids. J Supercomputing 6: 117–137, 1992.

Vinokur M. On one-dimensional stretching functions for finite-difference calculations. J Comput Phys 50: 215–234, 1983.

von Karman T, Trefftz E. Potential-stromung um gegebene Tragflachenquerschnitte, Z. Flugtechnische Moforluftsch 9: 111–116, 1918.

White AB, Jr. On selection of equidistributing meshes for two-point boundary-value problems. SIAM J Numer Analysis 16: 472–502, 1979.

5

Inlet, Duct, and Nozzle Flows

Charles E. Towne
NASA Lewis Research Center, Cleveland, Ohio

5.1   INTRODUCTION

The use of computational fluid dynamics (CFD) for the design and analysis of flow through inlets, ducts, and nozzles has increased greatly in the last few years. Several factors have contributed to this growth, including (1) complex geometric design requirements, leading to flow phenomena that are outside our established base of experience and may not be intuitively predictable; (2) high fuel costs, leading to potentially large cost savings for even small performance improvements; (3) high cost and/or lack of facilities for extensive experimental testing; (4) continued development and improvement of sophisticated numerical algorithms for solving the complex equations governing fluid flow; and (5) tremendous improvements in computational power.

It is interesting to note, though, that some of the analysis methods used in CFD predate the computer. In 1929 Prandtl and Busemann applied the method of characteristics using a graphical technique to design a two-dimensional supersonic nozzle. A few years later they designed a nozzle for the first practical supersonic wind tunnel (Anderson, 1982). Not surprisingly, though, the real growth of CFD as a practical way of solving real-world problems began with the introduction of the digital computer in about the mid-1960s.

Today CFD is being applied by industry, government, and universities to duct flow problems in a wide variety of areas. The principal users are probably in the aeronautics and space areas, for flows in a variety of ducts, such as jet engine inlets and nozzles, fuel lines and storage tanks, wind tunnels, and rocket nozzles. It is also being used in many other areas, including the automotive industry to design air ducts, cooling lines, and hydraulic lines; the heating and air conditioning industries to design air ducts and to study room airflows; and medicine to study blood flow.

Although CFD is also being widely used for compressors, turbines, and combustors, the focus in this chapter is on nonrotating and nonreacting flows. It should also be noted that the author's area of interest is inlets, ducts, and nozzles for airbreathing propulsion systems, and some of that bias will no doubt be apparent in the following discussion.

In this chapter we will first describe, in general terms, the steps involved in applying CFD to an inlet, duct, or nozzle problem. Two examples will then be presented to show how CFD is being applied to real problems in the aerospace field today. The first deals with using CFD to help design a subsonic inlet for a new general aviation jet aircraft, with the focus on lowering the total pressure distortion levels at the duct exit. This example is described in quite a bit of detail to clearly illustrate the steps involved in a real-world application. In the second example, CFD is being used as a critical tool in the reduction of data from a hypersonic engine experiment, allowing thrust data to be obtained that could not be obtained in any other practical way. We will conclude with a discussion of the current status of CFD for inlet, duct, and nozzle applications, and what needs to be done if CFD is to become more widely used for routine real-world design work.

5.2 CFD SOLUTION PROCESS

5.2.1 Gather Information/Choose Analysis Method and Flow Models

The first step in any CFD study is to gather the necessary information about the problem. The amount of detail required will depend to some extent on how the results will be used. For example, a calculation being done to quantify the various sources of total pressure loss in an air duct will likely require a high level of fidelity in simulating the problem, with a detailed description of the actual geometry, incoming flow profiles, boundary conditions, etc. Conversely, if a parametric study is being done to determine which of several ducts has the lowest losses, the absolute loss level may not be critical, and a lower fidelity simulation may be acceptable.

The information gathered in this step will be used to determine (1) the type of CFD analysis and flow modeling (e.g., turbulence modeling) that is appropriate for the problem; (2) the particular code to be used, based on how well its features and capabilities match those needed to solve the problem; and (3) the input parameters that will be used when running the code.

Some knowledge, or at least an experience-based guess, about the type of physical phenomena expected in the flow is required at this stage. The more that is known about the flow, the better the CFD simulation is likely to be. Following are some of the things that should be considered at this point.

Will the Flow Be Compressible or Incompressible?

Depending on the application, compressibility effects may become important for Mach numbers above 0.3 or so. A CFD code designed to compute incompressible flow will be useless for a compressible problem. A compressible code may be used at low Mach numbers, but iterative methods may take longer, possibly much longer, to converge. Some compressible codes, however, use numerical techniques such as matrix preconditioning that allow them to be used at "incompressible" conditions without suffering from slower convergence rates.

Will the Boundary Layers Be Thin or Thick?

If viscous effects can be neglected, Euler or even potential flow methods are probably the ones to use. They should be faster, both because the viscous terms do not have to be computed and because fewer grid points are needed since there are no shear layers to resolve. If viscous effects are important, but the boundary layers are thin, a boundary layer method may be used in conjunction with an inviscid method to compute these effects. If the boundary layers are thick, a fully viscous method such as a parabolized or Reynolds-averaged Navier-Stokes analysis will be necessary.

Are Regions of Separated Flow Possible?

Some parabolized codes include approximations that allow them to compute flows with small separation bubbles, but larger regions of flow separation can only be computed by solving the Navier-Stokes equations.

Will the Flow Be Laminar, Turbulent, or Transitional?

If the flow is turbulent or transitional, a turbulence model must be used. There are a wide variety of turbulence models in use today, and most CFD codes for viscous flow have more than one to choose from. In most cases, the choice will be between an algebraic model, a one-equation model, and a two-equation model.* Unfortunately, there is no single turbulence model that works well for all types of flow, and experience is invaluable when choosing an appropriate model for a particular problem. As one might expect, algebraic, or zero-equation, models are the simplest and fastest, and are most suitable for relatively simple attached flows. Two-equation models are more complicated and slower, but are more suitable for complex flows with shear layer interactions and/or flow separation. One-equation models fall somewhere in between. Some, but by no means all, of the models may be able to predict laminar-turbulent transition, but this capability is even less mature than the prediction of fully turbulent flows. Turbulence modeling is currently a very active research area, and will probably remain so for at least several more years.

---

* The terminology "one-equation" and "two-equation" refers to the number of differential equations that are solved in the model. There are even more complex models, called Reynolds stress models, that are being used, but in general these are not yet practical for most engineering applications. There are also codes that actually compute, as opposed to model, the turbulent eddies. These codes are *very* long-running, even for simple configurations, and this is still very much a research area.

Will Shock Waves Be Present?

Shock waves are always a possibility in transonic or supersonic flow. Some CFD codes for inviscid flows do shock fitting, in which the shock waves are computed separately and "fitted" into the solution as infinitely thin discontinuities. Most codes, though, do shock capturing, in which the shocks are automatically "captured" during the solution of the governing flow equations. With these codes, the discontinuity is smeared, typically across 3–5 grid points.

Will Three-Dimensional Effects Be Important?

All real flows are three-dimensional, but if the three-dimensional effects can be neglected, using a 2-D code will be significantly faster than using a 3-D code. Be aware though, that three-dimensionality may be important in a nominally "2-D" flow. One example is flow through a rectangular cross-sectioned supersonic inlet, where the sidewall boundary layers, and their interaction with the shock waves from the ramp, can play a critical role in determining the distribution of flow in the inlet.

Will Real-Gas Effects Be Significant?

Most CFD codes assume that the fluid is thermally perfect, i.e., that it satisfies the thermal equation of state

$$p = \rho RT$$

Some codes also assume a calorically perfect gas, i.e., one for which the ratio of specific heats $\gamma = c_p/c_v$ is constant. The molecular viscosity and thermal conductivity laws that are needed should also be determined, at least for laminar flows. (For turbulent flows, the turbulent values will likely overwhelm the molecular values, except very near solid surfaces where the turbulent values approach zero.) For the molecular viscosity coefficient, most CFD codes use either a power-law approximation, such as

$$\frac{\mu}{\mu_r} = \left(\frac{T}{T_r}\right)^{0.67}$$

which for air is valid at temperatures between about 300 and 900°R (167 and 500 K), or Sutherland's law:

$$\mu = C_1 \frac{T^{3/2}}{T + C_2}$$

where C_1 and C_2 are constants, which is valid between about 180 and 3400°R (100 and 1889 K) (Ames Research Staff, 1953). The thermal conductivity coefficient may be computed using similar equations, or related to μ through the Prandtl number $\mathrm{Pr} = c_p\,\mu/k$.

At hypersonic Mach numbers, however, temperatures may be high enough that the assumption of a perfect gas is no longer valid, and real-gas effects become important. Under these conditions, the specific heats are not constant, and the perfect gas equation of state no longer holds. An equilibrium air model is sometimes used, which generally consists of an empirical curve fit or table of gas properties as a function of pressure and temperature. For hypersonic nozzle flows where chemical reactions may occur, a nonequilibrium finite-rate chemistry model may be required (Numbers, 1994).

What Are the Appropriate Reference Conditions?

While the mechanics of this will vary from code to code, reference conditions must be specified to define the state of the flow. Three critical parameters are the Mach number M_r, the Reynolds numbers $\mathrm{Re}_r = \rho_r u_r L_r/\mu_r$, and the Prandtl number $\mathrm{Pr}_r = (c_p)_r \mu_r/k_r$. For a duct flow, these parameters are typically based on values at the duct entrance, or at some critical location like the throat in an inlet. The Prandtl number is a function of temperature, but is approximately constant for most gases, and is often assumed to be constant in CFD codes.

What Are the Appropriate Boundary Conditions?

The specification of boundary conditions is very important. After all, since the equations governing fluid flow are the same for every problem (i.e., the Navier-Stokes equations), the boundary conditions are really what determine the solution. Again, the mechanics of specifying the boundary conditions will vary from code to code. All methods, however, require information at the inflow boundary, which may range from a complete description of the flow profiles, to just a specification of the mass flow rate. Euler and Navier-Stokes codes also require information at the outflow boundary, such as the static pressure or mass flow rate. At solid wall boundaries, inviscid methods require the velocity to be tangent to the boundary, and viscous methods generally use the no-slip (i.e., zero velocity) condition. Either the wall temperature or heat transfer rate will also usually be required. Specialized boundary conditions may also be needed, such as the capability to specify bleed flow rates.

The analysis method used to solve a problem will, ideally, depend on answers to all of the questions above. For most internal flow problems the choice today is between a Euler parabolized Navier-Stokes (PNS) or Reynolds-averaged Navier-Stokes method, although potential flow methods have also been used for some applications. In practice, the choice of a specific code, and to some extent the analysis method and flow models to be used, is also determined by the computer resources available, the availability of the code, the level of user expertise, and the level of confidence in the code within the organization.

5.2.2   Define the Geometry

The geometry may be initially provided in a variety of ways. In many modern designs, the geometry is the output from a computer-aided design (CAD) system. For relatively simple configurations, it may also be defined analytically. Another option is simply to define the surfaces by a series of points in some coordinate system. In any case, it must be defined in sufficient detail to allow the CFD simulation to meet the goals of the study.

There is often a compromise that must be made at this point, though, as alluded to earlier. An exact representation of the real geometry would potentially yield the best solution, but may require a more sophisticated CFD method than would otherwise be needed, and use prohibitively large amounts of computer resources. As an extreme example, in a curved cooling duct there may be rivet heads or other protuberances that locally affect the flow. Including these in the CFD calculation may yield a very good simulation of the real-world flow, but would likely require a Navier-Stokes analysis and a very dense grid to resolve the flow details. But if the objective of the calculation is only to compare the flow profiles at the exit for two different turning angles, small details like this should not be included, as they are unlikely to affect the results.

Sometimes the effects of small geometric features like this may be modeled instead of actually computed. One example might be including the effect of rivet heads on the flow, without actually including them in the CAD geometry description, by specifying a surface roughness factor in the turbulence model. Another example is described as part of the Paragon *Spirit* inlet case in section 5.3, in which the effects of vortex generators on the flow are modeled.

5.2.3   Generate the Grid

The next step is to generate the computational grid. To at least some extent, the exact type of grid that is used will be dictated by the choice of CFD code used to compute the flow. For internal flows, body-fitted grids are generally used, which map the irregularly shaped physical flow domain into a rectangular computational domain. The surfaces of the geometry, and the inflow and outflow boundaries, become boundaries in the computational domain, which greatly simplifies the application of the numerical boundary conditions.

If the CFD code has multiblock capability (and most modern ones do), it may be desirable to divide the flow domain into blocks. A grid is generated for each block, and each block is solved separately by the CFD code, with interface boundary conditions used to pass information between blocks. This can greatly simplify the grid generation process for complex geometries. It may also be more efficient in the flow solution step by allowing parallel execution, with all the grid blocks being solved simultaneously on separate processors.

Grid generation usually proceeds by first constructing grids on the computational boundaries, then filling the interior of the flow field. For relatively simple geometries it may be possible or even desirable to write a customized program to generate the grid algebraically. This allows a great deal of user control over the distribution of grid points and is especially useful when the flow will be computed several times with systematic variations in the grid. More complex geometries will require the use of a generalized grid generation program. These typically generate the grid by solving a set of elliptic partial differential equations. The CFD analyst does not necessarily need to be an expert in the inner workings of the grid generation program, but it helps to be proficient in using the program. For complex geometries, the grid generation step may well be the most time-consuming one in the entire solution process.

There are several points to keep in mind when generating the grid for an application. First, grid points should be concentrated in regions where large flow gradients are expected, to adequately resolve the flow there. These high-gradient regions include boundary layers near solid surfaces, shear layers between adjacent streams or in wakes or jets, and shock waves. In addition, abrupt changes in grid spacing can lead to numerical problems, so the change in grid spacing should be smooth. The distribution of grid points may be controlled in a variety of ways, depending on the particular grid generation program being used. In the future, adaptive gridding techniques may allow this to be done automatically as the flow solution is being computed, but current CFD production codes do not generally have this capability.

While the grid must be dense enough to allow an accurate solution, using too many grid points is wasteful of computer resources. Determining the degree of resolution necessary is somewhat of an art, learned through experience with a particular code and type of flow. In general, though, for turbulent boundary layer calculations the grid point adjacent to the wall should have a y^+ value below 1.0.* As noted earlier, shock waves in most CFD codes are captured, and smeared across 3–5 grid points. Ideally, then, the grid should be dense enough around the shock that this amount of smearing is acceptable.

Second, highly nonorthogonal grids should be avoided if possible. Even though many CFD codes solve the governing equations in generalized non-orthogonal coordinates, excessive grid skewness (i.e., nonorthogonality), especially near boundaries, may adversely affect the solution. For example, if "normal-gradient" boundary conditions have been implemented by simply setting the value at the wall equal to the value at the adjacent interior grid point, an error will be introduced if the grid is nonorthogonal at the boundary.

Third, the grid should be smooth. Sharp changes in slope and/or curvature of the grid lines will cause sharp changes in the metrics of the transformation between physical and computational space, and can lead to numerical difficulties.

Traditionally, CFD codes have used structured grids, which consist of curvilinear sets of points whose coordinates are specified by 3-D arrays in the three spatial directions. The discussion in this section, at least in part, has assumed that a structured grid would be used. However, some newer CFD codes use unstructured grids, which usually consist of triangular or tetrahedral cells whose coordinates are specified on a point-by-point basis. In general, unstructured grids may be created more quickly than structured grids, especially for complex geometries. To date, they have been used mostly with Euler solutions, and some question their suitability for viscous flows with thin shear layers. This is an active research area, though, and the use of unstructured grids in CFD is likely to become more widespread.

5.2.4   Compute the Flow Field

For configurations and flow conditions that are within the user's established experience base for the code being used, the CFD calculation itself is often the easiest step in the process. For the most part, it is a matter of setting up the necessary input file(s), specifying the necessary job control information for the computer system being used (e.g., linking files to the

---

* y^+ is the inner region coordinate in the boundary layer law of the wall, given by $y^+ = u_\tau y/v_w$, where the friction velocity $u_\tau = (v \partial u/\partial y)_w^{1/2}$ and y is the distance from the wall.

proper Fortran I/O units), and starting the job interactively or submitting it to a batch queue. Long-running calculations are usually done in steps, with each successive computer run restarting the calculation where the previous one left off. For iterative methods, the results should be examined after each computer run to check the convergence status, and to identify physically unrealistic results that indicate a problem with the program input or the computational mesh. See the following section for more information about analyzing the computed results.

For configurations and/or flow conditions that are outside the established experience base, however, the CFD calculation may take significantly longer. Most codes have a variety of input options for things like the choice of artificial viscosity model and the magnitude of artificial viscosity; the solution algorithm and iteration control parameters; the form of the boundary conditions to be used; and the choice of turbulence model and all of its various input parameters. The proper values to use for a new case are often not obvious. The mesh density and quality required may also not be known. Under these conditions, it may take several "iterations"—running a case, examining the results, changing the input and/or mesh, and rerunning the case—to get the first good result for a new case. Unfortunately, successfully running a CFD code is, at least for some applications, a combination of art and science.

5.2.5 Analyze the Results

After each intermediate computer run, as noted above, the results should be examined to check the convergence status, as well as to identify physically unrealistic results that indicate problems. One way to check convergence is to examine the L_2 norm of the residual for each equation. Ideally, the residuals would all approach zero at convergence. In practice, however, for real-world problems they often drop a certain amount and then level off. Continuing the calculation beyond this point will not improve the results. A decrease in the L_2 norm of the residual of three orders of magnitude is sometimes considered sufficient. Convergence, however, is in the eye of the beholder. The amount of decrease in the residual necessary for convergence will vary from problem to problem, and will depend on how the computed results are to be used. For some problems, it may be more appropriate to measure convergence by some flow-related parameter, such as the pressure or skin friction distribution along a surface. Determining when a solution is sufficiently converged is, in some respects, a skill best acquired through experience.

CFD codes are capable of generating a tremendous amount of flow-field data. While some meaningful output may be created by the CFD code itself, examining the computed results generally requires the use of some sort of postprocessing routine. A postprocessing routine manipulates the results generated by the CFD code, and presents them to the user in a form that is meaningful for the problem being studied. In general, this means looking at the results graphically.

There are a variety of graphics packages available, both from commercial vendors and in the public domain, that may be used to display the results from a CFD calculation. These range from simple 2-D x-y plotting programs to fully 3-D interactive graphics systems. With the 3-D systems available today, results may be displayed and examined in almost any form imaginable. Pressures may be plotted on the surface of a diffuser, for example, in the form of colored contour lines, or as filled and shaded polygons. Velocity vectors may be displayed showing the development of secondary flow vortices in a curved duct. Animations may be created showing unsteady phenomena or tracking streamlines through a flow field. These interactive postprocessing systems can be tremendously useful in identifying problem areas and in understanding the critical physics of the flow.

Generalized postprocessors like these may not compute all the parameters of interest for a specific application, however. Things like the compressor-face distortion values in an inlet, for example, or the integrated thrust in a nozzle, are not usually computed by general-purpose CFD codes or postprocessing packages. It is sometimes necessary, therefore, to write a special-purpose code, or locate one that someone else has written, that reads the output files created by the CFD code and computes the needed values. When writing a postprocessing code like this, it is often tempting to do a "quick and dirty" job, that works for the specific problem at hand but no others. For all but the most unique situations, however, it will pay off in the long run if the code is written in as general a form as possible, to facilitate its application to other problems.

5.3 EXAMPLE—PARAGON *SPIRIT* INLET

In 1995 Paragon Aircraft Corporation asked NASA Lewis Research Center for help in the design of the subsonic inlet for the *Spirit*, a new six-passenger general aviation jet aircraft they were developing. The inlet (Fig. 5.1) has an S-shaped centerline and a slightly elliptical cross section that transitions to a circle at the compressor face. The area ratio is 1.14. The objective of the study was to determine the baseline performance of the inlet and to design a vortex generator system for the inlet that would ensure that it would meet the distortion criteria for the aircraft's medium bypass ratio commercial

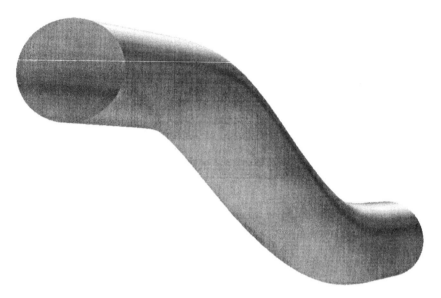

Figure 5.1   Paragon *Spirit* inlet.

turbofan engine.* In particular, values for total pressure recovery and distortion at the exit of the inlet were required.

5.3.1   Gather Information

Three operating points were of interest, as defined in Table 5.1. In the table, M_∞ and M_{thr} are the freestream and throat Mach numbers, respectively; $w\sqrt{\theta}/\delta$ is the corrected engine airflow, where w is the actual engine airflow, θ is the ratio of the engine face average total pressure to standard sea level pressure, and δ is the ratio of the freestream total temperature to standard sea level temperature; and Re_R is the Reynolds number based on throat

---

* Vortex generators are small airfoil-shaped devices mounted on a solid surface, usually in groups containing several pairs, that project up into the boundary layer. A vortex is shed from the trailing edge of each generator and propagates downstream. This enhances the mixing between the high-velocity core flow and the low-velocity boundary layer flow, and can thus delay or prevent flow separation and lower total pressure distortion levels.

Table 5.1 Operating Points for *Spirit* Inlet Calculations

Operating point	Altitude [ft (m)]	M_∞	M_{thr}	$w\sqrt{\theta}/\delta$ [lb$_m$/sec (kg/sec)]	Re$_R$
High-speed cruise	36,000 (10,973)	0.70	0.660	75.3 (34.1)	1.14×10^6
Best cruise	40,000 (12,192)	0.50	0.542	66.4 (30.1)	0.74×10^6
Takeoff	Sea level	0.20	0.468	59.4 (26.9)	2.25×10^6

conditions and the equivalent throat radius (i.e., the radius of a circle with the same area as the elliptical throat).

It was known that the flow in the inlet would be subsonic but compressible, with turbulent boundary layers. It was clearly a three-dimensional problem, probably with relatively thick boundary layers and strong pressure-driven secondary flows due to the S-shaped centerline curvature. Streamwise flow separation was a possibility, at least for the baseline configuration without vortex generators. However, it was not necessary to accurately compute any separated flows, since the mere existence of separation would be enough to invalidate the design. Thus, knowing that is separated would be enough. For the operating points of interest, the flow at the entrance to the inlet was expected to be relatively uniform. The principal elliptic effects would be caused by the effect of the S-shaped curvature on the pressure distribution.

Based on the above information, the *RNS3D* parabolized Navier-Stokes code was chosen for this problem. A Reynolds-averaged Navier-Stokes code could also have computed this flow, of course, but would have been much more expensive and time-consuming to run. One particularly nice feature of the *RNS3D* code that made it appropriate for this problem is the fact that it includes the capability to *model* vortex generators by adding streamwise vorticity to the flow at the location of the generators. Without this capability, it would have been necessary to include the small vortex generators in the actual geometry, greatly complicating and enlarging the computational grid. A Navier-Stokes code would also probably have been required to compute the shedding of the vortex from the generator trailing edges.

The original version of the *RNS3D* code was called *PEPSIG* and was developed in the late 1970s and early 1980s. Since then, additional modifications have been made and the code has been renamed. Like other spatial marching codes, it neglects the viscous and thermal diffusion terms in the streamwise direction. In addition, special treatment is required for the pressure gradient term in the streamwise momentum equation to suppress its elliptic behavior. In *RNS3D*, the pressure p in that equation is written as

$$p(x, y, z) = P(x, y, z) + p'(x) + p''(y, z)$$

where x is the marching direction. Here $P(x, y, z)$ is a known estimate for the pressure field, from a potential flow solution for example; $p'(x)$ is a one-dimensional correction computed during the marching solution using global mass flow conservation as a basis; and $p''(y, z)$ is a two-dimensional correction in the cross section, also computed during the marching solution.

The details of the derivation of the equations and the solution procedure are beyond the scope of this chapter. The basic analysis is described by Briley and McDonald (1979); Levy et al. (1980); Levy, Briley, and McDonald (1983); and Briley and McDonald (1984). Several reports and papers have been published presenting results of validation studies and applications using the *PEPSIG/RNS3D* code (Towne and Anderson, 1981; Towne, 1984; Vakili et al., 1984; Towne et al., 1985; Towne and Schum, 1985; Anderson, 1986; Kunik, 1986; Povinelli and Towne, 1986; Tsai and Levy, 1987; Anderson, 1991; Anderson and Gibb, 1992; Anderson et al., 1992; Anderson and Towne, 1993).

5.3.2 Geometry Definition

The *Spirit* inlet shown Fig. 5.1 is an S-shaped duct with elliptical cross sections perpendicular to the centerline. The cross-section shape at a particular streamwise station can thus be described by

$$\left(\frac{x_1}{a}\right)^2 + \left(\frac{x_2}{b}\right)^2 = 1$$

where a and b are the semimajor and semiminor axes of the elliptical cross section, and x_1 and x_2 are local Cartesian coordinates perpendicular to the centerline.

One of the input options in *RNS3D* is to specify the geometry in terms of polynomials in some marching parameter. For this case, the geometric values to be specified were the Cartesian x and y coordinates of the duct centerline and the semimajor and semiminor axes a and b.

The centerline coordinates were supplied as cubic splines: thus,

$$x_{CL} = a_0 + a_1 t + a_2 t^2 + a_3 t^3$$
$$y_{CL} = b_0 + b_1 t + b_2 t^2 + b_3 t^3$$

where t was the streamwise marching parameter.

To best fit an actual configuration, *RNS3D* allows the centerline to be split into sections, with different polynomial coefficients in each section. The equations solved by *RNS3D* require second derivatives of the coordinates, which are computed numerically. The geometry description should therefore be smooth from one section to another. In theory, second derivatives of the geometric parameters should be at least continuous. Note that this formally applies to the cross-section parameters a and b, as well as the centerline coordinates. Experience with the code, though, shows that in practice getting the centerline smooth is more critical.

The centerline for the *Spirit* inlet was divided into 15 sections. The smoothness was checked by plotting y_{CL} and $\partial y_{CL}/\partial x$ versus x (Fig. 5.2a,b). The symbols in Fig. 5.2a indicate the boundaries between the cubic spline sections defining the shape. Note the slope discontinuity in the plot of $\partial y_{CL}/\partial x$ near the end of the duct. Initially this caused some concern, but preliminary calculations indicated that it did not appreciably affect the numerical stability of the CFD calculation or the computed viscous results, so no additional work was done to eliminate this discontinuity.

The cross-section semiaxes a and b were supplied in the form of a table of values versus the marching parameter t. A curve-fitting routine was used to develop polynomials defining a and b. Like the centerline coordinates, the values of a and b computed from the curve fits were also plotted, both to determine how well the curve fits matched the supplied values and to examine the smoothness of the result. These plots are shown in Figs 5.3 and 5.4. The symbols in Figs 5.3a and 5.4a are the supplied tabular values of a and b, and the line is the fitted polynomial. Note that a and b have a discontinuous slope at $t = 14$. This is due to a short, constant-area section at the end of the actual duct. In theory this is probably not a "good thing," but again, preliminary calculations indicated that it did not adversely affect the computed viscous results.

Besides specifying the geometry in terms of polynomials, another input option in *RNS3D* is to read a 3-D file containing a grid of points defining the surfaces. This option is somewhat more robust and flexible, especially when defining the locations of vortex generators in a duct. The polynomials describing the centerline and cross-section axes were thus used to define the boundaries for some preliminary calculations with *RNS3D*. The resulting surface grid was saved in a file and used as input for subsequent calculations to define the geometry.

5.3.3 Computational Grid

The interior grid point distribution in *RNS3D* is defined at run time via input parameters. Thus, a separate grid generation step was not needed for these calculations. The computational grid used for most of the calculations is shown in Fig. 5.5. Note that, since the centerline is two-dimensional and

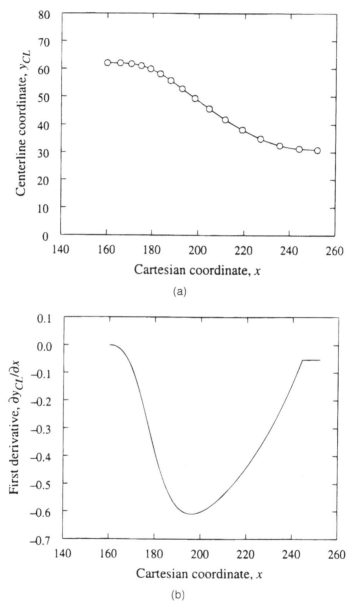

(a)

(b)

Figure 5.2 *Spirit* inlet centerline geometry: (a) centerline coordinate, y_{CL}; (b) first derivative, $\partial y_{CL}/\partial x$.

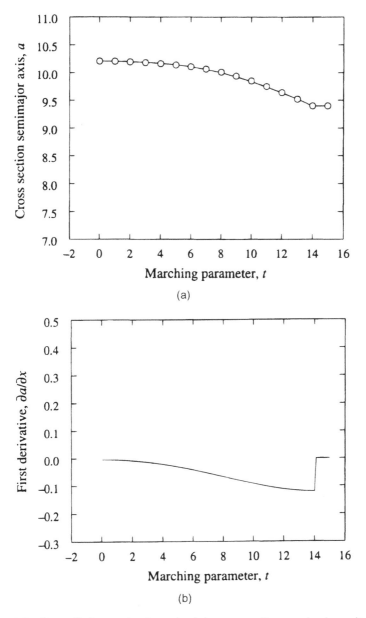

Figure 5.3 Curve fit for semimajor axis: (a) cross section semimajor axis, a; (b) first derivative, $\partial a/\partial x$.

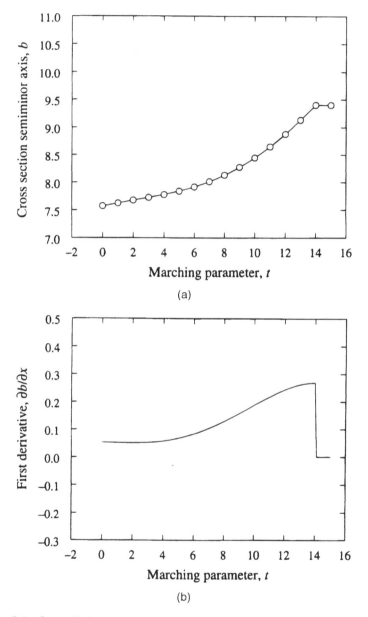

Figure 5.4 Curve fit for semiminor axis: (a) cross section semiminor axis, b; (b) first derivative, $\partial b/\partial x$.

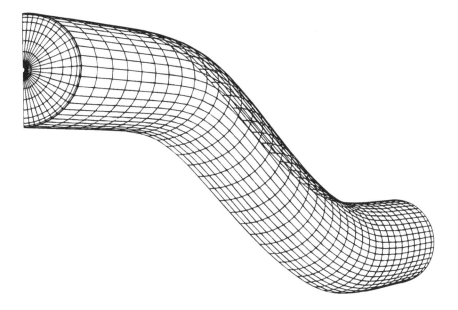

Figure 5.5   Computational grid for the *Spirit* inlet.

the inlet flow is uniform, the flow will be symmetric about the 0°–180° line, and only half the cross section needs to be computed.

A 49 × 49 mesh was used in the 180° cross section, with 151 streamwise stations. Grid points were packed in the radial direction near the outer boundary to resolve the boundary layers there. Uniform grid spacing was used in the circumferential and streamwise directions. For clarity, the grid in Fig. 5.5 has been thinned by a factor of 4 in the radial direction and 2 in the circumferential and streamwise directions.

5.3.4   Potential Flow Solution

Running a case with *RNS3D* is a two-step procedure. The first step is a potential flow run to compute the pressure estimate $P(x, y, z)$. This potential flow pressure field is saved in a file and used as input in the second step, the actual viscous marching calculation. Note that the potential flow calculation has to be done only once for a given geometry. Changes in flow conditions, initial profiles, grid density, etc., for the viscous calculation can be made without affecting the prestored potential flow pressure file.

Since there are no boundary layers to resolve in the potential flow, a coarser mesh may be used. For the *Spirit* inlet, the mesh size was

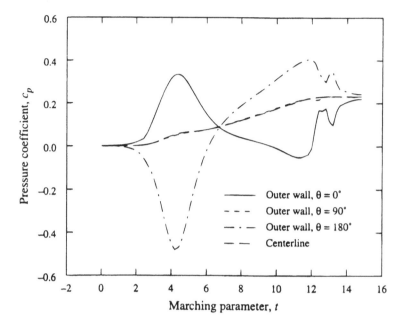

Figure 5.6 Potential flow pressure distribution.

$19 \times 20 \times 75$. The computed potential flow pressure coefficients, along the outer boundary at $\theta = 0°$, $90°$, and $180°$, and along the centerline, are shown in Fig. 5.6.

The outer wall values at $\theta = 90°$ and the centerline values are essentially identical and reflect the increase in cross-sectional area. The outer wall values at $\theta = 0°$ and $180°$ have the typical shape for an S-duct, with higher pressure on the outside of the bend and lower pressure on the inside. The wiggles at the downstream end are probably due to the discontinuous slope in the cross section axes a and b, described previously. As noted earlier, they had no significant effect on the viscous solution.

5.3.5 Distortion Criteria

Before describing the viscous calculation, it is useful to discuss the method used to quantify the amount of distortion. The distortion descriptors are based on compressor face total pressure values that normally would be measured in an experiment by a standard 40-probe rake, (Fig. 5.7). To get analogous results from the CFD calculation, the computed total pressures

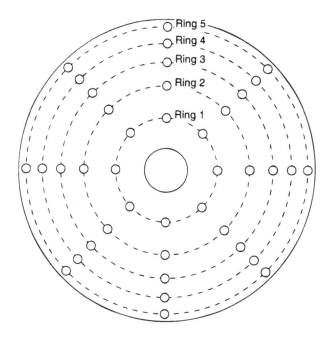

Figure 5.7   Standard 40-probe compressor face rake.

were interpolated from the much denser computational grid to the probe locations of a 40-probe rake.

For this study, a simplified stability assessment procedure supplied by the engine manufacturer was used. Four distortion descriptors are computed from the compressor face total pressure values, quantifying various aspects of the radial and circumferential distortion. These descriptors are then combined with empirical parameters, which are functions of corrected engine airflow rate, to define DLP(core) and DLP(fan), the distortion limit parameters for the core compressor and the fan tip. Both of these distortion limit parameters must be below 1.0 for stable engine operation.

5.3.6   Viscous Flow Solution/Analysis of Results

The next step was to compute the viscous flow in the inlet without vortex generators. This was done for all three operating conditions listed in Table 5.1. In addition, since no experimental data was available to determine the boundary layer thickness at the throat (the initial station in the marching analysis), each operating condition was run with three different initial

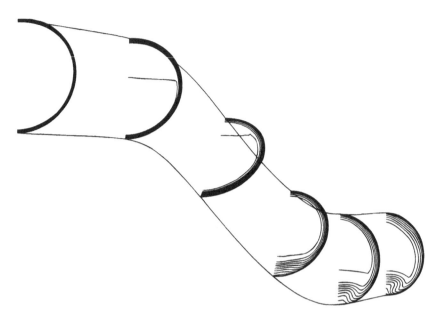

Figure 5.8 Total pressure contours, high-speed cruise condition, without vortex generators.

boundary layer thicknesses—$\delta/R = 0.025, 0.05$, and 0.10. The development of the flow through the inlet is illustrated in Fig. 5.8, in the form of computed total pressure contours at selected streamwise stations, for the high-speed cruise condition with $\delta/R = 0.05$. The horseshoe-shaped pattern at the exit is typical of S-duct flows. The curved centerline causes transverse pressure gradients to be set up in the cross section, as shown by the potential flow results in Fig. 5.6. Pressure-driven secondary flow vortices appear, which drive the low-energy boundary layer flow to the bottom of the duct.

The computed recoveries and distortion limit parameters are shown in Table 5.2 for all three operating conditions and initial boundary layer thicknesses. The total pressure recovery is satisfactory for all operating conditions and inlet boundary layer thicknesses. In addition, both DLP(fan) and DLP(core), the fan tip and core compressor distortion limit parameters, are well below the critical value of 1.0 at the best cruise and takeoff conditions, as is DLP(core) at the high-speed cruise condition. But, DLP(fan) at the high-speed cruise condition is clearly too high. This operating point was thus used to design the vortex generator system for the inlet.

Based on earlier experience in the use of vortex generators in subsonic inlets (Anderson and Gibb, 1992; Anderson et al., 1992), a system was

Table 5.2 Distortion Limit Parameters without Vortex Generators

Operating condition	Initial δ/R	Recovery	DLP(fan)	DLP(core)
High-speed cruise	0.025	0.982	2.156	0.008
	0.05	0.979	2.510	0.076
	0.10	0.972	2.300	0.307
Best cruise	0.025	0.985	0.238	0.024
	0.05	0.982	0.259	0.105
	0.10	0.978	0.264	0.279
Takeoff	0.025	0.992	0.087	0.002
	0.05	0.990	0.107	0.003
	0.10	0.988	0.137	0.037

designed for the *Spirit* inlet with 11 pairs of counterrotating generators distributed around the 360° cross section a short distance downstream of the throat, as shown schematically in Fig. 5.9. The design variable examined was the generator height h. Cases were run using *RNS3D* with $h/R = 0.04$ to 0.07 in increments of 0.005 for all three initial boundary layer thicknesses. The resulting values for the fan tip distortion limit parameter are shown in Fig. 5.10. Based on these results a generator height of $h/R = 0.05$

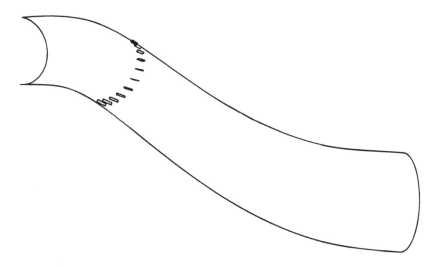

Figure 5.9 Vortex generator installation in the *Spirit* inlet.

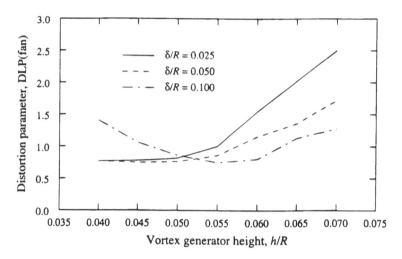

Figure 5.10 Effect of generator height on fan tip distortion limit parameter.

was selected as the optimum choice, given the uncertainty in the actual boundary layer thickness.

The computed total pressure contours for the $h/R = 0.05$ and $\delta/R = 0.05$ combination are shown in Fig. 5.11. By comparison with the results shown in Fig. 5.8, it can be seen that the effect of the vortex generators in this case is to split the large region of low total pressure at the compressor face into two smaller regions and to shift them slightly in the circumferential direction.

Additional computations were also performed to confirm that the distortion levels at the other two operating conditions were still acceptable with vortex generators installed. The results for the three operating conditions and initial boundary layer thicknesses are summarized in Table 5.3. By comparison with Table 5.2, it can be seen that all the distortion levels were lowered by the use of vortex generators. The values for DLP(fan) are all below 1.0, the limit set for the candidate engine, but they are still uncomfortably high at the high-speed cruise operating point. The total pressure contours at the compressor face for this condition (see the exit station in Fig. 5.11) are actually very similar to the results for the best cruise condition (not shown). The high DLP(fan) values are a result of the large corrected weight flow value of 75.3 lb_m/sec (34.1 kg/sec) at the high-speed cruise condition (Table 5.1). In the stability assessment procedure used in this study, the DLP(fan) values increase rapidly when the corrected engine airflow increases above 68 lb_m/sec (30.8 kg/sec).

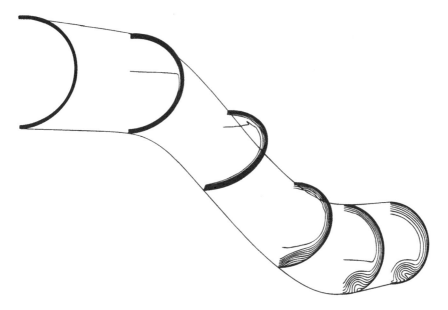

Figure 5.11 Total pressure contours, high-speed cruise condition, with vortex generators.

After discussing these results with the inlet designers and the engine manufacturer, it was determined that a lower throat Mach number of 0.61 should have been used for these calculations. At this Mach number, the corrected weight flow is 71.8 lb$_m$/sec (32.6 kg/sec). The high-speed cruise cases were therefore rerun with the lower throat Mach number, both with

Table 5.3 Distortion Limit Parameters with Vortex Generators

Operating condition	Initial δ/R	Recovery	DLP(fan)	DLP(core)
High-speed cruise	0.025	0.984	0.818	0.008
	0.05	0.982	0.762	0.007
	0.10	0.975	0.862	0.026
Best cruise	0.025	0.989	0.124	0.006
	0.05	0.986	0.144	0.007
	0.10	0.981	0.189	0.038
Takeoff	0.025	0.991	0.095	0.003
	0.05	0.991	0.102	0.004
	0.10	0.988	0.117	0.006

Table 5.4 Distortion Limit Parameters, High-Speed Cruise, $M_{thr} = 0.61$

	Initial δ/R	Recovery	DLP(fan)	DLP(core)
Without vgs	0.025	0.985	0.448	0.007
	0.05	0.981	0.541	0.052
	0.10	0.976	0.564	0.218
With vgs	0.025	0.987	0.259	0.007
	0.05	0.985	0.271	0.006
	0.10	0.979	0.301	0.045

and without vortex generators, and the resulting performance parameters are listed in Table 5.4. At the lower throat Mach number, the distortion levels are below 1.0 even without vortex generators, although a conservative designer may feel they are still too high. With vortex generators, though, the levels are well below 1.0.

Prior experience with *RNS3D* has shown that having a sufficiently dense mesh, especially in the streamwise direction, is required to get quantitatively accurate predictions of the development of secondary flow vortices. Some additional runs were therefore made for the high-speed cruise condition with vortex generators to investigate the effects of streamwise mesh density. One run was also made doubling the mesh in both cross-flow directions. The results are listed in Table 5.5. The initial boundary layer thickness δ/R for these cases was 0.05.

Using four times as many cross-section points increased the computed value of DLP(fan) only slightly. Going from 151 to 601 streamwise points, though, increased the value by over 50%. As the number of points was increased further, DLP(fan) dropped slightly and appeared to asymp-

Table 5.5 Effect of Mesh Density on Distortion Limit Parameters

Mesh	Recovery	DLP(fan)	DLP(core)
49 × 49 × 151	0.985	0.271	0.006
49 × 49 × 301	0.982	0.344	0.004
49 × 49 × 601	0.979	0.411	0.004
49 × 49 × 1201	0.980	0.379	0.004
49 × 49 × 2401	0.981	0.363	0.004
97 × 97 × 151	0.983	0.310	0.023

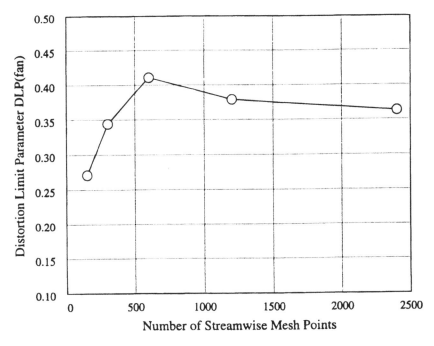

Figure 5.12   Effect of streamwise mesh density on DLP(fan).

totically approach a value of about 0.35, as shown more clearly in Fig. 5.12. This value is still well below the limit of 1.0 for the candidate engine.

Based on these CFD results, Paragon, in consultation with the engine manufacturer, concluded that the *Spirit* inlet would meet the performance criteria for the candidate engine and that an experimental test program that had been planned could be eliminated with minimal risk. They plan to proceed directly to a flight test with an instrumented inlet installed on the new aircraft. Thus, the careful use of the CFD in this project has resulted in significant savings in both cost and time.

5.4   EXAMPLE—STRUT-JET ENGINE

Propulsion systems for missiles, reconnaissance aircraft, and single-stage-to-orbit vehicles must operate efficiently at flight conditions ranging from takeoff to hypersonic cruise. Because a specific propulsion cycle is more efficient at one flight condition than others, a new family of combined cycle engines is being studied. These engines combine two or more different propulsion cycles into an integrated system for better overall performance throughout the flight envelope.

One such system currently being studied at the NASA Lewis Research Center is the strut-jet. This engine is based on the rocket-based combined cycle (RBCC) concepts of Escher, Hyde, and Anderson (1995). It combines a high-specific-impulse low-thrust-to-weight air-breathing engine with a low-specific-impulse high-thrust-to-weight rocket engine. From takeoff to high supersonic speeds (about Mach 3) the system operates as an air-augmented rocket. At approximately Mach 3 the rockets are shut down, and the system becomes a dual-mode ramjet. At very high Mach numbers (above about Mach 8) and high altitude, the air-breathing system may not provide adequate thrust, and the rockets would then be turned back on.

Demonstration tests of the strut-jet engine were run at NASA Lewis in 1996. Thrust measurements were made at several fuel flow conditions. The measured experimental thrust, however, includes aerodynamic forces on various pieces of attached external hardware, such as the instrumentation and model support system, and therefore does not represent the true thrust of the propulsion system. The true thrust may thus be written as

$$T_{sys} = T_{exp} - T_{ext}$$

where T_{sys} is the true internal thrust of the propulsion system, T_{exp} is the measured thrust in the experiment, and T_{ext} is the (negative) thrust due to external hardware. If the same experiment is run without fuel flow to the engine, we get

$$(T_{sys})_{nf} = (T_{exp})_{nf} - T_{ext}$$

where $(T_{sys})_{nf}$ is the internal force on the propulsion system, and $(T_{exp})_{nf}$ is the measured force in the experiment. Subtracting, we can write

$$T_{sys} = (T_{sys})_{nf} - \Delta T$$

where $\Delta T = T_{exp} - (T_{exp})_{nf}$ is the measured increment in thrust for a given fuel flow condition. The internal force $(T_{sys})_{nf}$ cannot be measured experimentally because the force T_{ext} due to external hardware cannot be determined independently. $(T_{sys})_{nf}$ can be computed using CFD, however, allowing the true thrust of the propulsion system to be determined.

To compute this internal force, the NPARC code was used. NPARC is a multiblock Navier-Stokes code being developed and supported by the NPARC Alliance, a partnership between the NASA Lewis Research Center and the USAF Arnold Engineering Development Center (NPARC Alliance, 1994). It solves the Reynolds-averaged, unsteady compressible Navier-

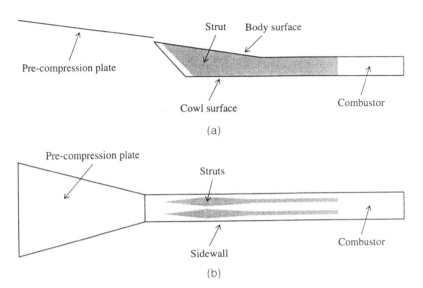

Figure 5.13 Rocket-based combined cycle engine: (a) sidewall removed; (b) cowl removed.

Stokes equations in generalized nonorthogonal body-fitted coordinates. Several turbulence models are available in the code; for this application, the Chien low-Reynolds-number k-ε model was used (Chien, 1982). Spatial derivatives in NPARC are represented using central difference formulas, and explicit boundary conditions are used. Jameson's artificial dissipation model is used for stability, and to smooth pre- and postshock oscillations and to prevent odd-even point decoupling (Jameson, Schmidt, and Turkel, 1981). The equations are solved by marching in time using an ADI algorithm derived using the Beam-Warming approximate factorization scheme (Beam and Warming, 1978).

After the NPARC CFD calculations were completed, the solution was postprocessed to obtain the internal force $(T_{sys})_{nf}$. Two calculation methods were used. The first was a simple momentum balance, subtracting the integrated momentum at the duct entrance from the integrated momentum at the exit. The momentum was computed by numerical integration over the computational grid. The second method integrated the pressure and skin friction forces on the internal surfaces of the configuration to obtain the internal force. Ideally, these two methods give identical results. However, several sources of error can contribute to a discrepency. Incomplete mass continuity and difficulty in calculating accurate skin friction are the two most common problems.

This methodology was first tested on a subscale model of the strut-jet engine that was tested in the NASA Lewis 1 × 1 ft (0.3 × 0.3 m) supersonic wind tunnel (Fernandez et al., 1996). A simplified schematic of the model is shown in Fig. 5.13. The strut-jet configuration that was analyzed consisted of a rectangular cross-sectioned inlet with swept leading-edge sidewalls. Two struts, also with swept leading edges, were installed in the inlet. In the actual engine test that will be run in the Hypersonic Test Facility, the rockets will be installed in the base of these struts. A precompression plate upstream of the inlet was used to simulate the effect of the vehicle forebody.

The computational grid was created using GRIDGEN, a grid generation package widely used for CFD applications (Steinbrenner, Chawner, and Fouts, 1990). Six grid blocks were used, as listed in Table 5.6, with a total of 1,400,319 points. Note that the configuration is symmetric, and thus only half the duct was computed, from the sidewall to the center symmetry plane between the two struts.

For these calculations, the freestream Mach number, static pressure, and static temperature were 6.0, 14.98 lb_f/ft^2 (717.2 N/m^2), and 93.13°R (51.74 K), respectively. The resulting Reynolds number was $3.80 \times 10^6/ft$ ($12.5 \times 10^6/m$). Convergence was achieved after the L_2 norm of the residual was reduced at least three orders of magnitude in each grid block, and when no discernable change in the static pressure and mass flow distributions were observed over at least 1000 iterations. Representative results are shown in Fig. 5.14a and b, where the computed and experimental static pressures distributions are plotted along the body and cowl centerlines.

The internal force values computed using the two calculation methods were within 1.5% of each other, as shown in Table 5.7. Positive values indicate thrust, and negative values indicate drag.

This example is an illustration of how CFD can be used to solve a problem that would be very difficult and expensive to solve any other way. Experimentally, there is no practical way to separate the true propulsion system thrust from the measured thrust at these hypersonic conditions.

Table 5.6 Grid Blocks for Strut-Jet Engine Computation

Block no.	Grid size	Description
1	22 × 76 × 30	Inlet entrance center duct
2	111 × 57 × 30	Center duct
3	111 × 57 × 51	Side duct
4	22 × 76 × 52	Inlet entrance side duct
5	59 × 95 × 80	Forebody
6	51 × 57 × 104	Combustor section

Table 5.7 Internal Force Balance for Strut-Jet Engine

	Momentum balance		Force integration	
Boundary	Momentum $[lb_f$ (N)]	Surface	Pressure $[lb_f$ (N)]	Skin friction $[lb_f$ (N)]
Inflow	−32.3999 (−144.122)	Body	−1.3563 (−6.033)	−0.4740 (−2.108)
Spillage	0.8108 (3.607)	Cowl	0.0000	−0.3226 (−1.435)
Outflow	28.3773 (126.228)	Strut	0.4575 (2.035)	−1.1703 (−5.206)
		Sidewall	0.0000	−0.6499 (−2.891)
		Base	0.2566 (1.141)	0.0000
Total	−3.2118 (−14.287)	Total	−3.2590 (−14.497)	

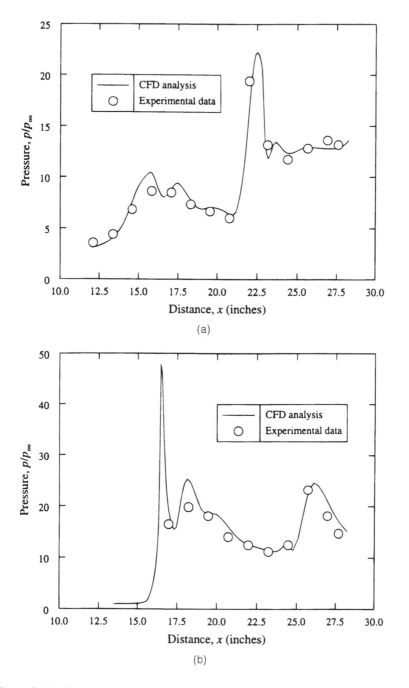

Figure 5.14 Pressure distribution in strut-jet engine: (a) along body centerline; (b) along cowl centerline.

Using CFD, however, allows the true thrust of the propulsion system to be determined.

5.5 CURRENT STATUS AND FUTURE DIRECTIONS

Over the last 10 years or so, several papers and journal articles have appeared describing the status of CFD for applications. By their very nature, of course, publications like these become outdated fairly quickly. This section presents this author's perception of the current status of CFD for inlet, duct, and nozzle applications and indicates possible future directions that would make CFD more useful in the real world (i.e., industry). Much of the material here has been influenced by the authors of two recent papers on the use of CFD in the aerospace industry (Cosner, 1994; Paynter, 1994).

There are several issues, sometimes overlapping, that are inhibiting the widespread routine use of CFD, especially in the design environment. Some of these are modeling issues, resulting from our lack of understanding of some of the basic but complex flow physics in many real-world applications. Some are numerical issues, dealing with how the equations are solved. Others are more procedural in nature, related to how the various steps involved in a CFD analysis are currently being accomplished, and to how CFD codes are written, tested, and evaluated.

5.5.1 Modeling Issues

For nonreacting flow through inlets, ducts, and nozzles, the principal flow modeling problems remaining today are the following.

Turbulence

As noted earlier, a universal turbulence model that works well for all types of flow does not yet exist. In general, guidelines based on experience must be used when choosing the model to use for a particular problem. This is an active area of research, and new turbulence models, or variations on existing ones, seem to be proposed weekly. This rapid growth makes it difficult to evaluate new models, however. The situation would improve with the development and acceptance of standards for software interfaces, validation, and documentation. (See below.)

Laminar-Turbulent Transition

Our capability to predict laminar-turbulent transition is even less mature than that for fully turbulent flows. In many CFD codes the flow must be either fully laminar or fully turbulent. Those that are capable of predicting transition generally use fairly crude models based on correlations with experimental data for simple flows. Better models are needed for use in computing realistic 3-D flows in engineering applications. See the recent papers by Simon (1993) and Simon and Ashpis (1996) for an overview of the current research in this area.

Boundary Conditions

CFD is typically used to study the flow through a component of a larger physical system, such as the inlet in a jet engine. At some types of boundaries, such as a simple no-slip solid wall or a supersonic inflow boundary, choosing appropriate boundary conditions is fairly straightforward. For many other types of boundaries, the situation is more complicated. The conditions specified at the outflow boundary of an inlet, for example, must properly model the influence of the compressor on the flow in the inlet. Porous wall boundary conditions are normally used to represent the flow through a bleed region in a supersonic inlet. These specialized boundary conditions must also be able to model unsteady interactions between components, such as the reflection at the compressor face of a disturbance in the inlet flow. Additional research is required to develop satisfactory boundary conditions for specialized applications like these.

5.5.2   Numerical Issues

The numerical algorithms being used in modern CFD codes to solve the governing equations are generally pretty fast. While faster algorithms are always desirable, other numerical issues are also of critical importance.

Computational Platform

The computational power available to the CFD user has increased dramatically over the last 10–15 years. In the not-too-distant past, CFD codes were almost always run on large mainframe computers, but today Navier-Stokes analyses for relatively simple 2-D problems can be run on desktop PCs and workstations. Even some 3-D problems are being run on midrange to high-end workstations. Parallel processing software has been developed that allows CFD codes to use multiple processors, either on a single computer with multiple CPUs or on a cluster of computers, with each processor computing a part of the problem.

This rapid and continual growth in the capability of the hardware has in many respects been the determining factor in the growth of CFD. The computer speed and memory that is available to the CFD user influences, for example, the size of the grid and the sophistication of the turbulence model. As the hardware continues to improve, CFD simulations will also continue to improve.

Other advancements are possible in CFD, besides those related to the raw speed and memory of the computer. New solution algorithms that are designed specifically to take advantage of parallel processing capabilities should be investigated. Faster algorithms may also be developed by taking advantage of other features present in a specific type of computational architecture. The disadvantage to this, and it is a big one, may be lack of portability between platforms. For long-term use in a design environment, the trade-off is probably not worthwhile.

Unsteady Flows

While many CFD codes are at least theoretically able to compute unsteady flows, not much emphasis has been placed on their numerical accuracy. Interest in unsteady flows is growing, however, and this area will see increasing activity in the future.

5.5.3   Procedural Issues

Another reason, perhaps the main reason, that CFD is not more widely used in routine design work is that the process is still too difficult and time-consuming, especially for non-CFD experts. There are several, sometimes interrelated, factors involved.

Ease of Use

The computer codes used in CFD, from the preprocessors used to define the geometry and generate the grid, through the postprocessors used to analyze the results, need to be made simpler to use. Until fairly recently CFD was basically a research area, with much effort being put into the development of faster and more accurate solution algorithms. Ease of use for the nonexpert user was not a high priority for the code developer. However, the situation is changing. Solution algorithms are now pretty good, as noted above, and ease of use is becoming more important.

Part of the solution will require closer coupling between the various steps in the CFD process, and the development of various standards will help, as discussed below. There are other things that should also be done, however, to make the individual steps in the solution process easier.

For complex configurations, grid generation is currently one of the more time-consuming steps. In particular, setting up the various blocks for a multiblock analysis and linking the grid blocks can be very labor intensive (Cosner, 1994). Automating this step as much as possible would be very beneficial. Cosner suggests an expert system type of approach, in which the key geometric features would be identified, and, along with the expected flow conditions, used as input to a system that would recommend the layout of the grid blocks, plus the grid size and grid point distribution within the blocks. A corollary to this idea is the use of adaptive grid techniques, in which the grid points are automatically redistributed to resolve high-gradient regions as the flow is being computed. Adaptive gridding is not a new idea, and has already been demonstrated for a variety of problems. However, it has not yet become a standard feature in most CFD analysis systems, perhaps because it requires close coupling between the grid generator and the flow solver. Adaptive gridding has the potential, though, to make the initial grid generation step much easier, to eliminate much of the manual iteration that is now sometimes necessary between the grid generator and flow solver, and to allow the best possible solution for a given number of grid points.

The flow solvers themselves can also be made easier to use. Some CFD codes are overly sensitive to things like nonorthogonal and nonsmooth meshes, the time step size, and the choice of artificial viscosity parameters. Getting good results (or in the most extreme cases, any results at all) from a CFD code may require "tweaking" the input until the "correct" value or combination of values is found. More robust solution algorithms, that are less sensitive to mesh and input irregularities, would help. Another improvement would be the development of an intelligent user interface, that could be used to help set up the input for a particular problem and to check it for inconsistencies.

Postprocessing systems, while generally very good, can still be improved. As noted earlier, with the interactive 3-D graphics packages available today, results may be displayed and examined in almost any form imaginable. While this can be tremendously useful, for the most part the user must visually examine the computed results. More automated methods should be developed to identify key flow features and problem areas. The capability to perform solution quality checks, similar to the grid quality checks already available in some grid generation codes, is also needed. These automated postprocessing capabilities, once available, should be used to examine the flow field as it is being computed, and recommend changes to the grid and/or input parameters where appropriate.

Standardization

The various steps in the solution process have historically been separate elements. As a result, too much of the time required to solve a problem is spent in between the elements, converting the output from one step to the input for the next step. In addition, there is often too much iteration required between steps (e.g., "change the grid and recompute"). Closer coupling between the various steps is needed, so that the entire solution process from the geometry specification through the analysis of the results becomes as seamless as possible. Using CFD for design requires, almost by definition, the ability to easily change the geometry and determine the effect of that change on the flow.

To accomplish this, standards need to be developed and accepted by the CFD community for the interfaces between the various steps. For example, the wide variety of CAD packages in use for geometry specification have resulted in a variety of formats for the CAD output, many of them not directly readable by popular grid generation programs. CFD flow solvers read grid files in a variety of formats, and there is no universal standard for the interface requirements and boundary conditions to be used between blocks in a multiblock grid. The *PLOT3D* format (Walatka et al., 1990) has become a de facto standard for the output from CFD codes and can be read by a variety of postprocessors. Unfortunately, this format does not include all the information necessary to fully represent some computed results, such as turbulence data.

A variety of more complete formats have been or are being developed, such as the NASA-IGES standard for CAD output (Blake et al., 1991) and the interface standards from the NASA-funded Complex Geometry Navier-Stokes (CGNS) project. Since there is no "governing body" in CFD, however, the development of a single accepted standard for the interface between each step in the solution process is unlikely, at least in the near future. Instead, several "standards" will probably coexist. Code developers should therefore strive to support directly as wide a variety of the proposed formats as possible, both for input and output. In addition, generalized interface routines should be developed to convert data between a variety of standard formats.

Besides the need for standard data formats between steps in the solution process, standard interfaces are needed between modules within the individual codes. This is especially true for the CFD flow solver itself, where standard subprogram interfaces would make the development and testing of new technology, such as improved turbulence models, much easier. The CGNS project is addressing this issue also.

Finally, research into multidisciplinary methods, such as a CFD analysis coupled with an elastic structure analysis, is increasing. For these methods to ultimately be useful in the real world, standards are required for sharing data between the multiple analyses involved.

Validation

CFD code validation has been the subject of much interest in recent years (e.g., Marvin, 1993; Mehta, 1995; Aeschliman, Oberkampf, and Blottner, 1995). While various terms, such as verification, certification, and validation, have been used to describe different aspects of the process, it basically refers to determining how well a CFD code is able to simulate reality.

In order to determine the strengths and weaknesses of a CFD code, cases should be run for a variety of geometric configurations and over a range of flow parameters. Computed results should be compared with benchmark-quality experimental data, well-accepted computational results, and/or analytic solutions. If CFD is to become an accepted tool for design, code validation must be emphasized. The code developer must demonstrate that his or her code is able to simulate reality accurately enough and quickly enough to be relied upon in a design environment.

Starting as far back as 1968, various organizations have developed databases containing standard sets of experimental data to be used for CFD validation for various types of flow (e.g., Coles and Hirst, 1968; AGARD, 1988; Settles and Dodson, 1991). As our capability to predict more complex flows increases, the need for high-quality experimental validation data for those flows also increases. Data are now needed for high-(i.e., flight level) Reynolds-number turbulent flows, low-Reynolds-number transitional flows, and unsteady flows. These data would be especially useful for evaluating newly proposed turbulence models.

Documentation

CFD codes are notoriously poorly documented. For many CFD codes the documentation, if it exists at all, consists only of a user's guide describing the input and output, with a few examples. It generally does not include a detailed description of the code itself, showing exactly how the various physical and numerical models involved have been implemented. These details are often not even described in comments within the code itself. Without this information, even a knowledgeable CFD researcher will have difficulty modifying the code to test hypotheses about the cause of any disagreement with experimental data in a validation study.

Papers presenting applications of CFD are also often poorly documented. In addition to describing the problem and the CFD method that was used, they should include at least brief descriptions of the turbulence model, any artificial viscosity that was used, the grid size and distribution, the boundary conditions, and, for iterative methods, the iteration method and convergence history. Without these details, it is difficult or impossible to assess the significance of the computed results.

ACKNOWLEDGMENTS

The author would like to acknowledge and thank his co-workers at NASA Lewis Research Center and in the NPARC Alliance for their valuable contributions to this chapter. Special thanks go to Bernie Anderson and Julie Dudek, who did most of the work on the Paragon *Spirit* inlet calculations;

to Jim DeBonis and Shaye Yungster, who supplied the material on the strut-jet engine; and to Ray Cosner and Jerry Paynter, who provided insight into the difficulties involved in using CFD for real-world design in an industrial environment.

NOMENCLATURE

a, b	semiaxes in *Spirit* inlet cross section
DLP(core)	core compressor distortion limit parameter
DLP(fan)	fan tip distortion limit parameter
h	vortex generator height
M_∞, M_{thr}	freestream and throat Mach numbers for *Spirit* inlet
p	static pressure
Pr	Prandtl number
R	equivalent throat radius in *Spirit* inlet
Re	Reynolds number
Re_R	Reynolds number based on throat conditions for *Spirit* inlet
t	streamwise marching parameter in *RNS3D* code
T_{exp}, T_{ext}, T_{sys}	measured thrust, thrust due to external hardware, and true system thrust
$w\sqrt{\theta}/\delta$	corrected engine airflow
x_{CL}, y_{CL}	centerline coordinates for *Spirit* inlet

Greek

δ	boundary layer thickness

Subscripts

nf	value at no-fuel-flow condition
r	reference quantity
w	wall value

REFERENCES

Aeschliman DP, Oberkampf WL, Blottner FG. A proposed methodology for computational fluid dynamics code verification, calibration, and validation. 16th International Congress on Instrumentation in Aerospace Simulation Facilities, Wright-Patterson AFB, OH, July 18–21 1995.

AGARD. Validation of computational fluid dynamics. Vol. 1. Symposium papers and round table discussion. AGARD-CP-437, 1988.

Ames Research Staff. Equations, tables, and charts for compressible flow. NACA report 1135, 1953.

Anderson BH. The aerodynamic characteristics of vortex ingestion for the F/A-18 inlet duct. AIAA paper 91-0130, 1991.

Anderson BH. Three-dimensional viscous design methodology of supersonic inlet systems for advanced technology aircraft. In: Murthy SNB, Paynter GC, eds. Numerical Methods for Engine-Airframe Integration. New York: AIAA, 1986.

Anderson BH, Gibb J. Application of computational fluid dynamics to the study of flow control for the management of inlet distortion. AIAA paper 92-3177, 1992.

Anderson BH, Huang PS, Paschal WA, Cavatorta E. A study on vortex flow control of inlet distortion in the re-engined 727-100 centre inlet duct using computational fluid dynamics. AIAA paper 92-0152, 1992.

Anderson BH, Towne CE. Application of computational fluid dynamics to inlets. In: Goldsmith EL, Seddon J, eds. Practical Intake Aerodynamic Design. Oxford: Blackwell and New York: AIAA, 1993.

Anderson JD. Modern Compressible Flow with Historical Perspective. New York: McGraw-Hill, 1982.

Beam RM, Warming RF. An implicit factored scheme for the compressible Navier-Stokes equations. AIAA J 16: 393–402, 1978.

Blake MW, Kerr PA, Thorp SA, Chou JJ. NASA geometry data exchange specification for computational fluid dynamics. NASA RP 1338, 1991.

Briley WR, McDonald H. Analysis and computation of viscous subsonic primary and secondary flows. AIAA paper 79-1453, 1979.

Briley WR, McDonald H. Three-dimensional viscous flows with large secondary velocity. J Fluid Mech 144: 47–77, 1984.

Chien K-Y. Predictions of channel and boundary-layer flows with a low-Reynolds-number turbulence model. AIAA J 20: 33–38, 1982.

Coles DE, Hirst EA. Computation of turbulent boundary layers—1968 AFOSR-IFP-Stanford Conference. Vols I, II. Stanford University, 1968.

Cosner RR. Issues in aerospace application of CFD analysis. AIAA paper 94-0464, 1994.

Escher WJD, Hyde EH, Anderson DM. A user's primer for comparative assessments of all-rocket and rocket-based combined-cycle propulsion systems for advanced earth-to-orbit space transport applications. AIAA paper 95-2474, 1995.

Fernandez R, Trefny CJ, Thomas SR, Bulman M. Parametric data from a wind tunnel test on a rocket based combined cycle engine inlet. NASA TM 107181, 1996.

Jameson A, Schmidt W, Turkel E. Numerical solutions of the Euler equations by finite volume methods using Runge-Kutta time-stepping schemes. AIAA paper 81-1259, 1981.

Kunik WG. Application of a computational model for vortex generators in subsonic internal flows. AIAA paper 86-1458 (also NASA TM 87327), 1986.

Levy R, Briley WR, McDonald H. Viscous primary/secondary flow analysis for use with nonorthogonal coordinate systems. AIAA paper 83-0556, 1983.

Levy R, McDonald H, Briley WR, Kreskovsky JP. A three-dimensional turbulent compressible subsonic duct flow analysis for use with constructed coordinate systems. AIAA paper 80-1398, 1980.

Marvin JG. Dryden lectureship in research, a perspective on CFD validation. AIAA paper 93-0002, 1993.

Mehta UB. Guide to credible computational fluid dynamics simulations. AIAA paper 95-2225, 1995.

NPARC Alliance. A User's Guide to NPARC Version 2.0, 1994.

Numbers KE. Survey of CFD applications for high speed inlets. WL-TR-94-3131, 1994.

Paynter GC. CFD status for supersonic inlet design support. AIAA paper 94-0465, 1994.

Povinelli LA, Towne CE. Viscous analyses for flow through subsonic and supersonic intakes. NASA TM 88831 (Prepared for the AGARD Propulsion and Energetics Panel Meeting on Engine Response to Distorted Inflow Conditions, Munich, Germany, Sept. 8–9, 1986), 1986.

Settles GS, Dodson LJ. Hypersonic shock/boundary-layer interaction database. NASA CR 177577, 1991.

Simon FF. A research program for improving heat transfer prediction for the laminar to turbulent transition region of turbine vanes/blades. NASA TM 106278, 1993.

Simon FF, Ashpis DE. Progress in modeling of laminar to turbulent transition on turbine vanes and blades. NASA TM 107180, 1996.

Steinbrenner JP, Chawner JR, Fouts CR. The GRIDGEN 3D multiple block grid generation system. WRDC-TR-90-3022, 1990.

Towne CE. Computation of viscous flow in curved ducts and comparison with experimental data. AIAA paper 84-0531 (also NASA TM 83548), 1984.

Towne CE, Anderson BH. Numerical simulation of flows in curved diffusers with cross-sectional transitioning using a three-dimensional viscous analysis. AIAA paper 81-0003 (also NASA TM 81672), 1981.

Towne CE, Povinelli LA, Kunik WG, Muramoto KK, Hughes CE. Analytical modeling of circuit aerodynamics in the new NASA Lewis altitude wind tunnel. AIAA paper 85-0380 (also NASA TM 86912), 1985.

Towne CE, Schum EF. Application of computational fluid dynamics to complex inlet ducts. AIAA paper 85-1213 (also NASA TM 87060), 1985.

Tsai T, Levy R. Duct flows with swirl. AIAA paper 87-0247, 1987.

Vakili A, Wu JM, Hingst WR, Towne CE. Comparison of experimental and computational compressible flow in an S-duct. AIAA paper 84-0033, 1984.

Walatka PP, Buning PG, Pierce L, Elson PA. PLOT3D User's Manual. NASA TM 101067, 1990.

Turbine Flows: The Impact of Unsteadiness

Om P. Sharma, Daniel J. Dorney,* Seyf Tanrikut, and Ron-Ho Ni

Pratt & Whitney Aircraft, East Hartford, Connecticut

6.1 INTRODUCTION

Although the importance of periodic unsteadiness, induced by the relative movements of adjacent airfoil rows in turbomachinery, has long been

---

**Current affiliation*: GMI Engineering & Management Institute, Flint, Michigan.

acknowledged by design engineers, the unavailability of established prediction methods has precluded an explicit impact on the hardware design. The effects of periodic unsteadiness, until recently, were accounted for through empiricism in correlations and criteria used in design procedures. Application of these procedures often yield nonoptimal designs requiring expensive and time-consuming development programs. Work has been done over the past 15 years to develop more rigorous procedures to account for these unsteady flow effects as discussed below (Sharma et al., 1994).

Experimental programs have been conducted to investigate the impact of periodic unsteadiness on the loss and heat load generation mechanisms in turbines. Highlights from these experiments indicate that losses and heat load in an unsteady flow environment are larger than those measured for the same airfoils in a steady flow environment, as shown in Fig. 6.1 (Hodson, 1983; Blair et al., 1988; Doorley et al., 1986; Sharma et al., 1988). Simple models (Doorley et al., 1986; Sharma et al., 1990; Speidel, 1957) are available to account for the effects of upstream wake-induced unsteadiness on the performance of the downstream airfoil. The model proposed in Sharma et al. (1988) showed that the change in the profile losses of an airfoil in an unsteady environment can be related to the "reduced frequency," as shown in Fig. 6.2. Reduced frequency in this figure is defined as a ratio of the "flow change period" (relative speed/pitch of the upstream airfoil row) to the "flow interaction period" (axial velocity/axial chord of the downstream airfoil). In addition to providing a good estimate of profile losses for an embedded row of a multistage turbine, the correlation in Fig. 6.2 can also be used to calculate time-averaged boundary layer properties

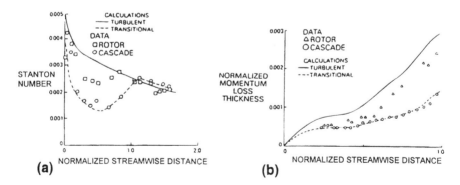

(a)   **(b)**

Figure 6.1 Measured streamwise distribution of time-averaged (a) Stanton number of Blair et al. (1988) and Sharma et al. (1988) and (b) boundary layer thickness of Hodson (1983) show larger values in an unsteady environment than in a steady cascade configuration.

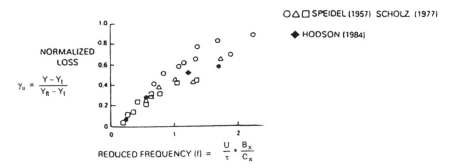

Figure 6.2 Additional time-averaged loss generated due to the unsteadiness induced by upstream wakes can be related to the frequency (Sharma et al., 1988).

including heat transfer as shown in Fig. 6.3. These calculations were conducted by assuming an increased level of intermittency factor in the airfoil suction surface boundary layer; the net increase in the intermittency factor is deduced from the correlation in Fig. 6.2.

Different models have been proposed (Scholz, 1977; Mayle and Dullenkopf, 1989; Hodson et al., 1992) to account for the effects of periodic unsteadiness on the airfoil boundary layer characteristics through the modification of the transition behavior, but results from these models are not found to be much different than those in Fig. 6.3. All of these models predict increases in both the loss and the heat load for downstream airfoils in the unsteady environment, provided the airfoils have attached boundary layers for steady incoming flows. In situations such as low-Reynolds-

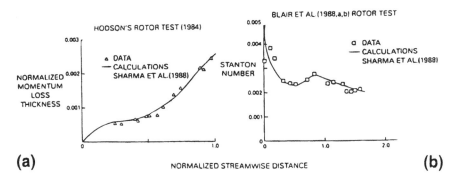

Figure 6.3 Sharma model (Sharma et al., 1988) yields good estimates of time-averaged (a) momentum loss thickness and (b) Stanton numbers from data acquired in an unsteady flow environment.

number operating conditions, where airfoils have extended regions of separated laminar flows (Fig. 6.4), interaction from upstream wakes can result in lower losses in an unsteady environment relative to the steady flow operating conditions. No reliable model is available to predict the behavior of airfoils operating at low Reynolds numbers.

It should be pointed out here that measured airfoil surface static pressure data in the above experiments could be well predicted by using steady flow codes, indicating that periodic unsteadiness has an insignificant effect on the airfoil loadings. Results in Fig. 6.3 were obtained by using time-averaged experimental data for the airfoil loadings. This implies that if airfoil loadings do not get affected by unsteadiness, fair estimates of time-averaged profile losses and heat loads can be obtained. Experience indicates that periodic unsteadiness has a relatively small effect on time-averaged loadings for turbines operating with moderate axial spacing between adjacent airfoils and operating at subsonic flow speeds.

Recent calculations (Rangwalla et al., 1991), conducted by using an unsteady Reynolds-averaged Navier–Stokes (RANS) code for the mean section of the first stage of a transonic turbine, indicated that the time-averaged loadings on the upstream stator are strongly affected by the axial gap between the stator and the rotor. These results, plotted in Fig. 6.5, clearly show that the time-averaged diffusion on the stator is significantly

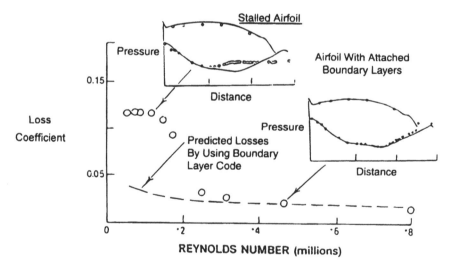

Figure 6.4 High loss levels measured for airfoils at low Reynolds numbers. Boundary layer separation is evident on airfoil surfaces.

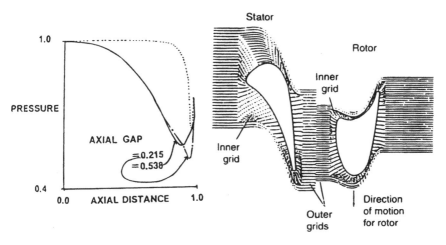

Figure 6.5 Diffusion on the upstream airfoil affected by the axial gap between rows. This effect is not accounted for in classical "design systems."

reduced as the axial gap between the airfoil rows is increased. Experimental verification of this numerical result is needed to enhance confidence in the predictive capabilities of the CFD codes. The predicted time-averaged loadings for the above-mentioned stator at the largest axial gaps were found to be quite close to those predicted by using steady CFD codes currently used in the Pratt & Whitney (P&W) design system.

The results discussed show limitations of the current prediction methods in accounting for the effects of periodic unsteadiness on airfoil loadings, which in turn affect performance and heat transfer coefficient distributions. The main focus of this chapter is to identify flow situations which can be adequately predicted by CFD codes used in the current design process and those flow situations which cannot be well predicted. This chapter also points out research work required to enhance the predictive capabilities of the current CFD codes to model these flow situations.

In the next section, the results obtained by analyzing data from a number of experimental turbine programs are described in order to provide an improved understanding of the flow physics. The impact of upstream wakes, hot streaks, and secondary flow vortices on the performance, heat load, and flow distribution through downstream airfoil rows are also discussed.

Advancements made in the flow simulation capabilities for multistage turbines through the use of unsteady CFD codes are discussed in section 6.3. It is shown that these unsteady codes provide realistic predictions of flow features through multistage turbines by using relatively simple models for the effects of turbulent viscosity.

Implications of the information are discussed in section 6.4. A clarification on situations where steady and unsteady CFD codes are needed in the design process is given.

Conclusions are outlined in section 6.5.

6.2 UNSTEADY FLOW EFFECTS IN TURBINES: EXPERIMENTAL DATA

All airfoil rows in turbines encounter spatial and temporal flow distortions generated by upstream airfoil rows and combustors, resulting in unsteady flows. Flow in the first-stage rotor passage is influenced by temporal distortions from adjacent stators. These temporal distortions consist of wakes, vortices, and entropic (temperature) disturbances from upstream stators, and potential (acoustic) waves from the upstream and downstream stators. Flow in the second-stage stator is affected not only by the temporal distortions generated by the adjacent rotor airfoils but also by the spatial distortions generated by the first-stage stator. The effects of temporal distortions, termed "upstream row-induced effects," are discussed in section 6.2.1. The effects of spatial distortions, termed "upstream stage-induced effects," are discussed in section 6.2.2.

6.2.1 Upstream Row-Induced Effects

The circumferential variations in the velocity field downstream of the first-stage stators in turbines are normally generated by the drag on the airfoil and endwall surfaces which cause reduced velocities and increased turbulence. For airfoil rows downstream of a combustor, high-velocity jets are found to exist due to large circumferential gradients in temperature. The effects of these upstream velocity variations can be simply illustrated through the use of velocity triangles (Fig. 6.6, Butler et al., 1989; Kerrebrock and Mikolajczak, 1970). This figure shows that the low-velocity fluid has a slip velocity toward the suction side of the downstream airfoil (for the compressor, the slip velocity is toward the pressure side) indicating that the high-turbulence, low-velocity fluid from the upstream airfoil wake will migrate toward the suction side of the airfoil. In a similar manner, high-velocity (high-temperature) fluid will migrate toward the pressure side of the downstream airfoil. This preferential migration of fluid particles has three effects:

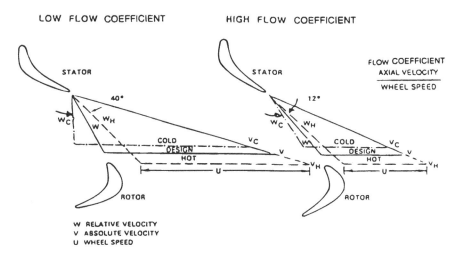

Figure 6.6 Rotor inlet gas temperature distortion causes large variation in blade incidence angle. Simple calculations conducted for hot-to-cold temperature ratio of 1.7 indicate angle variations of 12° and 40° for typical high- and low-flow-coefficient configurations, respectively.

1. Alterations in the boundary layer characteristics of the airfoil through its effect on the transition process. This effect is reasonably well accounted for in the turbine design process as outlined above.
2. Variations in the secondary flow generation in downstream passages.
3. Redistribution of the stagnation enthalpy (temperature).

Detailed discussions of the second and the third effects are given below.

Effect of Upstream Wakes on Secondary Flows

Total pressure loss data (Sharma et al., 1985) obtained by using high-response probes in the United Technologies Research Center (UTRC) large-scale rotating rig (LSRR) for the rotor as it passes through the upstream stator flow field, are shown in Fig. 6.7. This figure illustrates contours of the relative total pressure coefficient upstream and downstream of the rotor passage. In this figure, the residence time of the fluid particles in the rotor passage is accounted for in such a manner that the exit flow field corresponds to the given inlet flow field. Large variations in the rotor exit flow structures are seen in the figure for three different inlet conditions. These inlet conditions correspond to different positions of the upstream stators

$$\text{CPTR} = \frac{\text{RELATIVE TOTAL PRESSURE} - \text{REFERENCE PRESSURE}}{\text{DYNAMIC HEAD BASED ON WHEEL SPEED AT MID-SPAN}}$$

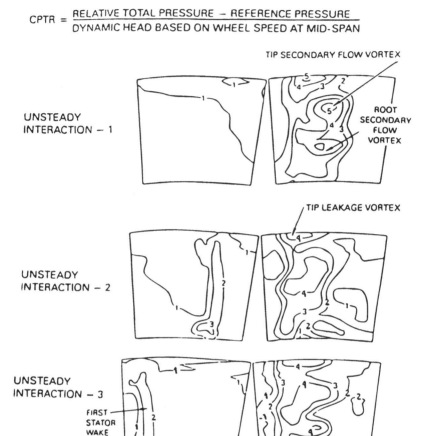

Figure 6.7 Total pressure loss contours and gap-averaged profiles at inlet and exit of the rotor in relative frame indicating the influence of unsteadiness.

relative to the rotor passage. When the inlet flow is circumferentially uniform due to the rotor passage being positioned between two adjacent stator wakes, the exit flow field shows three distinct vortices (Fig. 6.7a). The vortices are due to the hub and tip secondary flows, and the tip leakage effects. Without the tip leakage vortex, the flow field in Fig. 6.7a is similar to the one expected for this airfoil in a steady cascade environment. As the

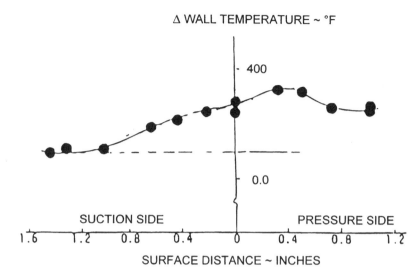

Figure 6.8   Hotter pressure sides indicated in turbine rotors. Temperature difference between pressure and suction surfaces of 250°F observed.

rotor passes through the upstream stator flow field, the tip leakage vortex shows little variation (Fig. 6.7a–c), indicating that the leakage phenomenon is not influenced by upstream circumferential distortions. The hub secondary flow vortex shows the largest variation, transforming from a distinct structure in Fig. 6.7a to a diffused structure in Fig. 6.7b, and becoming almost nonexistent in Fig. 6.7c. The overall variation in the size of the tip secondary flow vortex is smaller than that of the hub vortex but larger than the leakage vortex (Fig. 6.7c). This indicates that the secondary flow generation mechanisms, especially at the hub, are strongly influenced by the upstream circumferential distortions such as wakes.

The periodic oscillation in the size and strength of the secondary flow vortices observed in this experimental investigation equate to almost 40% variation in the secondary flow losses for the rotor passage. These data indicate that there is a potential to reduce secondary flow losses by manipulating the secondary flow vortices and increasing the unsteadiness in the rotor by increasing the number of upstream stator airfoils.

Effect of Upstream Temperature Streaks

Heat loads on turbine rotors are also affected by the migration of hot and cold air from the upstream stator and the combustor. Experimental data

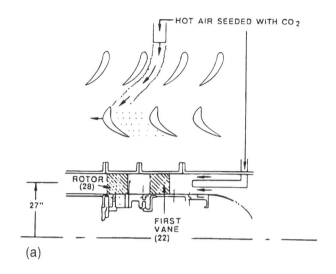

(a)

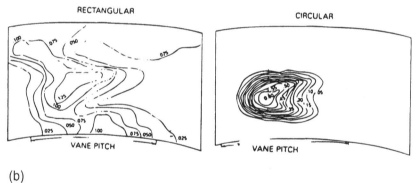

(b)

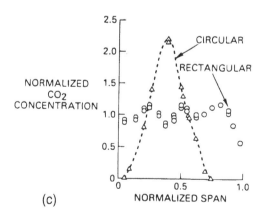

(c)

acquired in an engine indicate that pressure and suction sides of an airfoil can operate in different temperature regimes; the results plotted in Fig. 6.8 show that the differences can be on the order of 250°F. Extensive work has been done during the past 10 years to highlight the physical mechanisms responsible for this segregation of hot and cold air in turbine rotors and it is discussed here.

Results from an experimental investigation, conducted to quantify the influence of combustor induced hot streaks on the segregation of hot and cold air in turbine rotors, are discussed. In this investigation, the experimental data were acquired in the UTRC LSRR by introducing temperature streaks at inlet to the first-stage stator (Fig. 6.9). Two types of temperature profiles were generated upstream of the first-stage stator:

1. A hot streak generated in a circular pipe to yield temperature profiles both in the radial and the circumferential directions; some of the results from this investigation were reported in Sharma et al. (1990) and Butler et al. (1989). Data were acquired for two different flow coefficients by restaggering the stator and by increasing the speed of the rotor to maintain design incidence angle on the rotor airfoil.
2. A hot streak generated with a rectangular nozzle to yield a radially uniform profile that had temperature gradients in the circumferential direction. This experiment was conducted for the lower flow coefficient only.

The hot air in these experiments was seeded with carbon dioxide (CO_2) and the migration path of the hot streak through the turbine was deduced using static pressure taps and "sniffing" techniques as discussed in Butler et al. (1989). The temperature patterns at the exit of the first-stage stator for this test are given in Fig. 6.9b. The temperature patterns follow the Munk and Prim principle (Munk and Prim, 1947) and indicate little mixing in the stator passage. Measured concentrations of CO_2 on the rotor airfoil surfaces from the nozzle (rectangular hot streak) experiment are shown in Fig. 6.10. This figure shows higher levels of CO_2 on the rotor airfoil pressure

---

Figure 6.9 (a) Schematic of the experimental apparatus used to simulate the redistribution of a hot streak in a turbine rotor (Butler et al., 1989). (b) Contour plots of normalized CO_2 concentration downstream of the first-stage stator in the UTRC LSRR obtained with circular and rectangular hot streaks. High values indicate high temperature. (c) Spanwise distribution of normalized CO_2 concentration profiles (indicators of temperature) measured in the rotor frame from the circular and rectangular hot streaks.

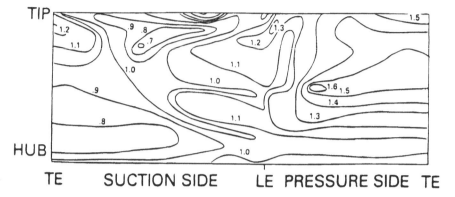

Figure 6.10   Larger time-averaged CO_2 concentration (temperature) measured on the pressure side of the rotor airfoil relative to the suction side indicates the segregation of hot and cold air.

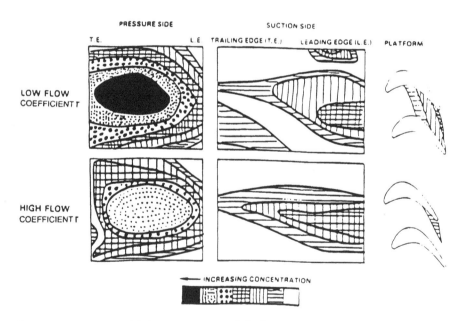

Figure 6.11   Effect of inlet temperature distortion can be reduced by increasing the flow coefficient.

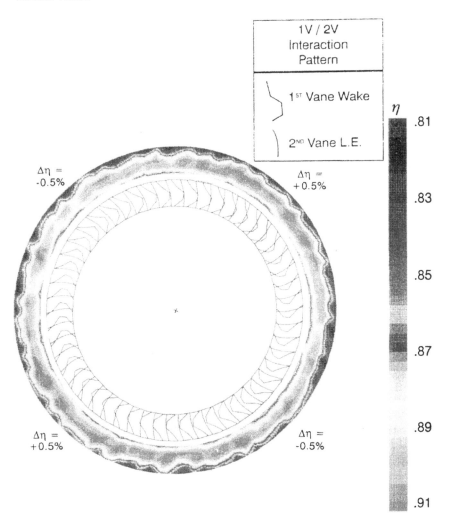

Figure 6.12 Original alternate turbopump design (ATD) turbine test article (TTA) measured efficiency contours.

side than on the suction side. These results, obtained with a radially uniform inlet profile (Fig. 6.9c), clearly demonstrate that the segregation of the hot and cold air in turbine rotors is mainly driven by two-dimensional mechanisms. Experimental data acquired on the rotor with a circular incoming hot streak are shown in Fig. 6.11 for two flow coefficients. These

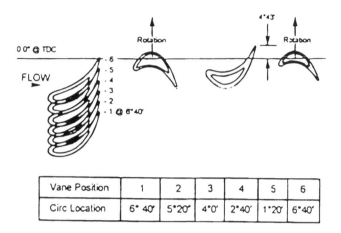

Figure 6.13   ATD TTA first vane clocking positions.

Vane Position	1	2	3	4	5	6
Circ Location	6° 40′	5°20′	4°0′	2°40′	1°20′	6°40′

data indicate that the rotor pressure side temperature is higher for the lower flow coefficient configuration.

Extrapolation of these data to an actual engine environment indicate that the pressure side of first-stage rotors can operate at temperatures between 100–700°F higher than the suction side. These temperature differences can cause significant durability problems for airfoils and endwalls. Large amounts of cooling air are required to accommodate these temperature levels, resulting in reduced efficiency of the cycle and increased specific fuel consumption of the engine.

6.2.2   Upstream Stage-Induced Effects

Experimental data show substantial variations in flow quantities measured downstream of a stage at stations where the second-stage airfoils are normally located. Indexing of second-stage airfoils relative to these incoming distortions can have a significant impact on the overall performance of the machine as discussed below.

The performance of the alternate turbopump development (ATD) turbine test article was evaluated at NASA Marshall Space Flight Center during 1991 (Gaddis et al., 1992). Measured turbine efficiencies downstream of the second stage (Fig. 6.12) showed a 2-cycle pattern on top of a 54-cycle pattern, the latter corresponding to the second-stage stator airfoil count.

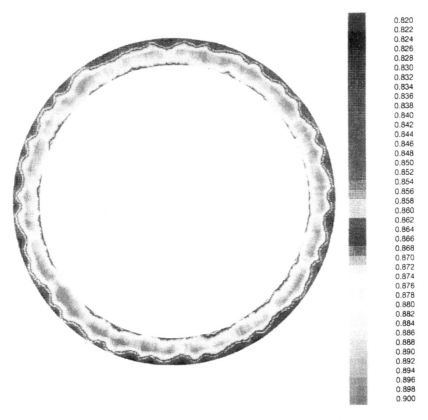

Figure 6.14   Measured efficiency contours for "clockable" turbine configuration.

The two-cycle pattern equated to a $\pm 0.5\%$ in the overall turbine efficiency, and was found to be due to the interaction between the first- and second-stage stators (i.e., dependent on where the wake fluid from the first-stage stator impinges on the second-stage stator). With airfoil counts for the first-and the second-stage stators of 52 and 54, respectively, a 2-cycle pattern will exist over the full annulus, as indicated in Fig. 6.12.

Experiments were conducted to assess the impact of indexing on the overall turbine performance and to establish whether this concept can be exploited for engine applications. Hardware was built to allow indexing of the two stators (Fig. 6.13). Both temperature and torque were measured to define the efficiency. Flow-field data were acquired over the full annulus. Results from these experiments (Huber et al., 1995) clearly indicate that the

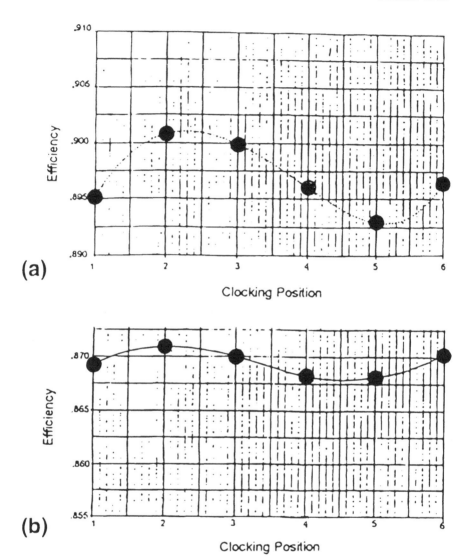

Figure 6.15 (a) Efficiency as a function of clocking position at midspan clearly indicates a minimum and a maximum. (b) Overall efficiency as a function of clocking position also indicates a minimum and a maximum.

performance of the turbine can be optimized by appropriately indexing first- and second-stage stator airfoils (Figs 6.14 and 6.15). Experimental data acquired over the range of incidence angles also showed large changes in performance due to indexing effects.

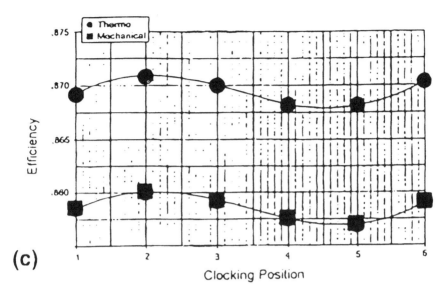

(c)

Figure 6.15 Continued. (c) Both thermodynamic and mechanical measurement of turbine efficiency in the ATD TTA confirm clocking effects and magnitude of the variations.

The above results indicate that a knowledge of the shapes and locations of wakes from upstream airfoils is required to optimize the design of downstream stage airfoils.

6.3   UNSTEADY FLOW EFFECTS IN TURBINES: NUMERICAL SIMULATIONS

Significant progress has been made over the past 20 years in developing flow prediction systems based on CFD codes. The application of these CFD codes in the turbine design process is shown in Fig. 6.16, where the interrelationships of the CFD codes with conventional design procedures are illustrated. The main contribution from CFD has come through the use of multistage flow analyses. Three different approaches are available to compute time-averaged flows through multistage machines.

1.   In the first approach, flow through each airfoil row in the machine is calculated for specified circumferentially uniform inlet and average exit boundary conditions. These boundary conditions are initially obtained from predicted or "data-matched" streamline curvature methods. Subsequently, these boundary conditions are deduced from circumferentially

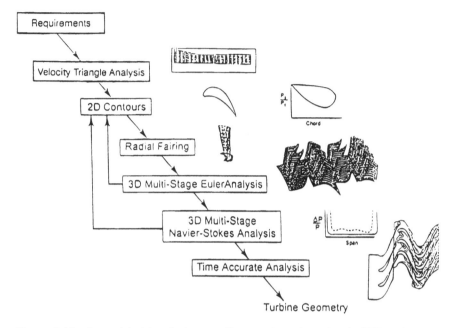

Figure 6.16   Current turbine design practices are based on steady CFD codes.

averaged mean flow quantities obtained from computations for adjacent airfoil rows. This approach (Ni and Bogoian, 1989; Denton, 1990; Dawes, 1990, 1990b), known as the "mixing plane" approach, was extensively used during the 1980s to establish spanwise distributions of airfoil loadings and flow profiles. An example of this is shown in Fig. 6.17. This approach relies on the solution of equations governing the conservation of mass, momentum, and energy, while the impact of periodic unsteadiness represented by "apparent-stress"-like terms is neglected. The effect of periodic unsteadiness has, therefore, not been accounted for in this approach.

2.   The second approach, termed the "average passage" approach, was developed (Adamczyk, 1985) to accurately simulate time-averaged flows through multistage machines. The effects of adjacent airfoil rows in this approach are accounted for through the use of body forces and "apparent stresses." Reliable models are not yet available to account for circumferential variations of "apparent stresses." These are currently assumed to be constant in the circumferential direction. The average passage approach has the potential to yield more accurate estimates of flow through multistage machines at off-design conditions than the mixing plane approach. Significant work, however, is needed to develop physics-based models to

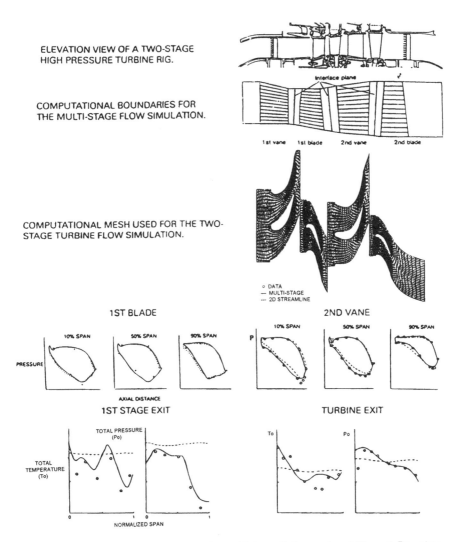

Figure 6.17 Three-dimensional steady multistage Euler code of Ni and Bogoian (1989) provides accurate estimates of airfoil loadings and total pressure as well as total temperature profiles in multistage turbines.

account for the radial and circumferential variations of apparent stresses and to enhance the predictive capabilities of codes based on the average-passage approach. In a number of situations the flow field may need to be computed over more than one airfoil passage in the machine; this implies a

Sharma et al.

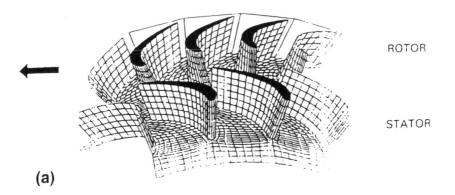

ROTOR

STATOR

(a)

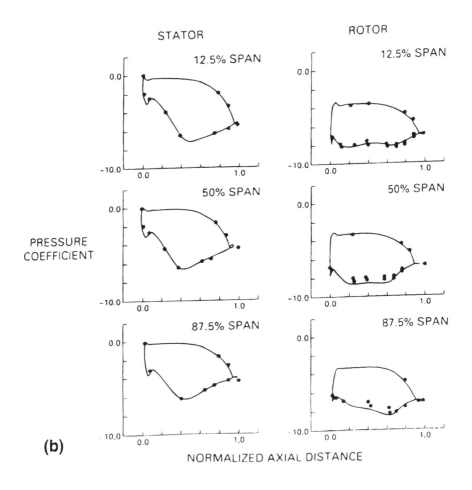

(b)

flow simulation over the most pertinent circumferential dimension that includes multiple airfoil passages. This requires minor modifications to mixing-plane and average-passage approaches. This point is further discussed in section 6.3.2.

3. Unsteady flow computations in the third approach utilize either the Euler (Ni et al., 1989, 1990) or RANS (Rai, 1989; Rai and Madavan, 1988; Rao and Delaney, 1990; Gundy-Burlet, 1991; Rhie et al., 1995; Hall, 1997; Weberand and Steinert, 1997) equations. These computations, conducted for actual airfoil counts, can yield very accurate results for the time-averaged flows. Simulations of actual airfoil counts in a multistage environment can require tremendous computational resources. The Euler codes, with approximate airfoil counts and with relatively coarse grids, have been extensively used in conducting design optimization studies. More recently the use of RANS codes has been incorporated into the design process. Numerical experiments conducted with unsteady RANS codes are providing information which indicates that their application to the design process needs to be accelerated as discussed below.

One of the cornerstones of the design process is the ability to predict airfoil surface static pressure distributions which provide information about work, losses, and heat loads. Once the pressure distribution is known, design criteria and boundary layer calculation methods are used to select high-performing airfoil sections. The main reason why Euler flow solvers were so easy to incorporate into the design process was that these methods produced reliable predictions of airfoil surface static pressure distributions in cascades and in multistage rotating rig environments, thus gaining the confidence of design engineers. Initial positive results (Huber and Ni, 1989; Huber et al., 1985) from the application of these codes encouraged turbine design engineers to look for further improvements in durability and performance through the use of more advanced codes.

Unsteady flow simulations of the UTRC LSRR model turbine have provided insight into the effects of upstream wakes and hot streaks on flow mechanics. These simulations (Ni et al., 1989; Takahashi and Ni, 1991) were conducted for single- and $1\frac{1}{2}$-stage configurations using a three-dimensional unsteady multistage Euler code. The airfoil count in these simulations was

---

Figure 6.18   (a) Computational mesh of Ni and Bogoian (1989) for the UTRC large scale rotating rig (LSRR). (b) Predicted time-averaged pressure distributions from the 3D Euler code of Ni and Bogoian (1989) show good agreement with measured data for the LSRR first stage. Both steady and unsteady multistage codes yield similar results for loadings.

three first-stage stators, four first-stage rotors, and four second-stage stators (instead of the experimental ratio of 22 first-stage stators, 28 first-stage rotors, and 28 second-stage stators) in order to contain computer requirements. The first-stage stators were scaled to maintain mass flow and pitch-to-chord ratio. Two sets of simulations were conducted, one with a uniform upstream total temperature profile (single stage) to simulate the flow conditions from Sharma et al. (1985) and the other with an upstream temperature streak ($1\frac{1}{2}$-stage configuration) to simulate flow conditions from Butler et al. (1989). A wall shear stress model (Denton, 1986) was used in the code to simulate the viscous flow effects. The tip leakage flow was not modeled in these calculations. The results from these simulations are discussed below.

6.3.1 Effect of Upstream Wakes on Rotor Secondary Flows

The computational mesh used in the single-stage simulation for the UTRC LSRR (Ni et al., 1989) is shown in Fig. 6.18a. A total of 70,000 grid points were used in the axial, radial, and tangential directions to discretize the flow field. The time-accurate flow solver and an interpolation method at the blade-row interfaces were used to obtain the unsteady periodic solution in time. Convergence was obtained at about 20,000 time steps or six rotor-passing cycles (a cycle is defined as the rotor through a distance equal to one stator airfoil pitch) and required about 10 CPU hours on a CRAY-XMP computer for each simulation. The predicted time-averaged loadings on the stator and the rotor airfoils from both simulations are compared to the experimental data in Fig. 6.18b for three spanwise locations. The predicted results are shown to be in good agreement with the experimental data. It should also be pointed out that these solutions are almost identical to those obtained by the steady multistage code (Ni and Bogoian, 1989), indicating that unsteadiness has a weak impact on the airfoil loadings in the UTRC LSRR.

The effects of upstream wakes on the secondary flow generation in the downstream rotor can be deduced through the review of unsteady total pressure contours downstream of the rotor from the simulation conducted with a uniform upstream total temperature profile. Contour plots of the computed instantaneous relative total pressure coefficients downstream of rotor, together with the 3-stator/4-rotor configuration, are shown in Fig. 6.19. These results indicate the unsteadiness has a strong effect on the rotor flow field. Organized flow structures pointed out in this figure indicate the existence of different secondary flow vortices in each rotor airfoil passage. The secondary flow vortices in the tip region are similar in all four rotor passages, which indicates that circumferential distortions generated by the upstream stator have relatively little effect on the flow in the tip region.

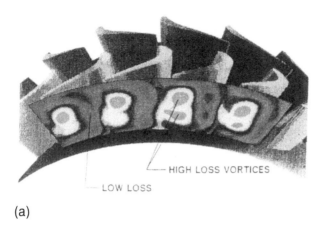

(a)

NUMERICAL SOLUTION ANIMATED

Relative Total Pressure Loss at Rotor Exit

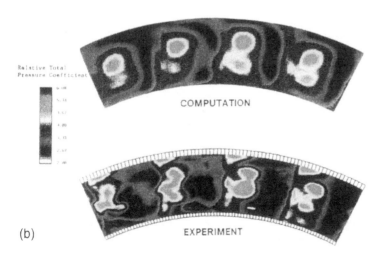

(b)

Figure 6.19 Relative total pressure loss at the exit of the rotor: (a) location of data plane, and (b) periodic disappearance of the root secondary flow vortex is in agreement with the experimental data.

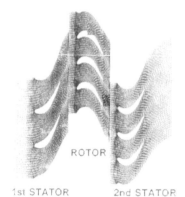

Computational mesh at midspan for 1 1/2
stage simulation

1st STATOR   2nd STATOR

ROTOR

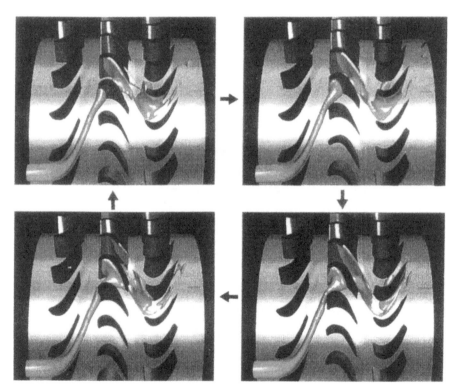

Figure 6.20 Four snapshots in time of an isotherm of one hot jet in a $1\frac{1}{2}$-stage flow simulation.

Significant passage-to-passage variations in the flow structures are observed in the root region. In particular, there are substantial reductions in the strength of the root secondary flow vortex as the rotor moves past the stator airfoils. This periodic disappearance of the root secondary flow vortex is in excellent agreement with experimental data (Sharma et al., 1985), as shown in Fig. 6.19. The comparison indicates that the unsteady flow simulation using a Euler code successfully predicted the unsteady and distorted flow features observed in the experimental data with primitive modeling of viscous effects.

Results from this numerical study clearly demonstrate that Euler codes can be used to obtain first-order effects of flow unsteadiness in turbines.

6.3.2   Hot Streak Migration through a $1\frac{1}{2}$-Stage Turbine

A three-dimensional hot streak simulation was conducted by using one hot streak, three first-stage stators, four first-stage rotors, and four second-stage stators in the geometric model. Four views of the isotherms associated with the hot streak as it migrates through the turbine are shown in Fig. 6.20. One hot streak is located in every third first stator passage; only one is shown in the figure for clarity. A complex interaction of the hot streak with the rotor and the second-stage stator is suggested by this figure. The hot streak is injected at the inlet midway between two first-stage stators and it is found to convect through the first-stage stator passage with only minor changes due to area variations. The hot streak next enters the rotor passage, where it is chopped up by the passing rotor airfoils into discrete eddies. The high-temperature eddies are convected into the second-stage stator passages, where they are further broken up.

The calculated time-averaged temperature distributions in four rotor passages are shown in Fig. 6.21. Note that each of the four rotor passages have identical time-averaged solutions because each rotor sees, over a periodic cycle, the identical inlet and exit boundary conditions. The hot gas tends to migrate toward the rotor pressure side (temperature segregation), and the rotor passage secondary flow transports the hot gas radially over the pressure surface and then over the endwalls (three-dimensional convection). Simultaneously, the hot gas on the rotor suction side appears to lift off the surface with increasing axial distance. An important observation from this figure is that maximum time-averaged temperatures downstream of the rotor leading edge are higher than the time-averaged maximum gas temperatures forward of the leading edge. These higher-than-inlet time-averaged temperatures are an indicator of the long residence time of the segregated hot gas in the rotor passage.

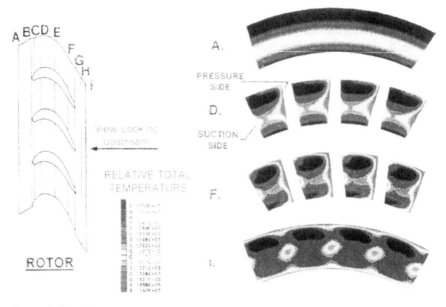

Figure 6.21 Time-averaged temperature in four rotor passages shows the hot jet segregation and migration.

A comparison of the temperature distribution in the rotor passage obtained by time-averaging the results of the unsteady simulations to those obtained from steady multistage (mixing-plane) simulations are shown in Fig. 6.22 in the form of contour plots at specified axial locations. This figure clearly shows that the maximum temperature from the steady multistage code is significantly lower than that obtained from the time-averaged results of the unsteady simulation. These results demonstrate that the time-averaged temperature on the rotor airfoil surfaces is strongly affected by periodic unsteadiness induced by combustor generated hot streaks. Conventional design procedures rely on axisymmetric rotor inlet temperatures to define the heat load on rotor airfoils; this has historically resulted in the underestimation of airfoil pressure surface temperatures and inaccurate estimates of cooling air requirements. The present work indicates that an unsteady Euler code can be used to provide a more accurate estimate of the gas temperature near the airfoil surface, which should result in a better cooling air estimate in the design process.

Two numerical experiments were conducted (McGrath et al., 1994) for the UTRC $1\frac{1}{2}$-stage LSRR turbine to assess the degree of complexity

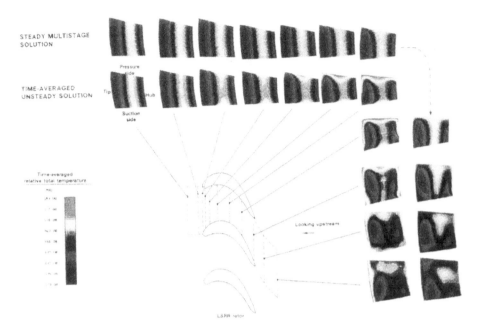

Figure 6.22 The maximum temperature from a steady solution is significantly lower than that from a time-averaged unsteady solution.

needed to model the effects of hot streak migration using steady multistage Euler codes. Since the flow in this problem is periodic over three first-stage stator passages, the simulation needed to be conducted using an average-period approach by utilizing three first-stage stators, four first-stage rotors, and four second-stage stators. In addition, the effects of periodic unsteadiness were accounted for through the use of apparent stresses. Distributions of apparent stresses were computed in the entire computational domain from the unsteady flow simulations discussed above. Circumferentially averaged values of these stresses, termed "axisymmetric apparent stresses", were used in the first numerical experiment, while the three-dimensional distribution of stresses (nonaxisymmetric) were used in the second experiment. Results from these numerical experiments are shown in Fig. 6.23 in the form of contours of relative total temperature at various axial stations in the rotor passage. These results clearly show that axisymmetric stresses are insufficient to explain the segregation of hot and cold air in rotor passages. Results obtained by using nonaxisymmetric stresses are, however, in excellent agreement with those obtained from unsteady simulations (Fig. 6.21).

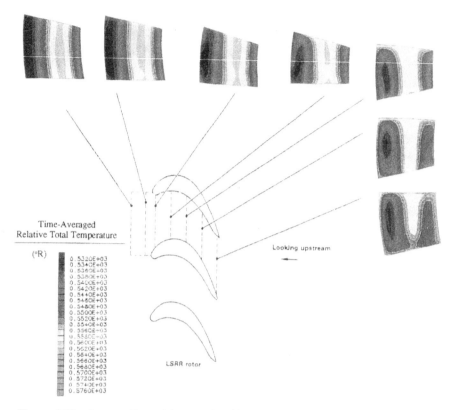

Figure 6.23 Segregation of hot and cold air is not predicted by an "average-passage" code. Relative total temperature contours in the rotor passage are in the same view as given in Fig. 6.22.

Time-averaged temperatures in the four second-stage stator passages are shown in Fig. 6.24. A number of observations can be made from the numerical results shown in this figure:

1. Hot gas in the second-stage stator passage is found to be confined to a small region of the entire 4-stator flow solution domain. This is in contrast to the results obtained for the rotor passages, which had identical time-averaged temperature fields.

2. The maximum time-averaged temperature levels in the second-stage stator passage are significantly higher than in rotor passage. This is mainly because the hot gas in the second stator passage is

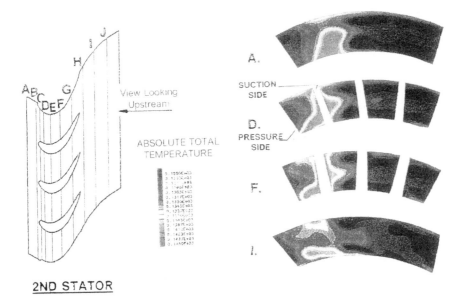

2ND STATOR

Figure 6.24 Time-averaged temperature in the four second-stage stator passages shows that hot gas is confined to a small region.

confined to a small region of the entire 4-stator flow solution regime.

3. The maximum downstream time-averaged temperatures are not higher than the maximum time-averaged temperatures at the second-stage stator inlet. This is opposite of what was observed in the time-averaged rotor flow field and indicates that hot gas is not lingering for a significant period of time on the second-stage stator surfaces.

4. There are no obvious signs of temperature segregation in the second stator passage; the hot gas does not have a tendency of preferential migration to either the pressure or the suction surface.

Figure 6.24 also shows that the hot streak has been split as it reaches the leading edge of a second stator airfoil. The hot gas that splits to the pressure side in passage 2 stays attached to the pressure surface, migrating radially along the surface toward the endwalls with increasing axial distance. This radial migration of the hot gas is similar to that observed on the rotor pressure side (Fig. 6.21) and indicates transport with classical secondary flows generated in the stator passage. The hot gas that splits toward the

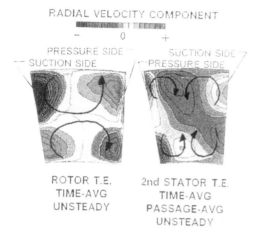

Figure 6.25 Time-averaged radial velocity contours show that there are four vortices in the second-stage stator passage and two vortices in the rotor passage.

suction side (passage 1 of Fig. 6.24) stays attached to the suction surface where it spreads radially toward both the inner and outer endwalls with increasing axial distance. This radial spreading on the suction side is opposite to the behavior observed on the rotor (Fig. 6.21), where the hot

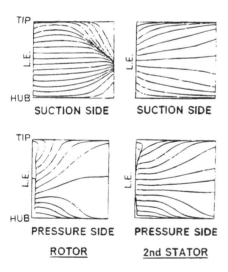

Figure 6.26 Streamlines on the rotor suction side converge to midspan at the trailing edge, whereas on the second-stage stator suction side the streamlines diverge toward the endwalls.

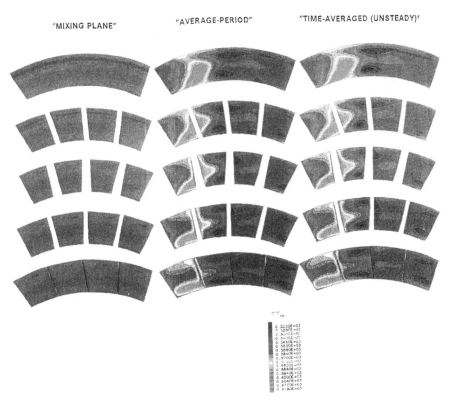

Figure 6.27 Computed temperature distributions in the UTRC LSRR second-stage stator.

gas on the rotor suction side migrated radially toward the midspan, transported by the secondary flow endwall vortices generated in the rotor passage. The suction surface hot gas in Fig. 6.24 is apparently being transported radially toward both endwalls by vortices near the suction side, for which analogous vortices did not appear in the rotor passage.

The second stator passages apparently contain four primary vortices as shown by the radial velocity component contours in Fig. 6.25, which are taken at a planar cut in passage 1 near the airfoil trailing edge. The mechanism believed to be generating the two extra suction side vortices is the interaction of vortices generated in the rotor passage with the downstream second-stage stator. Figure 6.25 also shows the radial velocity component at the trailing-edge plane of the upstream rotor. There appears to be only two primary vortices in the rotor passage, which are the classical counter-rotating endwall vortices generated in the rotor passage.

To further illustrate the differences between the rotor and second-stage stator secondary flows, the streamlines on the rotor and second-stage stator airfoil surfaces are shown in Fig. 6.26. The pressure side of both the rotor and second-stage stator show streamlines bending toward the endwalls with increasing axial distance from the leading edge, consistent with the existence of classical endwall secondary flow vortices on the pressure side. However, on the suction side the streamlines for the rotor bend toward the midspan region and the streamlines for the second-stage stator bend toward the end-walls. This observation is consistent with the existence of two primary vortices in the rotor and four in the second-stage stator.

Figure 6.27 shows total temperature contours at five axial planes in the second-stage stator passage obtained from

1. Time-averaged results from the unsteady Euler code
2. Steady multistage (mixing-plane) code
3. Steady multistage (average-period) code where the effects of periodic unsteadiness are accounted for through nonaxisymmetric distributions of the apparent stress.

The figure illustrates that the results obtained from the average-period approach are almost identical with those obtained from the unsteady code. Results obtained from the mixing-plane approach, however, are much different than those obtained from the unsteady code. This comparison indicates that current steady multistage codes, which solve for the flow through an average passage using either the mixing-plane or the average-passage approach, are insufficient to describe flow features having circumferential length scales larger than one airfoil passage. An average-period approach is required to describe these flows; in addition, the effects of periodic unsteadiness need to be accounted for through three-dimensional distributions of apparent stresses.

6.4 IMPLICATIONS OF RESULTS

Results from both experimental data and unsteady numerical simulations indicate that time-averaged loadings at subsonic Mach numbers on airfoil surfaces (and endwalls), with moderate axial gaps between adjacent airfoil rows, are not influenced by unsteadiness even in the presence of hot (temperature) streaks. Airfoil loadings are, however, affected by potential waves in turbines with small axial gaps between adjacent airfoil rows. The nonreflecting boundary conditions suggested in Giles (1988) allow this effect to be accurately accounted for in multistage steady CFD codes with the mixing-plane approach.

The pressure side of turbine rotors operate at higher temperatures than the suction side in the presence of circumferentially nonuniform temperature profiles. Currently available steady multistage CFD codes do not predict this phenomenon. Unsteady multistage Euler and Navier-Stokes codes can predict this flow phenomenon, and the computational resources have recently become available to allow the application of these codes in the turbine design process. Unsteady Euler codes by themselves, however, are not sufficient to provide accurate estimates of heat loads since viscous regions on airfoil surfaces also contribute toward establishing magnitudes of heat loads on airfoil surfaces. A viable technique may be to utilize a multistage steady RANS code based on the average-period approach discussed in section 6.3.2. The effects of periodic unsteadiness in this code may be accounted for through the use of apparent stresses computed from numerical simulations conducted by using the unsteady Euler code. Further work is needed to demonstrate the accuracy of utilizing this technique in the turbine design process.

Circumferential distortions generated by viscous flow mechanisms in first-stage airfoil passages have a significant effect on the performance of second-stage airfoils, as discussed in section 6.2.2. Significant improvements in the performance of multistage turbines may be achieved by indexing first- and second-stage airfoils. An improved physical understanding of the loss generation mechanisms, however, is needed to allow its routine application in the design process. Numerical simulations utilizing unsteady and/or average-period RANS codes should be able to provide this insight.

6.5 CONCLUSIONS

The following conclusions are drawn from the above discussions.

1. Three-dimensional steady multistage flow prediction codes provide accurate estimates of loadings for airfoil rows even in the presence of periodic unsteadiness and temperature distortions.

2. Turbine inlet circumferential temperature distortions result in hotter pressure sides and colder suction sides for rotor airfoils. This phenomenon is not predicted by the steady multistage codes which are currently used in turbine design procedures. Unsteady multistage Euler codes together with RANS codes based on the average-period approach are needed to provide accurate estimates of heat loads in a realistic turbine flow environment.

3. The efficiency of multistage turbines can be improved by indexing first- and second-stage airfoils. Improved understanding of the loss generation mechanisms, however, is needed to apply these concepts in a routine manner.
4. Continued development of unsteady RANS codes is needed to provide accurate estimates of the heat loads and losses in realistic turbine environments.

REFERENCES

Adamczyk JJ. Model equation for simulating flows in multistage turbomachinery. ASME paper #85-GT-226, 1985.

Blair MF, Dring RP, Joslyn HD. The effects of turbulence and stator-rotor interactions on turbine heat transfer. Part I. Design operating conditions. ASME paper #88-GT-125, 1988.

Butler TL, Sharma OP, Joslyn HD, Dring RP. Redistribution of an inlet temperature distortion in an axial flow turbine stage. AIAA J Propulsion Power 5: 64–71, 1989.

Dawes WN. A comparison of zero and one equation turbulence modeling for turbomachinery calculations. ASME paper #90-GT-303, 1990a.

Dawes WN. Towards improved throughflow capability: the use of 3D viscous flow solvers in multistage environment. ASME paper #90-GT-18, 1990.

Denton J. Calculation of three-dimensional viscous flows through multistage turbine. ASME paper #90-GT-19, 1990.

Denton JD. The use of a distributed body force to simulate viscous flow in 3D flow calculations. ASME paper #86-GT-144, 1986.

Doorley DJ, Oldfield MLG, Scrivener CTJ. Wake passing in a turbine rotor cascade. AGARD CP-390, paper No. 7, Bergen, Norway.

Gaddis SW, Hudson ST, Johnson PD. Cold flow testing of the space shuttle main engine alternate turbopump development high pressure fuel turbine model. ASME paper #92-GT-280, 1992.

Giles MB. Stator-rotor interaction in a transonic turbine. AIAA paper #88-3093, 1988.

Gundy-Burlet, KL. Computations of unsteady multistage compressors flows in a workstation environment. ASME paper #91-GT-336, 1991.

Hall EJ. Aerodynamic modeling of multistage compressor flowfields. Part I: Analysis of rotor/stator/rotor aerodynamic interaction. ASME paper #97-GT-344, 1997.

Hall EJ. Aerodynamic modeling of multistage compressor flowfields. Part 2: modeling deterministic stresses. ASME paper #97-GT-345, 1997.

Hodson HP. The development of unsteady boundary layers in the rotor of an axial-flow turbine. AGARD Proceedings No. 351, Viscous Effects in Turbomachines, 1983.

Hodson HP, Addison JS, Shepherdson CA. Models for unsteady wake-induced transition in axial turbomachines. J Phys III France 2: 545–574, 1992.

Huber FW, Rowey RJ, Ni R-H. Application of 3D flow computation to gas turbine aerodynamic design. AIAA-85-1216, 1985.

Huber FW, Ni R-H. Application for a multistage 3D Euler solver to the design of turbines for advanced propulsion system. AIAA paper #89-2578, 1989.

Huber FW, Johnson PD, Sharma OP, Staubach JB, Gaddis SW. Experimental investigation of vane wake clocking effects on turbine performance. ASME paper #95-GT-27, 1995; to appear in ASME J Turbomachinery.

Kerrebrock JL, Mikolajczak AA. Intra stator transport of rotor wakes and its effect on compressor performance. ASME J Eng Power Oct: 359–370, 1970.

Mayle RE, Dullenkopf K. A theory for wake-induced transition. ASME paper #89-GT-57, 1989.

McGrath D, Sharma OP, Ni R-H, Takahashi RT, Stetson GM, Staubach JB. Accurate simulations of flow through multistage turbomachines by using 3D steady CFD codes. Private communication, 1994.

Munk M, Prim RC. On the multiplicity of steady gas flows having the same streamline pattern. Proc Nat Acad Sci USA 33: 1947.

Ni R-H, Bogoian JC. Prediction of 3D multistage turbine flow field using a multiple-grid Euler solver. AIAA paper #89-0203, 1989.

Ni R-H, Sharma OP. Using 3D Euler flow simulations to assess effects of periodic unsteady flow through turbines. AIAA paper #90-2357, 1990.

Ni R-H, Sharma OP, Takahashi R, Bogoian J, 3D unsteady flow simulation through a turbine stage. Paper presented at the 1989 Australian Aeronautical Conference—Research and Technology—The Next Decade, Melbourne, Australia, Oct., 1989.

Rai MM. Three dimensional Navier-Stokes simulations of turbine rotor-stator interactions. AIAA J Propulsion Power 5: 307–319, 1989.

Rai MM, Madavan, NK. Multi airfoil Navier Stokes simulations of turbine rotor-stator interaction. AIAA paper 88-0361, 1988.

Rao K, Delaney R. Investigation of unsteady flow through transonic turbine stage. Part I: Analysis. AIAA paper #90-2408, 1990.

Rangwalla AA, Madavan NK, Johnson PD. Application of unsteady Navier Stokes solver to transonic turbine design. AIAA paper #91-2468, 1991.

Rhie CM, Gleixner AJ, Spear DA, Fischberg CJ, Zacharias RM. Development and application of a multistage Navier-Stokes solver. Part I: Multistage modeling using body forces and deterministic stresses. ASME paper #95-GT-342, 1995.

Scholz N. Aerodynamics of cascades. AGARD-AG-220, 1977.

Sharma OP, Butler TL, Joslyn HD, Dring RP. Three-dimensional unsteady flow in an axial flow turbine. AIAA J Propulsion Power 1: no. 1, 1985.

Sharma OP, Ni R-H, Tanrikut S. Unsteady flows in turbines—impact on unsteadiness. AGARD Lecture Series Turbomachinery Design Using CFD. Paper 195-5, May–June, 1994.

Sharma OP, Pickett GF, Ni R-H. Assessment of unsteady flows in turbines. ASME paper #90-GT-150, 1990.

Sharma OP, Renaud E, Butler TL, Milsaps K, Dring RP, Joslyn HD. Rotor-stator interaction in multistage axial turbines. AIAA paper #88-3013, 1988.

Speidel L. Beeinllussung der laminaren Grezschicht durch periodische Strorungen der Zustromung. Z Flugwiss 9: 5, 1957.

Takahashi R, Ni R-H. Unsteady hot streak simulation through $1\frac{1}{2}$ stage turbine. AIAA paper #91-3382, 1991.

Weber A, Steinert W. Design, optimization and analysis of a high-turning transonic tandem compressor cascade. ASME paper #97-GT-412, 1997.

7

Numerical Modeling of Materials Processing and Manufacturing Systems

Yogesh Jaluria
Rutgers University, New Brunswick, New Jersey

7.1 INTRODUCTION

In recent years, there has been a considerable interest and research activity
in manufacturing and materials processing. This is mainly due to increased
international competition and the critical need to optimize existing manufacturing systems and processes, improve product quality, reduce costs,
develop new processes, and produce a variety of new materials. Materials
processing generally refers to procedures and techniques that modify and
combine given materials to obtain desired characteristics in the product,
whereas manufacturing refers to the overall area of mass production of
useful items. An improved understanding of the physical mechanisms
underlying materials processing is needed for significant advancements in
this area. In addition, it is necessary to determine the dependence of
material characteristics on the process and operating conditions so that
these may be modified for improving product quality and reducing costs.
Mathematical modeling of the relevant thermal processes is the first step in
this direction. However, the governing equations are usually very complicated, and numerical modeling and simulation, with validation by analytical or experimental results, are needed for most practical systems and
processes.

     In a wide variety of manufacturing processes, heat transfer and fluid
flow considerations play a very important role. These include processes
such as casting, crystal growing, hot rolling, optical fiber drawing, soldering, welding, gas cutting, plastic injection molding and extrusion, metal
forming, and heat treatment. Table 7.1 presents different types of materials
processing operations along with a few important examples. Such manufac-

Table 7.1 Different Types of Materials Processing Operations, Along with Examples of Commonly Used Processes

Processes	Examples
1. Processes with phase change	Casting, continuous casting, crystal growing, drying
2. Heat treatment	Annealing, hardening, tempering, surface treatment, curing, baking
3. Forming operations	Hot rolling, wire drawing, metal forming, extrusion, forging
4. Cutting	Laser and gas cutting, fluid jet cutting, grinding, machining
5. Bonding processes	Soldering, welding, explosive bonding, chemical bond forming
6. Plastic processing	Extrusion, injection molding, thermoforming
7. Other processes	Chemical vapor deposition, composite materials processing, food processing, glass technology, optical fiber drawing, powder metallurgy, sintering, sputtering, microgravity materials processing

turing processes, in which the heat transfer and fluid flow aspects are of crucial importance in determining the properties of the final product, are also often categorized as thermal-based manufacturing or materials processing. A few typical thermal manufacturing processes are sketched in Fig. 7.1, including optical glass fiber drawing, continuous casting, mold casting, and plastic screw extrusion. A fairly wide range of important manufacturing techniques involve thermal processing of materials. Consequently, a tremendous research effort has been directed at the thermal transport that is of interest in such processes.

Several books are concerned with the area of manufacturing and materials processing. Most of these discuss the important practical considerations and manufacturing systems relevant to the various processes, without considering in detail the underlying thermal transport and fluid flow (Doyle et al., 1987; Schey, 1987; Kalpakjian, 1989). However, a few books have also been directed at the fundamental transport mechanisms in materials processing (Szekely, 1979; Ghosh and Mallik, 1986). Several other books consider specific manufacturing processes from a fundamental standpoint (Avitzur, 1968; Altan et al., 1971; Fenner, 1979; Easterling, 1983). In addition, there are several review articles and many symposia volumes on

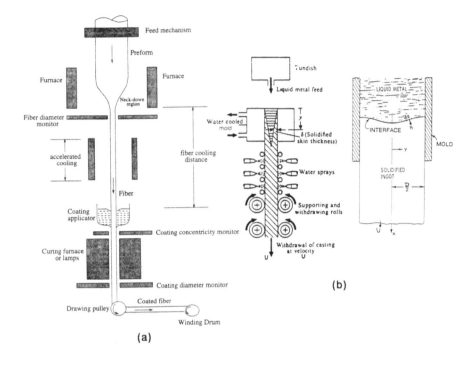

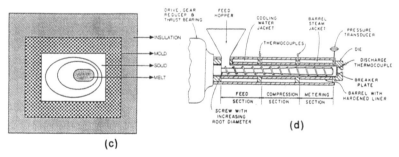

Figure 7.1 Sketches of a few common thermal manufacturing processes: (a) glass fiber drawing; (b) continuous casting; (c) mold casting; (d) plastics screw extrusion.

thermal aspects of materials processing that have appeared in the recent years (Hughel and Bolling, 1971; Kuhn and Lawley, 1978; Chen et al., 1983; Viskanta, 1985, 1988; Li, 1985). It is clear that a substantial amount of work has been done on the heat transfer and fluid flow phenomena underlying materials processing. This chapter discusses the relevant fundamental considerations and presents the numerical modeling of such manufacturing processes.

Many important considerations arise when dealing with the numerical modeling of materials processing (Table 7.2). Most relevant processes are time dependent, since the material often undergoes a given thermal variation in order to attain desired characteristics. Therefore, the variation with time is important, even though steady-state situations are also of interest in a few cases. Sometimes, a transformation of the variables in the problem can convert a time-dependent problem to a steady one. Most manufacturing processes involve combined modes of heat transfer. Conjugate conditions arise due to the coupling between conduction in the solid material and convection in the fluid. Radiation is frequently important in these processes. The material properties are often strongly temperature dependent, giving rise to strong nonlinearity in the energy equation (Lee and Jaluria, 1996a, 1996b). Also, the material properties may depend on the shear rate, as is the case for polymeric materials, which are generally non-Newtonian (Fenner, 1979; Jaluria, 1996). Therefore, material properties affect the transport processes and are, in turn, affected by the transport. This aspect often leads to considerable complexity in the mathematical modeling and in the numerical simulation. The material undergoing the thermal transport process may be moving, as in hot rolling or extrusion, or the thermal source itself may be moving, as in laser cutting or welding. Additional mechanisms

Table 7.2 Some Important Considerations in Materials Processing

1. Coupling of transport with material characteristics:
     different materials, resulting material structure,
     properties, behavior
2. Variable material properties
3. Complex geometries
4. Complicated boundary conditions
5. Interaction between different mechanisms:
     surface tension, heat and mass transfer,
     non-Newtonian flow, chemical reactions,
     free surface, powder, and particle transport,
     phase change, microstructure conversion
6. Inverse problems
7. Different energy sources:
     laser, chemical, explosive, gas, fluid jet, heat
8. System optimization and control

such as surface tension effects and chemical reactions are important in several cases. Complex geometry and boundary conditions are commonly encountered. Frequently, an inverse problem is to be solved to obtain the conditions that result in a given temperature field. Finally, the process is obviously linked with the manufacturing system design and operation.

All these considerations make numerical modeling of materials processing very involved and challenging. Special procedures and techniques are often needed to satisfactorily simulate the relevant boundary conditions and material property variations. However, the results obtained are important and interesting, since these are generally not available in the existing heat transfer and fluid mechanics literature. Numerical simulation results also provide appropriate inputs for the design and optimization of the relevant thermal system for the manufacturing process. Therefore, it is important to accurately model the process, mathematically and numerically. Let us first consider the basic conservation principles and the appropriate governing equations for these processes.

7.2 GOVERNING EQUATIONS

7.2.1 General Equations

The governing equations for heat transfer and fluid flow in materials processing are derived from the basic conservation principles for mass, momentum, and energy. For a pure viscous fluid, these equations may be written as

$$\frac{D\rho}{Dt} + \rho \nabla \cdot \mathbf{V} = 0 \tag{7.1}$$

$$\rho \frac{D\mathbf{V}}{Dt} = \mathbf{F} + \nabla \cdot \boldsymbol{\tau} \tag{7.2}$$

$$\rho C_p \frac{DT}{Dt} = \nabla \cdot (k\nabla T) + \dot{Q} + \beta T \frac{Dp}{Dt} + \mu \Phi \tag{7.3}$$

where D/Dt is the substantial or particle derivative, given in terms of the local derivatives in the flow field by $D/Dt = \partial/\partial t + \mathbf{V} \cdot \nabla$. Here, ρ is the fluid density, $\mathbf{V}$ the local velocity vector, $\mathbf{F}$ the body force acting per unit volume of the fluid, $\boldsymbol{\tau}$ the stress tensor, C_p the specific heat at constant pressure, k the thermal conductivity, β the coefficient of thermal expansion, μ the dynamic viscosity of the fluid, $\dot{Q}$ the energy source per unit volume, p the local pressure, Φ the viscous dissipation function, T the local temperature, and t the time.

Similarly, for a solid, the energy equation is written as

$$\rho C \frac{DT}{Dt} = \frac{\partial T}{\partial t} + \mathbf{V} \cdot \nabla T = \nabla \cdot (k\nabla T) + \dot{Q} \tag{7.4}$$

where C is the specific heat of the solid material. For a stationary solid, the convection term drops out and the particle derivative is replaced by the transient term $\partial/\partial t$. In a deforming solid, as in wire drawing, extrusion, or fiber drawing, the material is treated as a fluid, with the additional terms due to pressure work and viscous heating included. In the preceding equations, the material is taken as isotropic, with the properties, particularly the thermal conductivity k, assumed to be the same in all directions. For certain materials such as composites, the nonisotropic or orthotropic behavior must be taken into account.

The stress tensor in Eq. (7.2) can be written in terms of the velocity $\mathbf{V}$ if the material characteristics are known. For instance, if μ is taken as a constant for a Newtonian fluid, the relationship between the shear stresses and the shear rates, given by Stokes, are employed to yield

$$\rho \frac{D\mathbf{V}}{Dt} = \mathbf{F} - \nabla p + \mu \nabla^2 \mathbf{V} + \frac{\mu}{3} \nabla(\nabla \cdot \mathbf{V}) \tag{7.5}$$

Here, the bulk viscosity $K = \lambda + (2/3)\mu$ is taken as zero, where λ is a second viscosity coefficient. This approximation is usually employed for most problems of practical interest (Jaluria, 1980). However, for an incompressible fluid, $\rho = $ constant, $\nabla \cdot \mathbf{V} = 0$ from Eq. (7.1), and the last term in Eq. (7.5) drops out.

7.2.2 Buoyancy Effects

The body force $\mathbf{F}$ is also important in many manufacturing processes, such as crystal growing and casting where it gives rise to the thermal buoyancy term. The governing momentum equation is obtained, when thermal buoyancy is included, as

$$\rho \frac{D\mathbf{V}}{Dt} = -\mathbf{e}g\rho\beta(T - T_a) - \nabla p_d + \mu \nabla^2 \mathbf{V} \tag{7.6}$$

where T_a is the ambient temperature, $\mathbf{e}$ is the unit vector in the direction of the gravitational field, g is the magnitude of the gravitational acceleration, and p_d is the dynamic pressure, obtained after subtracting out the hydrostatic pressure p_a. Therefore, p_d is the component due to fluid motion, as discussed by Jaluria (1980) and Gebhart et al. (1988).

Boussinesq approximations, which neglect the effect of the density variation in the continuity equation, Eq. (7.1), yielding $\nabla \cdot \mathbf{V} = 0$, are employed here. Also, the density variation due to the temperature difference is approximated as linear. These approximations apply for small temperature differences, i.e., $\beta(T-T_a) \ll 1$. For large temperature and pressure differences, additional terms must be included. If the x-coordinate axis is vertically upward, i.e., in the direction opposite to gravity, then $\mathbf{e} = -\mathbf{i}$, where $\mathbf{i}$ is the unit vector in the x-direction, and the buoyancy term appears only in the x-component of the momentum equation. The energy equation remains Eq. (7.3) and the continuity equation is $\nabla \cdot \mathbf{V} = 0$. The governing equations are coupled because of the buoyancy term in Eq. (7.6) and must be solved simultaneously. This differs from the forced convection problem with constant fluid properties, for which the flow field is independent of the temperature distribution and may be solved for separately before solving the energy equation, as discussed in detail by Jaluria and Torrance (1986).

7.2.3 Viscous Dissipation

The viscous dissipation $\mu\Phi$ represents the irreversible part of the energy transfer due to the stress tensor. Therefore, it behaves like a thermal source in the flow and is always positive. For a Cartesian coordinate system, Φ is given by the expression

$$\Phi = 2\left[\left(\frac{\partial u}{\partial x}\right)^2 + \left(\frac{\partial v}{\partial y}\right)^2 + \left(\frac{\partial w}{\partial z}\right)^2\right]$$
$$+ \left(\frac{\partial v}{\partial x} + \frac{\partial u}{\partial y}\right)^2 + \left(\frac{\partial w}{\partial y} + \frac{\partial v}{\partial z}\right)^2 + \left(\frac{\partial u}{\partial z} + \frac{\partial w}{\partial x}\right)^2 - \frac{2}{3}(\nabla \cdot \mathbf{V})^2 \qquad (7.7)$$

where u, v, and w are the velocity components in the x-, y-, and z-directions, respectively. Similarly, expressions for other coordinate systems may be obtained. This term becomes important for high-viscosity fluids and at high speeds. The former circumstance is of particular interest in the processing of plastics, food, and other polymeric materials.

Equations (7.1)–(7.4) are written for variable properties. However, considerable simplification in the governing equations is obtained if constant properties are assumed, for instance, Eq. (7.5), which is written for constant viscosity μ. If the viscosity variation with temperature and/or shear rate is taken into account, the momentum diffusion terms become much more complicated. However, variable properties usually need to be included in the analytical/numerical modeling of manufacturing processes for an accurate simulation of the underlying heat transfer and fluid flow.

7.2.4 Processes with Phase Change

For materials processing problems that involve phase change, there are two main approaches for numerical simulation. The first one treats the two phases as separate, with their own properties and characteristics. The interface between the two phases must be determined so that conservation principles may be applied there and appropriate discretization of the two regions may be carried out (Ramachandran et al., 1981, 1982; Viskanta, 1985). This becomes fairly involved since the interface location and shape must be determined for each time step or iteration. The governing equations are the same as those given earlier for the solid and the liquid.

In the second approach, the conservation of energy is considered in terms of the enthalpy H, yielding the governing energy equation as

$$\rho \, \frac{DH}{Dt} = \rho \, \frac{\partial H}{\partial t} + \rho \mathbf{V} \cdot \nabla H = \nabla \cdot (k \nabla T) \tag{7.8}$$

where each of the phase enthalpies H_i is defined as

$$H_i = \int_0^T C_i \, dT + H_i^\circ \tag{7.9}$$

C_i being the corresponding specific heat and H_i° the enthalpy at 0 K. Then the solid and liquid enthalpies are given by, respectively,

$$H_s = C_s T, \qquad H_l = C_l T + [(C_s - C_l)T_m + L_h] \tag{7.10}$$

where L_h is the latent heat of fusion and T_m is the melting point. The continuum enthalpy and thermal conductivity are given, respectively, as

$$H = H_s + f_l(H_l - H_s), \qquad k = k_s + f_l(k_l - k_s) \tag{7.11}$$

where f_l is the liquid mass fraction, obtained from equilibrium thermodynamic considerations. The dynamic viscosity μ is expressed as the harmonic mean of the phase viscosities in the limit as $\mu_s \to \infty$; i.e., $\mu = \mu_l/f_l$. This model smears out the discrete phase transition. But the numerical modeling is much simpler since the same equations are employed over the entire computational domain and there is no need to keep track of the interface between the two phases (Bennon and Incropera, 1988; Viswanath and Jaluria, 1993). Figure 7.2 shows the grids employed for the two approaches outlined here for the numerical modeling of solidification processes, indicating the interface between the two regions for the two-phase approach and a single domain for the enthalpy method.

7.2.5 Numerical Solution

The governing equations given here are the usual ones encountered in fluid flow and heat transfer. Though additional complexities due to the geometry, boundary conditions, material property variations, etc., arise in materials processing, as mentioned earlier, the numerical solution of the

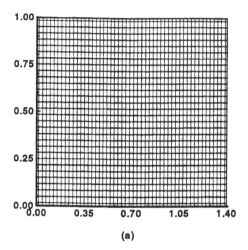

(a)

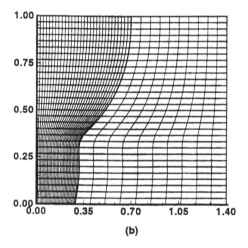

(b)

Figure 7.2 Grids used for (a) the enthalpy method (single region) and (b) the two-phase (two-region) method.

governing equations is based on the vast literature on computational fluid mechanics and heat transfer. Some of the special techniques needed due to complexities in materials processing are discussed later. Among the most commonly employed techniques for solving these equations is the SIMPLER algorithm, given by Patankar (1980), and several variations of this approach. The method employs the finite volume formulation with a staggered grid so that the value of each scalar quantity such as pressure and temperature is associated with the grid node and the vector quantities like velocity are displaced in space relative to the scalar quantities and generally located on the faces of the control volume. This grid system has an advantage in solving the velocity field since the pressure gradients that drive the flow are easy to evaluate and the velocity components are conveniently located for the calculation of the convective fluxes. A pressure correction equation is used during the iteration to converge to the solution.

For two-dimensional and axisymmetric problems, the governing equations are often cast in terms of the vorticity and stream function by eliminating the pressure from the two components of the momentum equation and by defining a stream function to take care of the continuity equation (Jaluria and Torrance, 1986). This reduces the number of equations by one and pressure is eliminated as a variable, though it can be calculated after the solution is obtained. The solution yields the stream function, which is used for obtaining the velocity field and drawing streamlines, the temperature, which is used for drawing the isotherms and calculating heat transfer rates, and the vorticity. Because the stream function is specified on the boundaries, convergence of the stream-function equation is usually faster, compared to the pressure equation, which involves gradient conditions. Thus, this approach is generally advantageous compared to the methods based on primitive variables of velocity, pressure, and temperature for two-dimensional and axisymmetric flows. The latter approach is more appropriate for three-dimensional circumstances.

Both transient and steady-state solutions are of interest, depending on the process under consideration. In the former case, time marching is used with convergence at each time step to obtain the time-dependent variation of the flow, temperature field, heat transfer rates, etc. For steady problems also, time marching may be used to obtain the desired results at large time. However, the problem can also be solved by iteration or by using false transients with large time steps (Mallinson and de Vahl Davis, 1973). Though central differences are preferred for all the approximations, numerical instability with the convection terms is often avoided by the use of upwind, exponential, or power-law differencing schemes (Patankar, 1980). Under-relaxation is often needed for convergence due to the strong nonlinearities that arise in these equations. Several methods are available to solve the vorticity transport and energy equations. The alternating direction implicit (ADI) method of Peaceman and Rachford (1955), as well as modifications of this time-splitting method, are particularly efficient for two-dimensional problems. Similarly, cyclic reduction, successive overrelaxation, and other standard methods may be used for the stream function or the pressure equation. Details on these techniques may be obtained from the references given earlier as well as from Roache (1976), Anderson et al.

(1984), Jaluria (1988a), and Minkowycz et al. (1988). A few typical results from the numerical modeling of manufacturing and materials processing systems are presented later.

7.3 NUMERICAL APPROXIMATION OF BOUNDARY AND INITIAL CONDITIONS

7.3.1 General Conditions

The boundary and initial conditions for the governing equations, presented in the preceding section, vary strongly with the thermal manufacturing process under consideration. Some of the general conditions are outlined here. Spatial boundary conditions generally arise from thermal conditions imposed at the solid surfaces, from ambient conditions far from the surfaces, and from symmetry. At the surface of a solid body undergoing a thermal process, such as heat treatment, the following three general types of boundary conditions may arise on the temperature:

$$T = f_1(\mathbf{x}), \quad \frac{\partial T}{\partial n} = f_2(\mathbf{x}), \quad a(\mathbf{x}) + b(\mathbf{x})\frac{\partial T}{\partial n} = f_3(\mathbf{x}) \tag{7.12}$$

where $\mathbf{x}$ is a vector denoting position on the boundary, $\partial/\partial n$ is the derivative normal to the boundary, and f_1, f_2, f_3, a, and b are arbitrary functions. These three types of boundary conditions are often referred to as Dirichlet (function), Neumann (gradient), or mixed boundary conditions, respectively. The first one is the easiest to simulate numerically, but practical processes typically involve the other two types of conditions. The conditions in materials processing often vary with time t; for instance, the temperatures of the heaters or the convective heat transfer coefficient in a furnace may vary with time. For such time-varying boundary conditions, the dependence on $\mathbf{x}$ in the above equations is replaced by dependence on $\mathbf{x}$ and t. These conditions may be nonlinear because of the temperature dependence of material properties or the presence of thermal radiation.

Frequently, heat flux is prescribed at a boundary, giving rise to Neumann conditions. Similarly, convective cooling at a surface yields mixed boundary conditions. Thus, typical boundary conditions will be of the form

$$-k\frac{\partial T}{\partial n} = q, \quad -k\frac{\partial T}{\partial n} = h(T - T_a),$$

$$-k\frac{\partial T}{\partial n} = h(T - T_a) + \varepsilon\sigma F(T^4 - T_a^4) \tag{7.13}$$

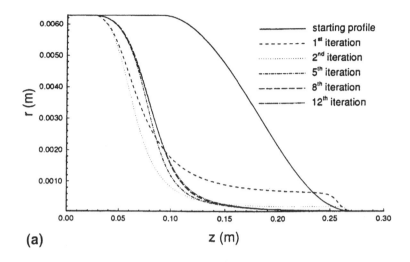

(a)

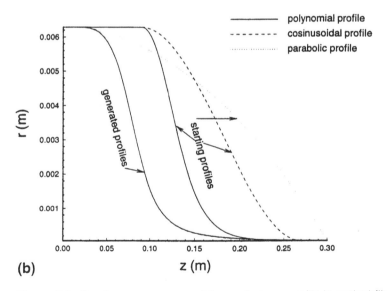

(b)

Figure 7.3   Iterative convergence of the neck-down profile in optical fiber drawing (Roy Choudhury et al., 1995).

where q is a heat flux lost at the surface, h the convective heat transfer coefficient, T_a the ambient temperature, ε the surface emissivity, σ the Stefan–Boltzmann constant, and F a geometric parameter for radiative transport. Similarly, boundary conditions may be written for the flow field. No-slip conditions at a stationary surface yield $\mathbf{V} = 0$ at the surface. On a moving surface, $\mathbf{V}$ is specified as the velocity of the solid surface.

7.3.2   Free Surfaces and Openings

At a free surface, the shear stress is often specified as zero, yielding a Neumann condition of the form $\partial\mathbf{V}/\partial n = 0$, if negligible shear is applied on the surface. If the shear stress exerted by the ambient fluid is significant, it replaces the zero in this equation. Basically, a balance of all the forces acting at the surface is used to obtain the interface. As considered in detail by Roy Choudhury et al. (1995), the free surface may be determined numerically by iterating from an initial profile and using the imbalance of the forces for correcting the profile at intermediate steps. Figure 7.3 shows such an iterative convergence for the neck-down profile of the glass preform in an optical fiber-drawing process, as sketched in Fig. 7.1a.

In a stationary ambient medium, far from the solid boundaries, the velocity and temperature may be given as $\mathbf{V} \to 0$, $T \to T_a$ as $n \to \infty$. However, frequently the condition $\partial\mathbf{V}/\partial n \to 0$ is used instead in order to allow for entrainment into the flow. The use of this Neumann condition generally allows the use of a much smaller computational domain than that needed for a Dirichlet condition imposed on the velocity $\mathbf{V}$ (Jaluria, 1992). The gradient conditions allow the flow to adjust to ambient conditions more easily, without forcing it to take on the imposed conditions at a chosen boundary. This consideration is very important for simulating openings in enclosures, such as the one in Fig. 7.4, where gradient conditions at the opening allow the flow to adjust gradually to the conditions outside the enclosure. Such conditions are commonly encountered in furnaces and ovens with openings to allow material flow.

7.3.3   Other Conditions

Several other boundary conditions that typically arise in materials processing may be mentioned here. The normal gradients at an axis or plane of symmetry are zero, simplifying the problem by reducing the computational domain. The temperature and heat flux continuity must be maintained in going from one homogeneous region to another. This yields the thermal conductivity at the interface being approximated as the harmonic mean of the conductivities in the two adjacent regions for one-dimensional transport (Jaluria and Torrance, 1986). Therefore, the conjugate conditions that arise

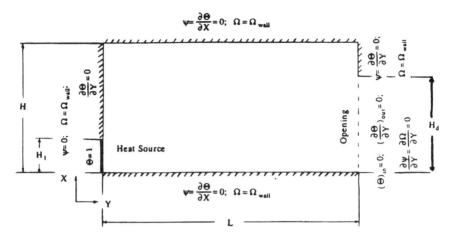

Figure 7.4 Numerical simulation of the boundary conditions at the opening of an enclosure. Here Ψ, Ω, and θ are the nondimensional stream function, vorticity, and temperature, respectively.

at a solid surface in heat exchange with an adjacent fluid are

$$T_s = T_f, \quad \left(-k\,\frac{\partial T}{\partial n}\right)_s = \left(-k\,\frac{\partial T}{\partial n}\right)_f \qquad (7.14)$$

where the subscripts s and f refer to the solid and the fluid, respectively. Figure 7.5 shows the grid in a typical conjugate transport problem, indicating the two regions and the boundary at which conditions such as those given in Eq. (7.14) are applied.

The initial conditions are generally taken as the no-flow circumstance at the ambient temperature, representing the situation before the onset of the process. However, if a given process precedes another, the conditions obtained at the end of the first process are employed as the initial conditions for the next one. For instance, if the material is preheated before entering a furnace, the temperature distribution in the material before the onset of thermal processing in the furnace provides the appropriate initial conditions. For periodic and steady-state processes, the initial conditions are arbitrary and may be taken as the ambient temperature.

7.3.4   Phase Change

If a change of phase occurs at the boundary, the energy absorbed or released due to the change of phase must be taken into account. Thus, the boundary conditions at the moving interface between the two phases (Fig. 7.1c) must also be given if a two-zone model is being used. This is not

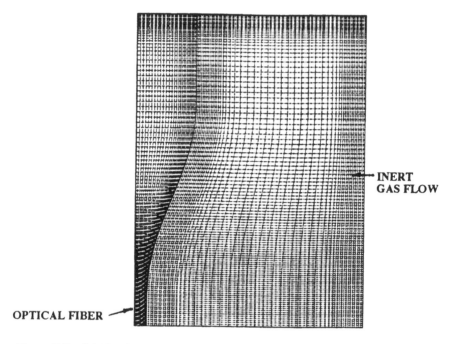

INERT
GAS FLOW

OPTICAL FIBER

Figure 7.5 Grid for the numerical modeling of the conjugate transport problem in optical fiber drawing.

needed in the enthalpy model given by Eqs. (7.8)–(7.11). For one-dimensional solidification, this boundary condition is given by the equation (Ramachandran et al., 1981)

$$k_s \frac{\partial T_s}{\partial y} - k_l \frac{\partial T_l}{\partial y} = \rho L_h \frac{d\delta}{dt} \tag{7.15}$$

where $y = \delta$ is the location of the interface. This implies that the energy released due to solidification is conveyed by conduction in the two regions. Similarly, for two-dimensional solidification, the boundary condition is written as (Ramachandran et al., 1982)

$$\left(k_s \frac{\partial T_s}{\partial y} - k_l \frac{\partial T_l}{\partial y}\right)\left[1 + \left(\frac{\partial \delta}{\partial x}\right)^2\right] = \rho L_h \frac{d\delta}{dt} \tag{7.16}$$

For a stationary interface, as shown in Fig. 7.1b, the boundary condition is (Siegel, 1978, 1984)

$$\left(-k \frac{\partial T}{\partial n}\right)_l + \rho U L_h \frac{dy}{ds} = \left(-k \frac{\partial T}{\partial n}\right)_s \tag{7.17}$$

where ds is a differential distance along the interface and n is the distance normal to it. Also, the temperature at the interface in all these cases is T_m. A few typical numerical results on this problem are given later.

7.4 GOVERNING PARAMETERS AND SIMPLIFICATIONS FOR NUMERICAL SIMULATION

7.4.1 Common Simplifications

The basic nature of the underlying physical processes and the simplifications that may be obtained under various circumstances can be best understood in terms of dimensionless variables that arise when the governing equations are nondimensionalized. The commonly encountered governing dimensionless parameters are the Strouhal number Sr, the Reynolds number Re, the Grashof number Gr, the Prandtl number Pr, and the Eckert number Ec. These are defined as

$$\text{Sr} = \frac{L}{V_a t_c}, \qquad \text{Re} = \frac{V_a L}{v}$$

$$\text{Gr} = \frac{g\beta(T_s - T_a)L^3}{v^2}, \qquad \text{Pr} = \frac{v}{\alpha}, \qquad \text{Ec} = \frac{V_a^2}{C_p(T_s - T_a)} \qquad (7.18)$$

where V_a is a characteristic speed, L a characteristic dimension, t_c a characteristic time, T_s the surface temperature, $v = \mu/\rho$ the kinematic viscosity, and $\alpha = k/\rho C_p$ the thermal diffusivity. It is often convenient to apply different nondimensionalization to the solid and the fluid.

The dimensionless equations may be used to determine the various regimes over which certain simplifications can be made. For instance, at small values of Re, the convection terms are small, compared to the diffusion terms and may be neglected. This approximation, often known as creeping flow, is applied to the flow of highly viscous fluids such as plastics and food materials. At large Re, boundary layer approximations can be made to simplify the problem. At very small Pr, the thermal diffusion terms are relatively large and yield the conduction-dominated circumstance, which is often applied to the flow of liquid metals in soldering and welding. A small value of Gr/Re^2 implies negligible buoyancy effects, for instance, in continuous casting where the effect of buoyancy on the transport in the melt region may be neglected. A small value of Ec similarly implies negligible pressure work effects, and a small value of Ec/Re can be used to neglect viscous dissipation. Finally, a small value of Sr indicates a very slow transient, which can be treated as a quasi-steady circumstance. Therefore, the numerical values of the governing parameters Re, Gr, Pr, Sr, and Ec can be employed to determine the relative importance of various physical mechanisms underlying the transport process. This information can then be used to simplify the relevant governing equations and the corresponding numerical modeling.

7.4.2 Approximations and Transformations

In the case of a moving cylindrical rod for extrusion or hot rolling (Fig. 7.6), the temperature T may be taken as a function of only the downstream distance x, assuming it to be uniform at each cross section. Such an assumption can be made if the Biot number Bi_R based on the radius R of the rod is small (i.e., $Bi_R = hR/k \ll 1.0$). Thus, for a thin rod of high thermal conductivity material, such an assumption would be valid. The governing energy equation is

$$\rho C\left(\frac{\partial T}{\partial t} + U\,\frac{\partial T}{\partial x}\right) = k\,\frac{\partial^2 T}{\partial x^2} - \frac{hP}{A}\,(T - T_a) \tag{7.19}$$

where P is the perimeter of the rod, A its cross-sectional area, and T_a the ambient temperature. The transient problem arises for small lengths of the rod, with steady flow arising for large lengths (Jaluria and Singh, 1983; Roy Choudhury and Jaluria, 1994). If the Biot number based on the surface area

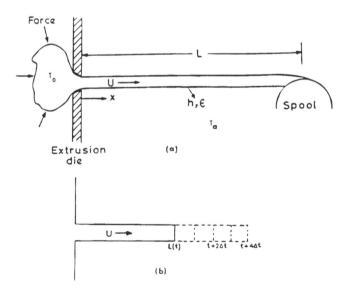

Figure 7.6 (a) Sketch of the extrusion process for a heated material; (b) moving material at different time intervals.

and volume of a given body is small, say, for a copper sphere, the entire body may be taken to be at a uniform temperature $T(t)$. Such an approximation can be made, for instance, regarding metal blocks being heat treated and the copper core of a plastic cable undergoing thermal processing (Jaluria, 1976).

The time-dependent nature of materials processing is another important consideration. By placing the coordinate system outside the moving material in Fig. 7.6, a steady problem is obtained if the edge of the rod is far from the inlet, $x = 0$. Then, the temperature $T(x, y, z)$ in a moving plate is governed by

$$\rho C U \frac{\partial T}{\partial x} = k\left(\frac{\partial^2 T}{\partial x^2} + \frac{\partial^2 T}{\partial y^2} + \frac{\partial^2 T}{\partial z^2}\right) \tag{7.20}$$

The boundary conditions in x may be taken as $T(0, y, z) = T_0$ and $T(\infty, y, z) = T_a$. For lumping in the y- and z-directions, an ordinary differential equation (ODE) is obtained from Eq. (7.20). Such an ODE may be solved analytically or numerically, for more complicated cases, by using shooting methods or the finite difference approach (Jaluria, 1988a).

In some cases, a transformation can be employed to convert a transient problem to a steady-state one. For instance, a moving thermal source at the surface of an extensive material gives rise to a transient circumstance if the coordinate system is fixed to the material. However, a steady-state situation is obtained by fixing the origin of the coordinate system at the source. If x is measured in the direction of the source movement from a coordinate system on the material surface and U is the location of the point source, the transformation used in $\xi = x - Ut$, which yields the governing equation

$$\frac{\partial^2 T}{\partial \xi^2} + \frac{\partial^2 T}{\partial y^2} + \frac{\partial^2 T}{\partial z^2} = -\frac{U}{\alpha}\frac{\partial T}{\partial \xi} \tag{7.21}$$

This transformation applies to processes such as welding and laser cutting. This steady-state problem is numerically solved, and the transformation is used to yield the time-dependent results.

In some manufacturing systems, the transient response of a particular component is much slower than the response of the others. The thermal behavior of this component may then be treated as quasi-steady, i.e., as a sequence of steady-state circumstances. For instance, in a heat treatment furnace, the walls and the insulation are often relatively slow in their response to the transport processes. Consequently, these may be assumed to be at steady state at a given time, with different steady states arising at different time intervals whose length is chosen on the basis of the transient response (Jaluria, 1988b).

For small Re, the creeping flow approximation is often employed, as mentioned earlier. For instance, the Reynolds number Re is generally smaller than 1.0 for plastics flow in a single-screw extruder and the inertia terms are usually dropped. Assuming the flow to be developing in the down channel (z) direction and lumping across the flights (i.e., velocity varies only with z and distance y from the screw root toward the barrel) (Fig. 7.7), the governing momentum equations become (Karwe and Jaluria, 1990)

$$\frac{\partial p}{\partial x} = \frac{\partial \tau_{yx}}{\partial y}, \qquad \frac{\partial p}{\partial y} = 0, \qquad \frac{\partial p}{\partial z} = \frac{\partial \tau_{yz}}{\partial y} \tag{7.22}$$

where the pressure terms balance the viscous forces. The coordinate system is generally fixed to the rotating screw and the channel straightened out mathematically, ignoring the effects of curvature. Then the complicated flow in the extruder is replaced by a pressure and shear-driven channel flow, with shear arising due to the barrel moving at the pitch angle over a stationary screw (Fig. 7.7). This substantially simplifies the numerical model. A few typical results on this problem are presented later, following the consideration of the material properties.

Several other such simplifications and approximations are commonly made to reduce the computational effort in the numerical simulation of thermal manufacturing processes. For instance, dy/ds may be taken as unity in Eq. (7.17) for many continuous casting processes that use an insulated mold, which gives rise to a fairly planar interface. Also, for slow withdrawal

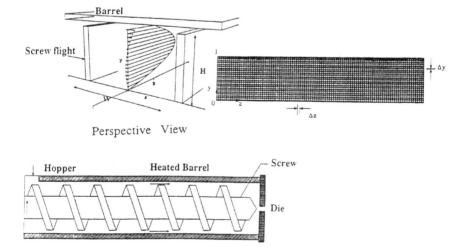

Figure 7.7 Screw channel and computational domain for a single-screw extruder.

rates, the heat transfer due to convection is small compared to that due to conduction within the moving material and may be neglected. If the extent of the material undergoing, say, thermal processing at the surface, is large, it may often be assumed to be semi-infinite, simplifying both the analysis and the numerical simulation. Similarly, the boundaries are often approximated as planar to simplify the imposition of the boundary conditions there. Clearly, the preceding discussion is not exhaustive and presents only a few common approximations and simplifications.

7.5 NUMERICAL COUPLING DUE TO MATERIAL PROPERTIES

7.5.1 Variable Properties

The properties of the material undergoing thermal processing play a very important role in the mathematical and numerical modeling of the process. As mentioned earlier, the ranges of the process variables, particularly temperature, are usually large enough to make it necessary to consider material property variations. Then, properties such as density (ρ), thermal conductivity (k), specific heat (C or C_p), coefficient of thermal expansion (β), and viscosity (μ) are taken as variables. The governing equations are Eqs. (7.1)– (7.4), which are written for variable properties. Usually, the dependence of the properties on temperature T is the most important effect. Numerical curve fitting may be employed to obtain a given material property as a function of T, say, $k(T) = k_r[1 + a(T - T_r) + b(T - T_r)^2]$, where T_r is a reference temperature at which $k = k_r$. Thus, a continuous function $k(T)$ replaces the discrete data on k at different temperatures (Jaluria, 1988a). This gives rise to nonlinearity since

$$\frac{\partial}{\partial x}\left[k(T)\frac{\partial T}{\partial x}\right] = \frac{\partial k}{\partial x}\frac{\partial T}{\partial x} + k\frac{\partial^2 T}{\partial x^2} = \frac{\partial k}{\partial T}\left(\frac{\partial T}{\partial x}\right)^2 + k\frac{\partial^2 T}{\partial x^2} \qquad (7.23)$$

Similarly, the data for other material properties may be represented by appropriate curve fits. Because of the resulting nonlinearity, due to property variations, the numerical solution of the equations becomes more involved than that for the constant property circumstance. Iterative numerical procedures are often required to deal with such nonlinear problems, as discussed by Jaluria and Torrance (1986). Due to these complexities, average constant property values are frequently employed to simplify the solution. Though this approach is satisfactory for small ranges of the process variables, most manufacturing processes require the solution of the full variable property problem for accurate predictions of the resulting transport. For convective transport in common fluids, a reference temperature is often employed at which the fluid properties are determined and the constant property model is considered. The film temperature $T_f = T_s - 0.5(T_s - T_a)$, with $\beta = 1/T_a$ for gases, is frequently used, where T_s is the surface temperature. Other reference temperatures, with 0.5 in the preceding equation being replaced by 0.38, 0.46, etc., have also been used for different cases (Gebhart et al., 1988).

7.5.2 Viscosity Variation

The variation of viscosity μ requires special consideration for materials such as plastics, polymers, food materials, and several oils that are of interest in a variety of manufacturing processes. Most of these materials are non-Newtonian in behavior, implying that the shear stress is not proportional to the shear rate. The viscosity μ is a function of the shear rate and, therefore, of the velocity field. Figure 7.8 shows the variation of the shear stress τ_{yx} with the shear rate du/dy for a shear flow such as the flow between two parallel plates with one plate moving at a given speed and the other held stationary. The viscosity is independent of the shear rate for Newtonian fluids like air and water, but increases or decreases with the shear rate for shear thickening or thinning fluids, respectively. These are viscoinelastic (purely viscous) fluids, which may be time dependent or time independent, the shear rate being a function of both the magnitude and the duration of shear in the former case. Viscoelastic fluids show partial elastic recovery on the removal of a deforming shear stress. Food materials are often viscoelastic in nature.

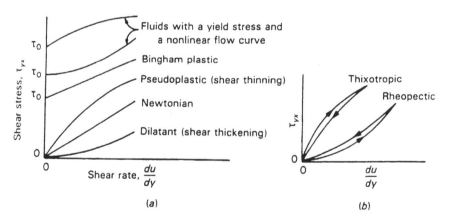

Figure 7.8 Plots of shear stress versus shear rate for viscoinelastic non-Newtonian fluids: (a) time-independent and (b) time-dependent fluids.

Various models are employed to represent the viscous or rheological behavior of fluids of practical interest. For instance, time-independent viscoinelastic fluids without a yield stress are often represented by the power-law model, given by (Tadmor and Gogos, 1979)

$$\tau_{yx} = K_c \left|\frac{du}{dy}\right|^{n-1} \frac{du}{dy} \qquad (7.24)$$

where K_c is the consistency index and n the power-law fluid index. Note that $n = 1$ represents a Newtonian fluid. For $n < 1$, the behavior is pseudoplastic (shear thinning) and for $n > 1$, it is dilatant (shear thickening). The viscosity variation may be written as (Tadmor and Gogos, 1979)

$$\mu = \mu_0 \left(\frac{\dot{\gamma}}{\dot{\gamma}_0}\right)^{n-1} e^{-b(T-T_0)} \qquad (7.25)$$

where $\dot{\gamma} = [(\partial u/\partial y)^2 + (\partial w/\partial y)^2]^{1/2}$, $\tau_{yx} = \mu(\partial u/\partial y)$, $\tau_{yz} = \mu(\partial w/\partial y)$ for a two-dimensional flow, with u and w varying only with y.

Similarly, expressions for other two- and three-dimensional flows may be written. Here $\dot{\gamma}$ is the shear strain rate, the subscript 0 denotes reference conditions, and b is the temperature coefficient of viscosity. For food materials, the viscosity is also a strong function of the moisture concentration c_m, which is often represented as

$$\mu = \mu_0 \left(\frac{\dot{\gamma}}{\dot{\gamma}_0}\right)^{n-1} e^{-b(T-T_0)} e^{-b_m(c_m - c_{m0})} \qquad (7.26)$$

where b_m is the moisture coefficient of viscosity and c_{m0} is the concentration in the ambient fluid. The temperature dependence is also often represented more accurately by an Arrhenius type of variation, i.e.,

$$\mu = \mu_0 \left(\frac{\dot{\gamma}}{\dot{\gamma}_0}\right)^{n-1} e^{B/T} \qquad (7.27)$$

where B is the corresponding temperature coefficient. Other models, besides the power-law model, are also employed to represent different materials (Tadmor and Gogos, 1979; Pearson and Richardson, 1983; Jaluria, 1996).

The non-Newtonian behavior of the material complicates the viscous terms in the momentum and the energy equations. For instance, the viscous dissipation term for the two-dimensional flow given by Eq. (7.25) is

$$\Phi = \tau_{yx} \frac{\partial u}{\partial y} + \tau_{yz} \frac{\partial w}{\partial y} \qquad (7.28)$$

where the variation of μ with $\dot{\gamma}$ and, therefore, with the velocity field is taken into account. Similarly, the viscous force term in the momentum equation yields $\partial(\tau_{yx})/\partial y$ in the x-direction, requiring the inclusion of the non-Newtonian behavior of the fluid (Karwe and Jaluria, 1990). Similarly, other flow circumstances may be considered for the flow of non-Newtonian fluids. Viscous dissipation effects are generally not negligible in these flows because of the large viscosity of the fluid. However, as mentioned earlier, because of the relatively small Reynolds numbers in most cases, the convection, or inertia, terms can be neglected, simplifying the equations.

Glass is another very important, though complicated, material. The viscosity varies very strongly with temperature. In optical fiber drawing, for instance, the viscosity changes through several orders of magnitude in a relatively short distance. This makes it necessary to employ very fine grids and substantial underrelaxation to achieve convergence. Even a small change of a few degrees in temperature in the vicinity of the softening point, which is around 1600°C for fused silica, can cause substantial changes in viscosity and thus in the flow field and the neck-down profile (Lee and Jaluria, 1996a, 1996b; Roy Choudhury et al., 1995).

7.5.3 Other Aspects

There are several other important considerations related to material properties. Constraints on the temperature level in the material, as well as on the spatial and temporal gradients, arise due to the nature of the material. In thermoforming, for instance, the material has to be raised to a given temperature level, above a minimum value T_{min}, for the process to be carried out. However, as shown in Fig. 7.9, the maximum temperature T_{max} must not be exceeded to avoid damage to the material. In polymeric materials, $T_{max} - T_{min}$ is relatively small and the thermal conductivity k is also small, making it difficult to design a process which restricts the tem-

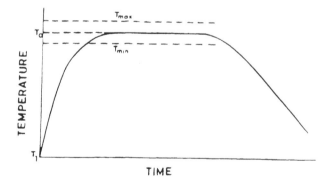

Figure 7.9 Typical temperature cycle for a heat treatment process.

perature to T_{max} while raising the entire material to above T_{min} (Jaluria, 1976).

Similarly, constraints on $\partial T/\partial t$, $\partial T/\partial x$, etc., arise due to thermal stresses in the material undergoing thermal processing. Such constraints are particularly critical for brittle materials such as glass and ceramics. Then the design of the thermal manufacturing process is governed by the material constraints.

In several circumstances, the thermal conductivity is not the same in all directions because of the nature of the material or because of the configuration. For anisotropic materials, such as wood, asbestos, composite materials, cork, etc., the conduction flux vector $\mathbf{q}$ may be written as $\mathbf{q} = -\mathbf{k}\,\nabla T$, where $\mathbf{k}$ is the conductivity tensor, with nine components k_{ij} obtained by varying i and j from 1 to 3 to represent the three directions. For orthotropic materials, the coordinate axes coincide with the principal axes of the conductivity tensor and the energy equation for a stationary material, in the Cartesian coordinate system, is

$$\rho C \frac{\partial T}{\partial t} = \frac{\partial}{\partial x}\left(k_x \frac{\partial T}{\partial x}\right) + \frac{\partial}{\partial y}\left(k_y \frac{\partial T}{\partial y}\right) + \frac{\partial}{\partial z}\left(k_z \frac{\partial T}{\partial z}\right) + \dot{Q} \qquad (7.29)$$

Similarly, the equations for other coordinate systems may be written. In the annealing of coiled steel sheets, the thermal conductivity k_r in the radial direction is often much smaller than k_z in the axial direction, due to gaps within the coils, and the governing conduction equation may be written taking this effect into account (Jaluria, 1988b). The preceding discussion brings out the importance of material properties in a satisfactory mathematical and numerical modeling of thermal manufacturing processes. The basic characteristics of the material undergoing thermal processing must be known and should be appropriately modeled to accurately predict the resulting transport and thermal field.

7.6 ADDITIONAL CONSIDERATIONS AND TYPICAL NUMERICAL RESULTS

We have considered the governing equations and boundary conditions for several different types of manufacturing processes, focusing on the concerns and difficulties commonly encountered in such problems. Several simplifications that may be employed to reduce the complexity of the governing equations and the associated boundary conditions have also been presented. However, as mentioned earlier, despite these simplifications, analytical solutions are very rarely possible because of the transient, variable-property, complex, and multidimensional nature of the transport that is of interest in most practical manufacturing processes. Therefore, numerical procedures are generally needed to obtain the important characteristics of the process and to provide inputs for the design, operation, control, and optimization of the system. Some of the computational techniques, approximations, and approaches that may be employed for materials processing have been outlined earlier. In this section, we shall present a few additional and more specific considerations in the numerical modeling of manufacturing processes, as well as typical results for a few important problems to illustrate the use of numerical modeling in this area.

7.6.1 Basic Aspects

Because of the complexities associated with materials processing, several specialized techniques have been developed for numerical simulation. A few of these have been discussed, and it is not possible to discuss all the other approaches in detail here. However, several books (Patankar, 1980; Anderson et al., 1984; Jaluria and Torrance, 1986; Minkowycz et al., 1988) present various computational techniques that may be employed to simulate a wide range of manufacturing processes. Several examples have been considered by Jaluria and Torrance (1986) to outline important considerations in the numerical modeling of the thermal processing of materials, considering processes such as heat treatment, surface treatment, melting, solidification, hot rolling, and thermal forming.

The finite element method (FEM) has been used most extensively in recent years to simulate manufacturing processes and systems, due to the complexity of the computational domain and of the boundary conditions (Huebner and Thornton, 1983). However, the finite difference method (FDM) is still used for a wide variety of processes, since it is often much easier to use and the numerical modeling of convective processes is well understood and documented. Therefore, unless the complexity of the region demands the finite element approach, the finite difference, or the similar finite volume approach, is quite satisfactory. Even for complicated geometries, a coordinate transformation is often employed to simplify the geometry, as, for instance, in solidification where complicated domains arise due to the interface between the solid and the liquid (Ramachandran et al., 1982; Siegel, 1984). The finite difference approach is particularly useful when dealing with additional transport mechanisms, such as radiation in participating media, combined heat and mass transfer, surface tension effects, etc. The boundary element method (BEM) has also become an important method for the numerical simulation of these processes, particularly for processes where transport at the surface is of crucial importance, for instance, in laser cutting and welding (Brebbia, 1977).

7.6.2   Material Property Variation

The nonlinearity arising from variable material properties requires special treatment in the numerical approach. Linearization of the resulting algebraic equations is often achieved by using the values from the previous time step or iteration to treat the coefficients of the various terms as known and, thus, linearize the equations in terms of the unknowns. For instance, we may write

$$\frac{\partial}{\partial x}\left(k\,\frac{\partial T}{\partial x}\right) = \left[\frac{\partial k}{\partial T}\,\frac{\partial T}{\partial x}\right]\frac{\partial T}{\partial x} + [k(T)]\,\frac{\partial^2 T}{\partial x^2} \tag{7.30}$$

Then the two quantities within brackets may be evaluated at the previous iteration or time step in order to linearize the equation for the solution of T at the next step. Thus, the iteration process becomes

$$T_i^{(n+1)} = F_i[T_1^{(n+1)},\ T_2^{(n+1)},\ \dots,\ T_{i-1}^{(n+1)},\ T_i^{(n)},\ \dots,\ T_m^{(n)}] \tag{7.31}$$

where T_i at the $(n+1)$th iteration is evaluated in terms of the most recently computed temperatures, with i varying from 1 to m. F_i represents the functional dependence. A similar equation may be written for marching in time for transient problems. Also, other variable property circumstances, such as non-Newtonian fluids, are linearized by employing the upstream values or those at the preceding time to treat the coefficients as known, resulting in a system of linear algebraic equations.

An important manufacturing process whose formulation has been discussed in detail in the preceding sections is plastic screw extrusion, sketched in Figs. 7.1d and 7.7. The material characteristics and the governing equations for a relatively simple two-dimensional model are given by Eqs. (7.22) and (7.24)–(7.28). This is a fairly complicated problem because of the strong shear rate and temperature dependence of the viscosity, large viscous dissipation, and the resulting coupling between the energy and momentum equations. Figure 7.10 shows typical computed velocity and temperature fields in an extruder channel. Large temperature differences are seen to arise across the channel height because of the relatively small thermal conductivity of plastics. The flow is well layered, with little bulk mixing, due to the high viscosity of these fluids, the typical viscosity being more than a million times that of water at room temperature. Viscous dissipation causes the temperature to rise beyond the imposed barrel temperature. A lot of work has been done on this problem because of its importance to industry, as reviewed by Tadmor and Gogos (1979) and Jaluria (1996).

Another process considered in detail was optical fiber drawing (Fig. 7.1a). Again, the viscosity of glass, which is a subcooled liquid, is a very strong function of temperature. At its softening point, the viscosity is still

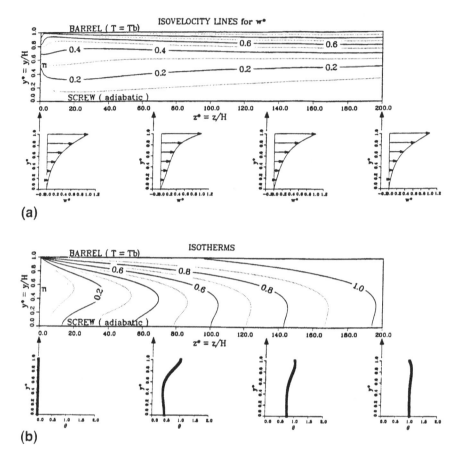

Figure 7.10 Calculated velocity and temperature fields in the channel of a single screw extruder at $n = 0.5$ (Jaluria, 1996).

very high, being of the same order as that of polymer melts. Thus, viscous dissipation is important, and the energy and momentum equations are coupled. Even small temperature differences are important because of the effect on the viscosity and, thus, on the flow characteristics. However, glass flow may be treated as Newtonian. In optical fiber drawing, the diameter of the cylindrical rod, known as a preform, typically changes from around 5 cm to about 125 μm in a distance of only a few centimeters. This places difficult demands on the grid as well as on the numerical scheme because of the large change in the surface velocity. The simulation of the free surface is another difficult problem, as discussed earlier. Typical computed results

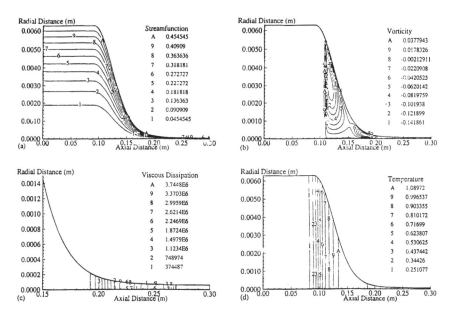

Figure 7.11   Calculated (a) stream function, (b) vorticity, (c) viscous dissipation, and (d) temperature contours in the optical fiber drawing process (Lee and Jaluria, 1996a).

in the neck-down region are shown in Fig. 7.11, indicating the stream function, vorticity, viscous dissipation and temperature contours. The flow is smooth because of the high viscosity. A typical temperature difference of 50–100°C arises across the fiber. Even this small difference is an important factor in fiber quality and characteristics. Viscous dissipation, though relatively small, is mainly concentrated near the end of the neck-down, in the small-diameter region, and plays an important role in maintaining the temperatures above the softening point. Further details on this problem may be obtained from Li (1985), Lee and Jaluria (1996, 1997), and Roy Choudhury et al. (1985).

7.6.3   Moving Boundary

A moving boundary results in another important complication in the numerical simulation of manufacturing processes. If the location of the moving boundary is known, as is the case for the circumstance of Fig. 7.6b, the continuous movement of the boundary may be replaced by steps so that the length L is held constant over a time increment Δt and the transient conduction problem is solved over this interval. The length L is then taken at the increased value for the next time interval, with the additional finite region adjacent to the base taken at temperature T_0, and the computation is carried out for this interval. The procedure is carried out until results are obtained over a given time interval or until the steady state circumstance is obtained (Jaluria and Singh, 1983). The corresponding initial and boundary conditions are

$$t = 0: \quad L(t) = 0$$

$$t > 0: \quad \text{at } x = 0, \ T = T_0; \quad \text{at } x = L(t), \ -k \frac{\partial T}{\partial x} = h_L(T - T_a) \quad (7.32)$$

where h_L is the heat transfer coefficient at the end of the moving rod. Typical results on this problem are shown in Fig. 7.12. The problem may be solved analytically or numerically, with the latter approach more appropri-

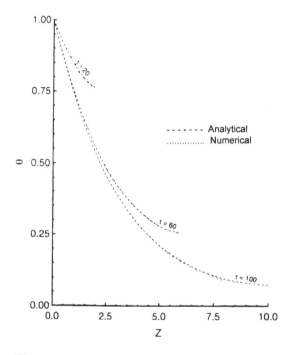

Figure 7.12 Transient temperature distribution in a moving cylindrical rod of radius R for one-dimensional conduction, at Peclet number $Pe_R = UR/\alpha = 0.2$ and Biot number $Bi_R = hR/k = 0.1$. Here, $\theta = (T - T_a)/(T_0 - T_a)$ and dimensionless time $\tau = \alpha t/R^2$. The analytical and numerical results essentially overlap (Roy Choudhury and Jaluria, 1994).

ate for two- and three-dimensional problems. It is seen that as time increases, the rod length L increases and the temperature at the end of the rod decreases. At large time, a steady-state distribution arises over the rod, and the temperature at the moving end reaches the ambient temperature. The problem may then be solved as a steady, continuously moving, infinite-rod case (Jaluria, 1992).

7.6.4 Solidification

If the location of the moving boundary is not known, it must be obtained from the solution, as is the case for mold casting shown in Fig. 7.1c. A coordinate transformation, such as the Landau transformation (Ramachandran et al., 1982), may be employed to make the computational domains rectangular or cylindrical, from the complicated ones shown. This considerably simplifies the numerical procedure by allowing a regular rectangular or cylindrical mesh to be used. Several other techniques have been developed to treat such moving-boundary problems and the complicated domains that arise. For the continuous casting problem of Fig. 7.1b, the interface between the solid and liquid is not known at the onset and an iterative procedure may be adopted to determine its shape and location. Again, body-fitted coordinates may be employed to approximate the irregularly shaped computational domains. Of course, if the enthalpy model, outlined in Eqs. (7.8)–(7.11), is employed, the entire region is treated as one, considerably simplifying the computational procedure.

Figure 7.13 shows the numerical results for melting in an enclosed region using the enthalpy model. Streamlines and isotherms are shown for four different times during the melting of pure gallium. This is a benchmark problem in which melting is initiated by a step change in the temperatures at the left and right boundaries, the left being at a higher temperature than the melting point and the right lower. The streamlines indicate the effect of thermal buoyancy which causes the interface between the solid and the liquid to bend rather than remain parallel to the vertical boundaries. The amount of material melted increases with time till it reaches a steady state for this problem. The recirculation in the liquid is clearly seen. These results are found to agree well with experimental results available in the literature. The two-region approach can also be used for modeling this problem. A transformed grid is generally used to model the complex domains, and the computational effort is much greater. However, for pure metals, the two-phase, two-region approach leads to more accurate results, whereas the enthalpy method is more useful for alloys and mixtures. A lot of work has been done on such melting and solidification problems, as reviewed by Viskanta (1985, 1988).

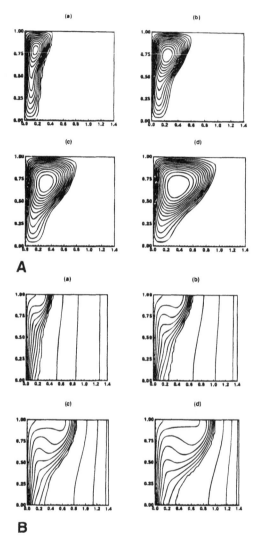

Figure 7.13 (A) Streamlines and (B) isotherms for melting of gallium in an enclosed region, with the left vertical boundary at a temperature higher than melting point, the right vertical boundary at a temperature lower than melting point, and the remaining two boundaries insulated. The enthalpy method is used and results are shown at different dimensionless time t following the onset of melting: (a) $t = 0.5248$, (b) $t = 1.0416$, (c) $t = 1.5622$, (d) $t = 1.9789$ (Viswanath and Jaluria, 1993).

The numerical results for continuous casting are shown in terms of isotherms in Fig. 7.14, again using the enthalpy method. The material is n-octadecane which starts as a liquid at the top and solidifies as it flows through a mold. The shaded region indicates the demarcation between pure liquid and pure solid. Therefore, the liquid fraction f_l is 1.0 at the top of the shaded region and zero at the bottom of this region. A value of 0.5 may be

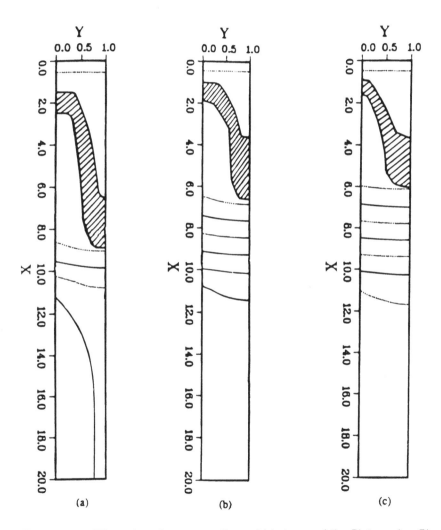

Figure 7.14 Effect of cooling rate at the mold in terms of the Biot number Bi on the solidification in continuous casting of n-octadecane, using the enthalpy method. (a) Bi = 0.05, (b) Bi = 0.1, (c) Bi = 0.15 (Kang and Jaluria, 1993).

taken to represent the liquid-solid interface, but the enthalpy method yields a finite region over which solidification is predicted to occur. It is seen that the material solidifies over a shorter distance at a larger value of the Biot number, as expected.

7.6.5 Conjugate Transport

Conjugate conditions are very frequently encountered in materials processing since the conductive transport in the material undergoing thermal treatment is coupled with the convective and radiative transport mechanisms at the surface. Figure 7.15 shows a sketch of a continuously moving material subjected to heat transfer at the surface. Such a situation arises for manufacturing processes like hot rolling, extrusion, metal forming, and fiber drawing. Conjugate conditions arise at the surface and the convective transport in the fluid must be solved in conjunction with conduction in the moving solid. This is an important problem and has received some attention in the literature (Jaluria, 1992). The region near the point of emergence of the material has large axial gradients and requires the solution of the full equations. However, far downstream, the axial diffusion terms are small, and a parabolic marching scheme may be adopted. This reduces the computational time compared to the solution of the elliptic problem over the entire computational domain.

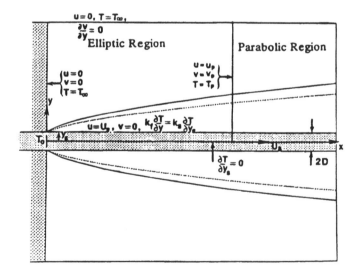

Figure 7.15 Sketch of a continuously moving material subjected to conjugate convective transport at the surface.

Figure 7.16 shows typical numerical results for this problem. A heated Teflon plate moves in quiescent water or air. The growth of the velocity and thermal boundary layer in the direction of the plate motion is clearly seen. The ambient fluid is entrained into the boundary layer flow which is driven by the moving plate due to viscous effects. The plate cools more rapidly in water, as expected. The conjugate transport is seen in the distribution of the

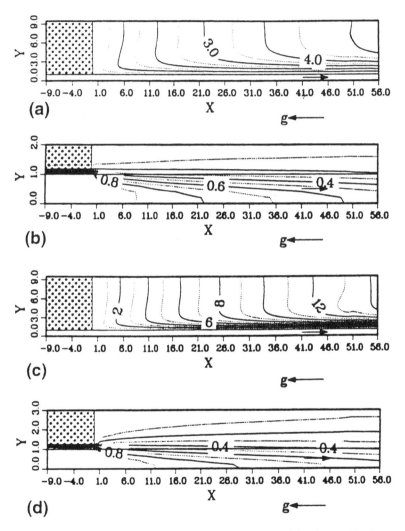

Figure 7.16 Calculated steady-state streamlines and isotherms for the conjugate problem of a heated Teflon plate moving in (a, b) water and (c, d) air (Kang et al., 1991).

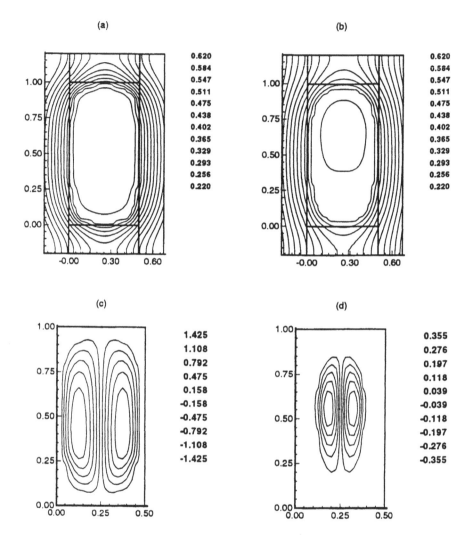

Figure 7.17 Isotherms and streamlines for solidification in a cavity with conjugate transport to the mold. (a, c) Dimensionless time $t = 0.05$ and (b, d) $t = 0.1$ (Viswanath and Jaluria, 1995).

isotherms in the solid and fluid. Many different fluids, materials, and flow conditions, including channel flows and the effect of buoyancy, have been considered for a wide variety of applications in materials processing, as reviewed by Jaluria (1992).

Conjugate transport is also important in other problems such as solidification in a mold for casting. Figure 7.17 shows typical numerical results when conduction in the mold is coupled with heat transfer in the liquid and the solid. With increasing time, the liquid region shrinks due to solidification, whereas the solidified region increases. The effect of the imposed conditions at the outer surface of the mold on the solidification process can be investigated by solving this conjugate problem, which yields the temperature field in the mold along with that in the solid and the liquid, as shown.

7.6.6 System Simulation

Another important aspect that must be mentioned here is the numerical simulation of the overall thermal system, since the thermal process undergone by the material is a consequence of the thermal exchange with the various components of the system. The numerical simulation of the system refers to the use of the numerical model to obtain a quantitative representation of the physical system and to characterize its behavior for a given design or set of operating conditions and for variations in these. Consider, for instance, a typical electrical furnace (Fig. 7.18), which consists of the heater, walls, insulation, inert gas environment, and the material undergoing heat treatment. The thermal transport mechanisms in all these components are coupled through the boundary conditions. Thus, each individual component may first be numerically simulated as uncoupled

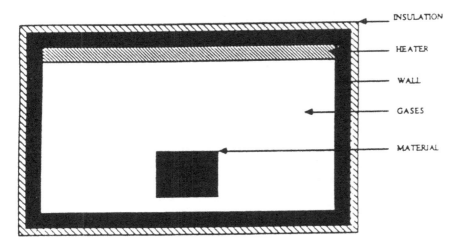

Figure 7.18 Sketch of a typical electrical furnace, indicating the various components of the system.

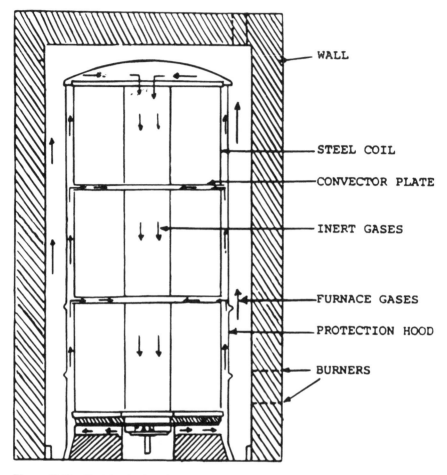

WALL

STEEL COIL

CONVECTOR PLATE

INERT GASES

FURNACE GASES

PROTECTION HOOD

BURNERS

Figure 7.19   Sketch of a batch annealing furnace for steel (Jaluria, 1988b).

from the others, by employing prescribed boundary conditions. Then these individual simulations are combined, employing the appropriate coupling through the boundary conditions. This procedure provides a systematic approach to the numerical simulation of the thermal system, which may be a simple one (Fig. 7.18) or a complicated practical one (Fig. 7.19) (Jaluria and Torrance, 1986; Jaluria, 1988b). Once the simulation of the system is achieved, the design and optimization of the thermal process as well as of the system may be undertaken. The results obtained from the simulation provide the necessary inputs for improving existing designs and developing new ones for improving the productivity and the product quality for a given manufacturing process.

7.7 CONCLUDING REMARKS

This chapter outlines the basic considerations in the numerical modeling of the heat transfer and fluid flow mechanisms that arise in the thermal processing of materials. Of particular interest are the numerical solution of the governing equations, numerical imposition of the relevant initial and boundary conditions, and simulating the material characteristics and the underlying physical mechanisms. The governing equations are presented in their general forms and are nondimensionalized to obtain the main governing parameters for these processes. The properties of the material undergoing thermal processing play a very important role in the mathematical modeling of these processes and are discussed in detail. Variable properties, including non-Newtonian fluid behavior, are considered and the corresponding mathematical and numerical modeling discussed.

An important consideration in such processes is the simplification of the governing equations and the associated boundary conditions in order to make these complicated problems tractable. Several typical processes are considered, and various simplifications that may be made on the basis of the values of the governing parameters and the transport regimes that arise are presented. The numerical simulation of a few important manufacturing processes is discussed, focusing on some of the special techniques needed for accurate and valid results. Numerical modeling of the process as well as of the thermal system is considered, and the basic approach for obtaining the inputs needed for design and optimization is outlined. Only a brief discussion of many of these aspects is given, with references provided for more detailed information. Heat transfer and fluid flow form a very important consideration in a wide variety of manufacturing processes, particularly in the thermal processing of materials, and this chapter presents the important aspects that need to be included for a numerical modeling of these processes.

ACKNOWLEDGMENTS

The support of the National Science Foundation, under grants CBT-88-03049 and DDM-92-13458, for some of the work reported here and for the preparation of this chapter is gratefully acknowledged. The author also acknowledges the work done by several students, as referenced here, that made it possible to present this review.

NOMENCLATURE

A	cross-sectional area
b, B, B_m	constants, defined in Eqs. (7.21)–(7.23)
c_m	concentration
C	specific heat of a solid
C_p	specific heat at constant pressure
$\mathbf{e}$	unit vector in the direction of gravitational force
Ec	Eckert number
f_l	liquid mass fraction
$\mathbf{F}$	body force vector
F_x, F_y, F_z	body force components in x-, y-, and z-directions, respectively
g	magnitude of gravitational acceleration
Gr	Grashof number, dimensionless
h	convective heat transfer coefficient
H	enthalpy
H_0	enthalpy at 0 K
$\mathbf{i}$	unit vector in x-direction
k	thermal conductivity,
k_x, k_y, k_z, k_r	thermal conductivity in x-, y-, z-, and r-directions, respectively
K	bulk viscosity
K_c	consistency index for non-Newtonian fluid, Eq. (7.20)
L	characteristic length
L_h	latent heat of fusion
n	power-law fluid index
p	local pressure
p_a	hydrostatic pressure
p_d	dynamic pressure due to fluid motion
P	perimeter
Pr	Prandtl number
q	heat flux
$\dot{Q}$	volumetric source
Re	Reynolds number
Sr	Strouhal number
t	time
T	temperature

u, v, w	velocity components in x-, y-, and z-directions, respectively
U_s	speed of a moving solid
$\mathbf{V}$	velocity vector
$\mathbf{x}$	position vector
x, y, z	coordinate distances
X, Y, Z	dimensionless coordinate distances

Greek

α	thermal diffusivity
β	coefficient of thermal expansion
$\dot{\gamma}$	strain rate
δ	location of interface between solid and liquid
ε	surface emissivity
θ	dimensionless temperature
λ	second viscosity coefficient
ν	kinematic viscosity
ξ	transformed coordinate distance
ρ	density
σ	Stefan–Boltzmann constant
τ	shear stress
$\boldsymbol{\tau}$	stress tensor
Φ	viscous dissipation function
Φ_v	viscous dissipation

Subscripts

0	reference
a	ambient
c	characteristic
l	liquid
m	melting point
s	solid

REFERENCES

Altan T, Oh SI, Gegel HL. Metal Forming Fundamentals and Applications. Metals Park OH: Amer. Soc. Metals, 1971.

Anderson DA, Tannehill JC, Pletcher RH. Computational Fluid Mechanics and Heat Transfer. New York: Hemisphere, 1984.

Avitzur B. Metal Forming: Processes and Analysis. New York: McGraw-Hill, 1968.

Bennon WD and Incropera FP. Developing Laminar Mixed Convection with Solidification in a Vertical Channel. J Heat Transfer 110: 410–415, 1988.

Brebbia CA. The Boundary Element Method for Engineers. 3rd ed. London: McGraw-Hill, 1977.

Chen MM, Mazumder J, Tucker CL, Eds. Transport Phenomena in Materials Processing. New York: Amer. Soc. Mech. Engrs., HTD 29, 1983.

Doyle LE, Keyser CA, Leach JL, Scharder GF, Singer MB. Manufacturing Processes and Materials for Engineers. Englewood Cliffs, NJ: Prentice-Hall, 1987.

Easterling K. Introduction to Physical Metallurgy of Welding. London: Butterworths, 1983.

Fenner RT. Principles of Polymer Processing. New York: Chemical Publishing, 1979.

Gebhart B, Jaluria Y, Mahajan RL, Sammakia B. Buoyancy-Induced Flows and Transport. New York: Hemisphere, 1988.

Ghosh A, Mallik AK. Manufacturing Science. Chichester UK: Ellis Horwood, 1986.

Huebner KH, Thornton EA. The Finite Element Method for Engineers. 2nd ed. New York: Wiley, 1983.

Hughel TJ, Bolling GF, Eds. Solidification. Metals Park, OH: Amer. Soc. Metal, 1971.

Jaluria Y. Computer Methods for Engineering. Englewood Cliffs, NJ: Prentice Hall, 1988a.

Jaluria Y. Heat and mass transfer in the extrusion of non-Newtonian materials. Adv Heat Transfer 28: 145–230, 1996.

Jaluria Y. Natural Convection Heat and Mass Transfer. Oxford, UK: Pergamon Press, 1980.

Jaluria Y. Numerical simulation of the transport processes in a heat treatment furnace. Int J Num Meth Eng 25: 387–399, 1988b.

Jaluria Y. Temperature regulation of a plastic-insulated wire in radiant heating. J Heat Transfer 98: 678–680, 1976.

Jaluria Y. Transport from continuously moving materials undergoing thermal processing. Ann Rev Heat Transfer 4: 187–245, 1992.

Jaluria Y, Singh AP. Temperature distribution in a moving material subjected to surface energy transfer. Comp Meth Appl Mech Eng 41: 145–157, 1983.

Jaluria Y, Torrance KE. Computational Heat Transfer. New York: Hemisphere, 1986.

Kalpakjian S. Manufacturing Engineering and Technology. Reading, MA: Addison-Wesley, 1989.

Kang BH, Jaluria Y. Thermal modeling of the continuous casting process. AIAA J Thermophys Heat Transfer 7: 139–147, 1993.

Kang BH, Jaluria Y, Karwe MV. Numerical simulation of conjugate transport from a continuous moving plate in materials processing. Numer Heat Transfer 19: 151–176, 1991.

Karwe MV, Jaluria Y. Numerical simulation of fluid flow and heat transfer in a single-screw extruder for non-Newtonian fluids. Num Heat Transfer 17: 167–190, 1990.

Kuhn HA, Lawley A, Eds. Powder Metallurgy Processing, New Techniques and Analysis. New York: Academic Press, 1978.

Lee SH-K, Jaluria Y. Effect of variable properties and viscous dissipation during optical fiber drawing. J Heat Transfer 118: 350–358, 1996.

Lee SH-K, Jaluria Y. Simulation of the transport processes in the neck-down region of a furnace drawn optical fiber. Int J Heat Mass Transfer 40: 843–856, 1997.

Li T, Ed. Optical Fiber Communications. Vol. 1. Fiber Fabrication. New York: Academic Press, 1985.

Mallinson GD, de Vahl Davis G. The method of false transient for the solution of coupled elliptic equations. J Comp Phys 12: 435–461, 1973.

Minkowycz WJ, Sparrow EM, Schneider GE, Pletcher RH. Handbook of Numerical Heat Transfer. New York: Wiley, 1988.

Patankar SV. Numerical Heat Transfer and Fluid Flow. New York: Hemisphere, 1980.

Peaceman DW, Rachford HH. Numerical solution of parabolic and elliptic differential equations. J Soc Ind Appl Math 3: 28–41, 1955.

Pearson JRA, Richardson SM, eds. Computational Analysis of Polymer Processing. London: Applied Science, 1983.

Ramachandran N, Jaluria Y, Gupta JP. Thermal and fluid flow characteristics in one-dimensional solidification. Int Comm Heat Mass Transfer 8: 69–77, 1981.

Ramachandran N, Gupta JP, Jaluria Y. Thermal and fluid flow effects during solidification in a rectangular enclosure. Int J Heat Mass Transfer 25: 187–194, 1982.

Roache PJ. Computational Fluid Dynamics. Albuquerque: Hermosa, 1976.

Roy Choudhury S, Jaluria Y. Analytical solution for the transient temperature distribution in a moving rod or plate of finite length with surface heat transfer. Int J Heat Mass Transfer 37: 1193–1205, 1994.

Roy Choudhury S, Jaluria Y, Lee SH-K. Generation of neck-down profile for furnace drawing of optical fiber. ASME Heat Transfer Div 306: 23–32, 1995.

Schey JA. Introduction to Manufacturing Processes. 2nd ed. New York: McGraw-Hill, 1987.

Siegel R. Shape of two-dimensional solidification interface during directional solidification by continuous casting. J Heat Transfer 100: 3–10, 1978.

Siegel R. Two-region analysis of interface shape in continuous casting with superheated liquid. J Heat Transfer 106: 506–511, 1984.

Szekely J. Fluid Flow Phenomena in Metals Processing. New York: Academic Press, 1979.

Tadmor Z, Gogos C. Principles of Polymer Processing. New York, Wiley, 1979.

Viskanta R. Heat transfer during melting and solidification of metals. J Heat Transfer 110: 1205–1219, 1988.

Viskanta R. Natural convection in melting and solidification. In: Kakac S, Aung W, Viskanta R, eds. Natural Convection: Fundamentals and Applications. New York: Hemisphere, 1985.

Viswanath R, Jaluria Y. A comparison of different solution methodologies for melting and solidification problems in enclosures. Numerical Heat Transfer 24B: 77–105, 1993.

Viswanath R, Jaluria Y. Numerical study of conjugate transient solidification in an enclosed region. Numer Heat Transfer 27: 519–536, 1995.

8

Passive Thermal Control of Electronic Equipment

Yogendra Joshi
University of Maryland, College Park, Maryland

8.1 INTRODUCTION

Electronic products constitute ever-increasing segments of the necessities and conveniences of modern society. Their current use spans computer, telecommunications, military/aerospace, industrial, instrumentation, con-

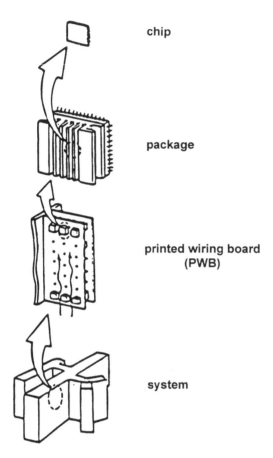

chip

package

printed wiring board
(PWB)

system

Figure 8.1 Hierarchy of electronic packaging.

sumer products, business/retail, and automotive markets. While the current
market share is dominated by the first three product groups, other markets,
such as automotive electronics, are poised for significant future growth.
Cost, size, and weight are often the primary constraints on the thermal and
physical design of electronic products.

While the use of electronic products spans many applications, their
internal physical construction or packaging usually follows a similar hierar-
chical structure. In most current systems, four structural levels can be identified (Fig. 8.1). At the smallest level, the electronic circuitry is contained
within the semiconductor chip (or die). The chip must be protected from
the environment, electrical signals must be routed to and from it, and

dissipation of the heat generated due to the electrical losses, with acceptable resulting temperature rises, must be ensured. This may be accomplished through the use of a chip package. If multiple chips are housed in a single package for enhanced electrical performance and compactness, the resulting configuration is called a multichip module. The next common structural level is a circuit card or board, which houses many individual packages and provides the appropriate electrical interconnections between them through a network of conducting traces. Such traces can be on single surface layer or within multiple layers of conducting planes to reduce size. Many boards are next housed within a chassis or box. The heat generated within the overall electronic system is rejected to the environment at this level. The box may be hermetically sealed, may have vents, or be equipped with fans for air circulation. One or more of the walls may contain heat exchanges for heat removal.

8.2   THERMAL ISSUES IN ELECTRONICS PACKAGING

Thermal engineering is needed in many aspects of the packaging of electronic products. These activities identified in Fig. 8.2 include the development of effective techniques for thermal management/control, as well as thermal characterization. Thermal control techniques may be passive, active, or a combination. Passive techniques, the focus of this chapter, are used for low-to-medium power dissipation in cost and weight-sensitive, high-reliability applications. These techniques involve no expenditure of external energy for heat dissipation from the package to the environment. Examples of these include combined natural convection and radiation in air, use of high-thermal-conductivity substrates, natural convection in dielectric liquids, heat pipes, and solid-liquid phase-change materials for transient applications. Active techniques are used for high-power dissipation applications. Examples include single-phase forced air or liquid convection, jet impingement, and forced convection boiling. In some applications, the heat removal from the electronic package may be by passive means but the transfer to the environment may require active thermal management. Examples of such passive/active schemes include heat pipes or pool boiling with forced-air-cooled condensers.

When considering the thermal control strategy for an entire electronic system, a combination of techniques is usually required. For example, in an avionics module, heat may be removed primarily by conduction and surface radiation at the package and board levels and rejected from the system boundaries to a convectively cooled heat exchanger plate. Thermal

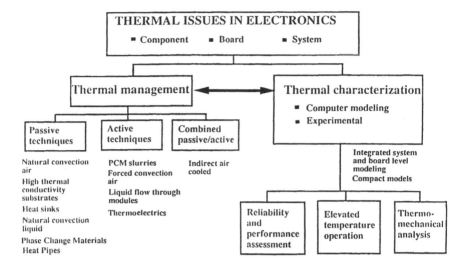

Figure 8.2 Thermal issues in the packaging of electronic equipment.

characterization is needed to ensure that the selected thermal management scheme(s) indeed provide the desired performance at each level of packaging. This is achieved through computational, experimental, or combined means.

Thermal characterization also forms the basis for several other important activities during the physical design, such as system reliability and performance assessments. Many failure mechanisms are dependent on temperature or its spatial or temporal gradients. Also, electrical characteristics such as signal propagation delay and many current and voltage parameters display a temperature dependence. For example, the operation of CMOS circuits at cryogenic temperatures for higher carrier mobility is currently being employed for processor speed gains for some high-end workstations and servers.

During design, the power dissipations, maximum allowable operating temperatures of the chips and the operating environmental thermal conditions are identified. Through thermal characterization, resulting temperatures throughout the system are determined. Packaging and interconnection materials must be carefully selected to withstand these temperatures. Thermal characterization is also a prerequisite for the determination of interfacial thermal stresses in packaging structures. Invariably, such dissimilar material interfaces are encountered in transitioning from one level of packaging to the next.

The challenges in the thermal design of electronic systems have traditionally resulted from a combination of two trends:

1. High heat fluxes and volumetric heat generation rates. For example, the surface heat flux on a 5-mm-by-5-mm chip dissipating 10 W is 2×10^5 W/m², which is only one order of magnitude below that on the sun's surface. Such large heat fluxes are encountered in logic chips in large mainframe computers. Recent projections (SIA Roadmap, 1994) show a moderate increase in heat fluxes with time for the higher-end computing equipment.

2. Upper operating temperature limit of the chip. For conventional electronic products, this is 70–125°C.

When considered together, the two requirements make electronics thermal control a very challenging problem. Electronic products also are unique in that the lifetime of a given packaging technology is rapidly decreasing. The typical product development cycle time has decreased from a few years to just a few months within a decade. To handle these rapid cycle times, product development increasingly relies on extensive computer-aided design. This has made thermal characterization through validated computer simulations an extremely important part of the overall design activity.

8.3 APPROACHES FOR THERMAL ANALYSIS OF ELECTRONIC EQUIPMENT

Heat transfer in electronic equipment usually takes place through a combination of modes. The heat generated in the chip is conducted across the package walls to the board or module. If the chip sits within an open cavity in the package, thermal radiation, and in some cases convection, adds to the transport. In some recent packaging schemes, such as direct-chip-attach, the package around the chip may be absent and the heat is conducted directly to the board, as well as convected and radiated from the chip. From the board, the heat transfer to the enclosure boundaries is dependent on the thermal management scheme employed. For air-cooled electronics, the heat transfer is by a combination of conduction within the board, convection to the air, and radiation to the enclosure boundaries. For other thermal management strategies, these modes can be altered. For example, when indirect liquid cooling at the enclosure boundaries is employed, the heat transfer takes place largely by conduction across the board to heat exchanger plates cooled by passing liquid through internal flow paths. For direct liquid immersion cooling, a combination of conduction in the packages and board, along with convection in the liquid, takes place.

Thermal analysis of electronic equipment is complicated not only by the numerous heat transfer paths but also by unique factors. For example, geometrical complexity often results in complex flow and heat transfer patterns within electronic systems. In most applications, it renders the conventional heat transfer correlations inapplicable. Also, the roughness associated with electronic components may promote earlier transition to turbulence. Very little information on the modeling of transitional and turbulent flows over electronic components currently exists.

Thermal analysis of electronic equipment has traditionally been done with heat-conduction-type models or thermal resistance networks. The former approach has been used for individual components or circuit board assemblies, while resistance networks are employed at all levels of packaging. Both conduction- and network-type models require specification of convection coefficients and radiative heat transfer information as inputs. As mentioned, the conjugate, effects and geometrical complexity invariably make the specification of this information difficult. The use of conventional correlations can result in large uncertainties in predicted temperatures.

Computational fluid dynamics/computational heat transfer (CFD/CHT) approaches have recently gained popularity in simulations of electronics cooling, as they allow the handling of conjugate processes and geometrical complexity. The specification of convection coefficients is not required once the conjugate effects are modeled. However, the computational effort required in these simulations is considerable, especially for natural convection flows. Also, the capability of modeling flows in transition or turbulence is currently rather limited.

In the following sections, two examples of the CFD/CHT approach for the analysis of passive electronics cooling techniques are presented. These illustrate the methodology used for considering conjugate transport effects. In each case, an enclosure containing the heat-dissipating electronic component is considered for laminar flow conditions. While considerably more complex electronic enclosures are typically found in applications, the modeling issues addressed in these examples are equally applicable to the more complex cases.

8.4 LIQUID IMMERSION COOLING OF A HEAT SOURCE IN AN ENCLOSURE

Thermal control of electronic components using dielectric liquids has recently received increased attention due to inherently high heat removal capabilities of liquids compared with air. Candidate liquids for immersion cooling must have high dielectric strength, be chemically inert, environmentally safe, and nontoxic. A family of such coolants is now commercially available as Fluorinerts (Product Manual, 1985) with a wide range of boiling points. Their availability has prompted a number of investigations of their heat transfer characteristics both for single-phase and phase-change schemes. A number of these are reviewed by Bergles and Bar-Cohen (1990) and Bar-Cohen (1991).

The fluid circulation in immersion cooling can be either forced or buoyancy induced. Since the passive liquid cooling systems employing buoyancy induced flow typically provide significant heat transfer enhancement over forced convection air cooling, along with design simplicity and resulting high reliability, they are of considerable current interest. Both experimental and computational investigations have been undertaken to characterize the resulting transport in configurations of interest in electronic packaging. Typically the electronic components in the available studies have been simulated as discrete heat sources flush mounted on, or protruding from, a substrate.

Experimental studies of natural convection liquid cooling have investigated arrays of simulated electronic components arranged (1) on vertical surfaces (Park and Bergles, 1987; Joshi et al., 1989a), (2) in vertical channels (Joshi et al., 1989b), and (3) in rectangular enclosures (Kelleher et al., 1987; Keyhani et al., 1988, 1990; Joshi et al., 1990; Heindel et al., 1995a, 1995b). Numerical computations have also been reported for the last category. Two-dimensional computations of transport for a protruding heat source in an enclosure were carried out by Lee et al. (1987) and for three flush heat sources in a liquid filled enclosure by Prasad et al. (1990). In both of these studies the heat sources were simulated as uniform flux regions and the substrate was considered adiabatic. The fluorocarbon coolants used in liquid immersion cooling are characterized by thermal conductivity values almost an order of magnitude smaller than conventional coolants such as water. This, in combination with the use of moderate thermal conductivity substrate materials such as ceramics makes the conductive spreading within the substrate an important and sometimes dominant mode of heat removal from the electronic component. These effects were examined for a substrate mounted protruding heat source in the two-dimensional computations by Sathe and Joshi (1991).

A 3 × 3 array of protruding heat sources on a vertical wall was investigated by Liu et al. (1987a, 1987a, 1987b). A uniform heat flux surface condition was applied to the protrusion faces. They found that there was little interference among chips and that the heat transfer was affected more by the stratified temperature field outside of the boundary layer region than by relative position within the boundary layer. Long-time solutions were characterized by temperature oscillations on the faces. Substrate conduction effects were included by Wroblewski and Joshi (1993) in their study of natural convection from a substrate-mounted protrusion in a dielectric-liquid-filled cubical enclosure. Comparisons of the computations with the experimental results of Joshi and Paje (1991) for the same configuration were also provided. This study is discussed next in some detail.

8.4.1 Liquid Immersion Natural Convection Cooling of an Electronic Component in an Enclosure

Figure 8.3a illustrates the configuration under study. A single protruding chip (or package hereafter), of dimensions $h_c \times l \times l$, is centered on a substrate of thickness h_s. The chip is assumed to have uniform properties, with internal heat generation rate of Q. The back of the substrate forms one of the vertical walls of a cubical enclosure of side L_e, filled with a dielectric liquid: The opposite vertical wall is maintained at a constant temperature of T_c, while all other walls of the enclosure, including the back of the substrate, are insulated. This particular geometry is similar to one of the experimental arrangements of Joshi and Paje (1991), with which the present computations are compared. Figure 8.3b shows the internal details of the package modeled for experimental comparisons.

Governing Equations and Nondimensional Parameters

The nondimensional governing equations for the three-dimensional problem, assuming laminar flow, constant properties, and the Boussinesq approximation, are as follows:

continuity:

$$\frac{\partial U}{\partial X} + \frac{\partial V}{\partial Y} + \frac{\partial W}{\partial Z} = 0 \tag{8.1}$$

x-momentum:

$$\frac{\partial(UU)}{\partial X} + \frac{\partial(VU)}{\partial Y} + \frac{\partial(WU)}{\partial Z} = -\frac{\partial P}{\partial X} + \left(\frac{\mathrm{Pr}}{\mathrm{Ra}}\right)^{1/2}\left(\frac{\partial^2 U}{\partial X^2} + \frac{\partial^2 U}{\partial Y^2} + \frac{\partial^2 U}{\partial Z^2}\right) \tag{8.2}$$

y-momentum:

$$\frac{\partial(UV)}{\partial X} + \frac{\partial(VV)}{\partial Y} + \frac{\partial(WV)}{\partial Z} = -\frac{\partial P}{\partial Y} + \left(\frac{\mathrm{Pr}}{\mathrm{Ra}}\right)^{1/2}\left(\frac{\partial^2 V}{\partial X^2} + \frac{\partial^2 V}{\partial Y^2} + \frac{\partial^2 V}{\partial Z^2}\right) + \theta$$

$$\tag{8.3}$$

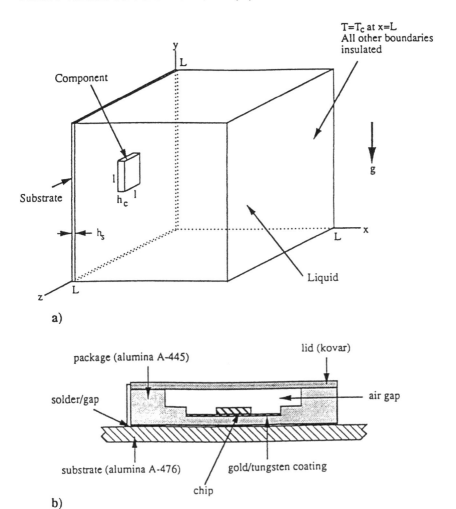

Figure 8.3   (a) Geometry of enclosure; (b) internal details of electronic package.

z-momentum:

$$\frac{\partial(UW)}{\partial X} + \frac{\partial(VW)}{\partial Y} + \frac{\partial(WW)}{\partial Z} = -\frac{\partial P}{\partial Z} + \left(\frac{Pr}{Ra}\right)^{1/2}\left(\frac{\partial^2 W}{\partial X^2} + \frac{\partial^2 W}{\partial Y^2} + \frac{\partial^2 W}{\partial Z^2}\right)$$

$$(8.4)$$

energy (fluid):

$$\frac{\partial(U\theta)}{\partial X} + \frac{\partial(V\theta)}{\partial Y} + \frac{\partial(W\theta)}{\partial Z} = \left(\frac{1}{\text{Pr Ra}}\right)^{1/2}\left(\frac{\partial^2\theta}{\partial X^2} + \frac{\partial^2\theta}{\partial Y^2} + \frac{\partial^2\theta}{\partial Z^2}\right) \tag{8.5}$$

energy (chip):

$$\left(\frac{\partial^2\theta}{\partial X^2} + \frac{\partial^2\theta}{\partial Y^2} + \frac{\partial^2\theta}{\partial Z^2}\right) + \frac{1}{R_C H_C} = 0 \tag{8.6}$$

energy (substrate):

$$\left(\frac{\partial^2\theta}{\partial X^2} + \frac{\partial^2\theta}{\partial Y^2} + \frac{\partial^2\theta}{\partial Z^2}\right) = 0 \tag{8.7}$$

The appropriate nondimensional parameters are $\text{Ra} = g\beta Ql^2/\alpha\nu k_f$, $\text{Pr} = \nu/\alpha$, $U = u/U_0$, $V = v/U_0$, $W = w/U_0$, $U_0 = (g\beta Q/k_f)^{1/2}$, $\theta = (T - T_c)/(Q/lk_f)$, $P = p/\rho U_0^2$, $X = x/l$, $Y = y/l$, $Z = z/l$, $H_c = h_c/l$, and $H_s = h_s/l$. It is noted that the normalization scales for the velocities and temperature involve Q, since it is the driving parameter for the transport. Since the normalized velocities and temperatures are functions of Q additionally through their dependence on Ra, the dimensional changes in the transport variables with Q need to be carefully interpreted.

The boundary conditions at the enclosure walls are as follows:

$$X = 0; \quad \frac{\partial\theta}{\partial X} = 0, \quad U = 0, \quad V = 0, \quad W = 0$$

$$X = X_L; \quad \theta = 0, \quad U = 0, \quad V = 0, \quad W = 0$$

$$Y = 0, X_L; \quad \frac{\partial\theta}{\partial Y} = 0, \quad U = 0, \quad V = 0, \quad W = 0$$

$$Z = 0, X_L; \quad \frac{\partial\theta}{\partial Z} = 0, \quad U = 0, \quad V = 0, \quad W = 0$$

where $X_L = L_e/l$.

In addition, the boundary conditions at the interfaces between two different materials are

$$R_i\left(\frac{\partial\theta_i}{\partial X_n}\right)_i = R_j\left(\frac{\partial\theta_j}{\partial X_n}\right)_j, \quad \theta_i = \theta_j, \quad U = 0, \quad V = 0, \quad W = 0$$

where X_n is the coordinate along the outward normal of the surface (i.e. $X_n = X$ if the surface is in the Y-Z plane), i and j refer to the two different materials (s for substrate, c for chip, and f for fluid), and $R_s = k_s/k_f$ and $R_c = k_c/k_f$. These are implicitly satisfied in the present computations through the use of the harmonic mean formulation for interface diffusivities, as discussed later.

For this investigation, $X_L = 5.1$, $H_s = 0.08$, and $H_c = 0.21$ were used to allow comparison with experimental data. The Prandtl number was chosen as 25, with $R_c = 2360$. This corresponds to the use of fluorinert liquid FC-75 (Product Manual, 1985) as the coolant with a silicon chip. The Rayleigh number was varied over a range from 10^3 and 10^9, assuming $R_s = 575$, which corresponds to an alumina ceramic substrate. A 1-cm^2 chip operated with a power level of 1.5 W in FC-75 corresponds to a Rayleigh number of approximately 10^9. To study the thermal spreading along the substrate, values of $R_s = 0.5$, 5, and 50 were also employed with Ra $= 10^6$.

Numerical Method

The governing equations are discretized using a control-volume approach as described by Patankar (1980). This approach uses control volumes for velocities that are staggered with respect to those for temperature and pressure; a power-law scheme for the differencing of dependent variables; a harmonic mean formulation for the interface diffusivities; and the SIMPLER algorithm for velocity-pressure coupling. The solution is obtained from an initial guess through an iterative scheme using a line-by-line tridiagonal matrix algorithm. The conjugate conduction in the chip and substrate is handled numerically by solving the same full set of momentum and energy equations throughout the entire enclosure, but with a large value of viscosity specified for the solid regions. The numerical solution is assumed converged when the maximum temperature change during successive iterations is less than 0.00001 times the maximum temperature for that iteration, and when overall energy balances on the enclosure, the chip, and the fluid are obtained within 1%.

The solution was obtained throughout the entire enclosure; no symmetry condition was imposed at $Z = X_L/2$. This prevented the forcing of a symmetric solution where one may not exist, as may be the case if the plume above the chip became unsteady. As it turned out, the solutions were indeed symmetric for all conditions examined. Most of the results presented here were obtained using a 23 × 22 × 22 or a 24 × 22 × 22 nonuniform grid. Several points were chosen near the hot and cold surfaces to capture the thermal boundary layers. These points were often adjusted slightly for various Rayleigh number ranges (i.e., more points closer to the wall for higher Rayleigh numbers). Below Ra $= 10^5$, the grid in the remainder of the flow was fairly uniform. At higher Rayleigh numbers, in which the boundary layer regions accounted for smaller portions of the enclosure, several intermediate grid points were located between the thermal boundary layer

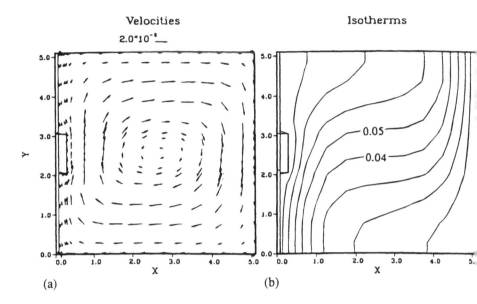

(a)   (b)

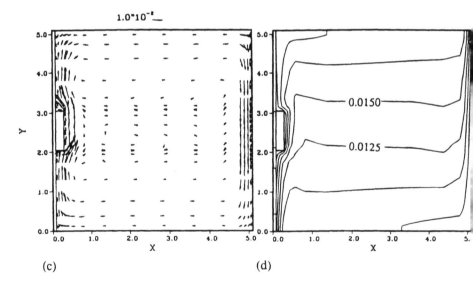

(c)   (d)

Figure 8.4 U-V velocity vectors and isotherms for the X-Y plane at $Z = 2.55$ for $R_s = 575$: (a, b) Ra $= 10^3$; (c, d) Ra $= 10^6$; (e, f) Ra $= 10^9$; (g, h) detail near chip for Ra $= 10^9$.

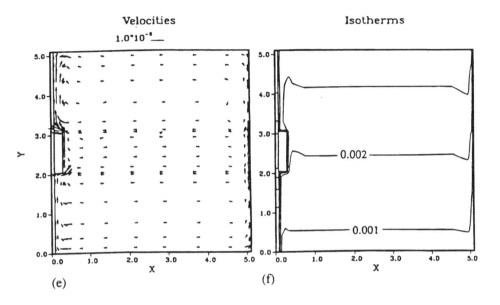

(e)   (f)

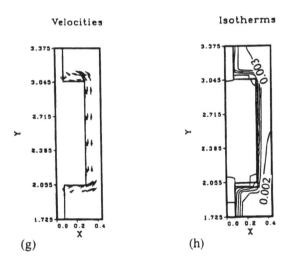

(g)   (h)

Figure 8.4 *Continued.*

and the core of the flow, to capture the outer region of the momentum boundary layer. The grid spacing within the core of the enclosure was relatively coarse.

Effect of Rayleigh Number ($R_s = 575$)

To describe the three-dimensional nature of the flow and heat transfer, data for a few selected planes are shown in Fig. 8.4 for Ra = 10^3, 10^6, and 10^9. Figure 8.4 shows the U-V velocity vectors and the isotherms for the X-Y plane at $Z = 2.55$ (i.e., the symmetry plane). At Ra = 10^3 (Fig. 8.4a), the primary flow is in the X-Y plane and it is characterized by a single cell that is disturbed only slightly near the protruding chip. The isotherms at Ra = 10^3 in Fig. 8.4b indicate that the relatively strong U velocities convect low-temperature fluid from the cold wall toward the hot wall in the lower portion of the cavity and high-temperature fluid from the hot wall toward the cold wall in the upper portion. Note that the chip is nearly isothermal, mainly as a result of the high R_c.

When Ra = 10^6, Fig. 8.4c, a cellular flow is also evident, but the bulk of the primary flow in the vertical direction is confined to boundary layers along the hot and cold surfaces rather than spreading throughout the enclosure, as seen at Ra = 10^3. These boundary layers thicken significantly near the lower and upper walls due to three-dimensional effects of the horizontal walls. The flows around the upper and lower corners of the chip lead to weak jets of fluid moving toward the substrate at the top of the chip and away from the substrate near the bottom of the chip. The effect of the flow on the isotherms in Fig. 8.4d is similar as that in Fig. 8.4b, but with the temperature gradients confined to thermal boundary layers along the hot and cold walls. The remainder of the core of the enclosure, between the boundary layers, can be characterized as well-stratified.

At Ra = 10^9 in Fig. 8.4e, the vertical primary flow is confined to even thinner boundary layers near the hot and cold walls. The vertical velocities are much greater in the region near and above the chip compared to that along the substrate or the cold wall. In particular, the velocities in the plume from the top of the chip are much higher than the rest of the flow. There is still a weak cellular structure to the primary flow. Like the Ra = 10^6 case, most of the horizontal flow between the hot and cold boundary layers occurs in regions near the top and bottom walls, but with velocities less than 10% of the maximum vertical velocity. The isotherms in Fig. 8.4f reveal very steep gradients near the hot and cold surfaces and a well stratified core. The cross convection is negligible everywhere. Details of the flow near the chip (Fig. 8.4g, h) present a clearer view of the growing thermal boundary layer along the chip face.

Velocity vectors and isotherms for the Y-Z plane at $X = 0.185$ (through the center of the chip) are shown in Fig. 8.5. These plots are composite drawings which take advantage of the symmetry about $Z = 2.55$, with isotherms shown on the right and velocity vectors shown on the left of each plot. The V velocity component near the substrate is fairly constant in

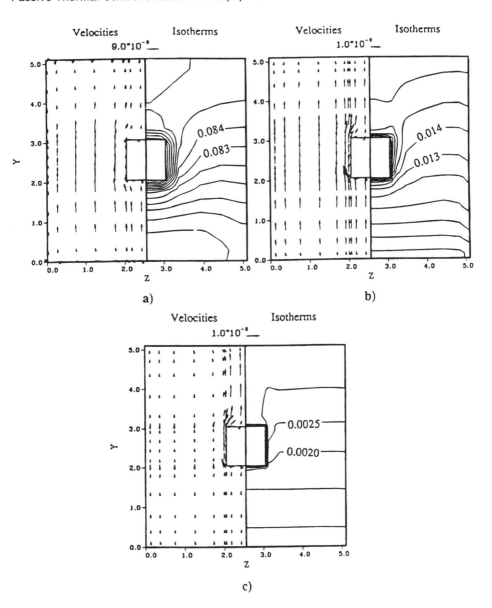

Figure 8.5 *W-V* velocity vectors and isotherms for the *Z-Y* plane at *X* = 0.185 for R_s = 575: (a) Ra = 10^3; (b) Ra = 10^6; (c) Ra = 10^9.

the Z-direction, except in the vicinity of the sidewalls for $Ra = 10^3$ in Fig. 8.5a. The W-velocity component near the substrate is outward, toward the sidewalls, near the bottom of the enclosure and inward near the top of the enclosure.

At $Ra = 10^6$ (Fig. 8.5b), the strength of the primary flow (in the Y-direction) again remains fairly constant across the Z-direction, with a smaller region of diminishing strength near the side walls. The one exception to this is the region above the chip, where the flow is slightly stronger than that along the rest of the substrate. This is a result of the emergence of a plume from the top of the chip. Near the upper edge of the substrate directly above the chip, the W velocity component is outward rather than inward as in Fig. 8.5a, as this plume encounters the solid top surface. At $Ra = 10^9$ (Fig. 8.5c) the plume grows to dominate the primary flow. When it reaches the top wall, it spreads horizontally outward toward the side walls, as well as toward the cold wall, creating a small circular region of very strong secondary velocities.

The isotherms in Fig. 8.5 reveal spreading of the heat away from the chip in a nonsymmetrical manner for all three Ra. Also seen is the formation of a thin thermal boundary layer type region around the chip with increasing Ra. The thermal stratification in the enclosure liquid below the chip becomes evident with increasing Ra.

Figure 8.6 presents the fractions of the net generated power that are lost from the various chip surfaces, the maximum chip temperature, and the temperature of the substrate at $Z = 1.125$ and $Y = 2.55$, as a function of the Rayleigh number. Because of the large value of the ratio of substrate-to-fluid thermal conductivity, conduction through the back of the chip to the substrate accounted for most of total heat loss from the chip—over 90% at $Ra = 10^3$ decreasing to 76% at $Ra = 10^9$. The heat loss from the front face was the next largest and increased with Rayleigh number as a result of the increase in the heat transfer coefficients. The other four surfaces of the chip accounted for almost negligible heat loss at $Ra = 10^3$ (insert in Fig. 8.6a), with substantial relative increases with increasing Rayleigh number.

The nondimensional chip temperature decreased by more than an order of magnitude from 0.101 for $Ra = 10^3$ to 0.00735 for $Ra = 10^9$. For comparison, the chip temperature for the conduction-only solution was 0.21. Figure 8.6b also shows how the reduction in substrate conduction at higher Rayleigh numbers led to a greater difference between the substrate and chip temperatures.

Effect of Substrate Thermal Conductivity ($Ra = 10^6$)

The decrease in the thermal conductivity of the substrate has a profound effect on the surface heat loss fractions and the maximum chip temperatures

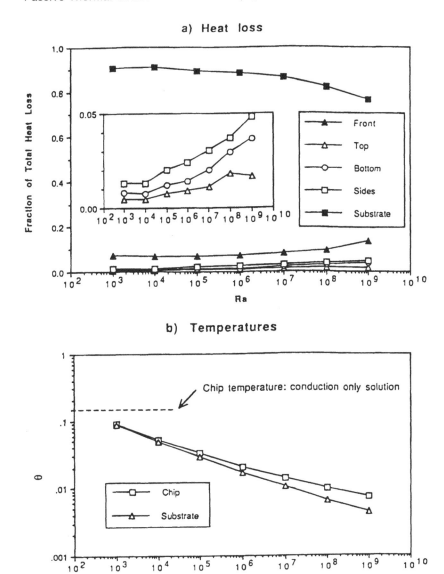

Figure 8.6   Effect of Rayleigh number on (a) fraction of heat loss from the hot surfaces (chip and substrate) and (b) maximum temperature of chip and temperature of substrate at $Z = 1.125$ and $Y = 2.55$.

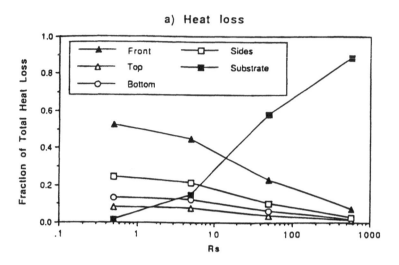

a) Heat loss

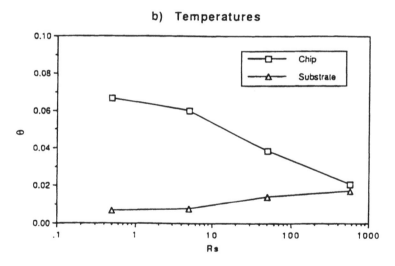

b) Temperatures

Figure 8.7 Effect of substrate thermal conductivity on (a) fraction of heat loss from the hot surfaces (chip and substrate) and (b) maximum temperature of chip and temperature of substrate at $Z = 1.125$ and $Y = 2.55$.

(Fig. 8.7a). As R_s is reduced, the fraction of loss through the substrate falls, while fractions from the various chip surfaces increase. At $R_s = 5$, the heat loss from the front face is the largest, while the substrate fraction has dropped to only 14%. At $R_s = 0.5$, the substrate loss is only 1%.

The chip temperature increases substantially with decreasing R_s. As seen in Fig. 8.6b, it increases from 0.02 for the baseline case to 0.06 for $R_s = 5$. For $R_s < 5$, further reductions of R_S lead to only moderate increases in chip temperature, since substrate conduction plays a minor role.

Comparison with Experiments

As mentioned, the geometry employed here was similar to one of the two configurations in the experiments performed by Joshi and Paje (1991). A key difference was that in the experiments the uniform heat generation was confined to the 1.52-mm square, 0.4-mm-thick silicon chip (or die as it is often called). This was located within a 20-pin "leadless chip carrier" package 8.9 mm square and protruding 1.9 mm from the substrate (Fig. 8.3b). The die was attached to the ceramic package through thin layers of gold and tungsten. An air space between the die and the lid of the package (made of kovar alloy) provided an additional internal resistance to heat transfer. The package itself was elevated slightly above the substrate, resting only on the 20 solder joints. Measurements of steady temperatures on the chip and selected substrate locations were reported in three dielectric liquids for a range of power levels.

The 3-D code was run for this particular package design with a power level of 1.84 W with FC-75 as the coolant. A slightly modified form of Eq. (8.6) was used to prescribe the heat generation only within the chip. Using properties evaluated at the mean temperature between the lid and the cold wall (from the experiments), this corresponded to Ra $= 1.15 \times 10^9$ and Pr $= 24$. Comparisons of measured substrate and chip temperatures with the computations are shown in Table 8.1. With the exception of the substrate temperature directly behind the package, the results are within $- 9\%$ and $+ 15\%$. Given the complicated design of the package, which could not be modeled in all its detail, and the uncertainty associated with the contact resistance between the package and the substrate, these results are satisfactory.

The numerical simulation revealed very large temperature gradients along the substrate in the Y- and Z-directions directly behind the package, the region where a large discrepancy (35%) was observed between the numerical and experimental results. In the experiments, three factors could have led to locally increased spreading in this region, which might account for the lower measured temperatures. First, a high thermal conductivity paste was used to attach the thermocouple to the substrate and the size of this connection was sufficient to cover a region where the numerical results showed large temperature variations. Second, behind the substrate was a

Table 8.1 Comparison of Numerical and Experimental Results for Leadless Chip Carrier Package

Location	θ experiment	θ numerical	% diff.
Chip	0.00197	0.0226	+15
Substrate[a]			
4. ($Z = 2.55$, $Y = 3.98$)	0.0043	0.0048	+12
5. ($Z = 2.55$, $Y = 2.55$)	0.0072	0.0097	+34
6. ($Z = 2.55$, $Y = 1.12$)	0.0032	0.0029	-9
7. ($Z = 1.12$, $Y = 3.95$)	0.0033	0.0037	+12
8. ($Z = 1.12$, $Y = 2.55$)	0.0035	0.0039	+11

[a] Numbers designate references to thermocouples used by Joshi and Paje (1991).

slab of plexiglass, which might also have acted as an additional thermal path for spreading. In addition, heat loss from the edges of the enclosure might have resulted in lower temperature throughout the entire enclosure. This may explain why most of the numerically predicted temperatures are above those measured experimentally.

8.4.2 Application of Solid-Liquid Phase-Change Materials for Passive Thermal Control of Electronic Packages

With increasing miniaturization of electronic products, there is a trend toward monochip packages with high power dissipation. For such packages, the use of phase-change materials (PCM) may provide thermal control for the entire duration of time the package is powered. These materials store thermal energy in the form of the latent heat of fusion, which can be rejected later to reverse the phase change. The PCMs to be used for thermal control purpose must meet several criteria. First, their melting point should be below the maximum allowable operating temperature range of the component. They should also be nonflammable, nonexplosive, and noncorrosive. For design purposes, their thermophysical properties must be available.

Bentilla et al. (1966) identified four organic paraffins that meet these criteria. An experimental and analytical study of PCM for thermal control of electronics was performed by Witzman et al. (1983). Their analytical model uses correlations for phase-change heat transfer. Their experiment showed the potential of using PCM for cooling high-power modules for a substantial amount of time. Snyder (1991) performed a two-dimensional analysis of a PCM cooling scheme for an electronic module. Melting of PCM was assumed to be driven by conduction only, and effects of natural

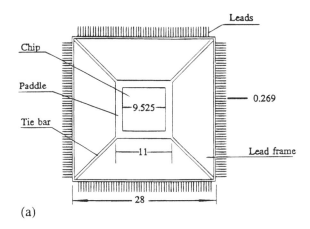

(a)

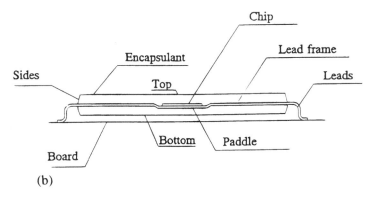

(b)

Figure 8.8 Details of the plastic quad flat package (dimensions in mm): (a) plan and (b) cross-sectional views.

convection in the melt pool were neglected. Ishizuka and Fukuoka (1991) used a metallic eutectic layer of Bi/Pb/Sn/In with a melting point of 57°C under a simulated electronic package. Their experimental data show that the operating temperature rise of the component can be arrested for a significant amount of time. They also performed a one-dimensional network analysis of their PCM cooling technique.

The computational studies of cooling of electronics in literature often consider simple geometries and boundary conditions. Heat transfer from discrete heat sources is often approximated by uniform heat flux from packages with internal details of the actual electronic package omitted. With these simplifications, the dominant heat transfer path from the package

cannot be ascertained. Information about relative distribution of heat flow
for these various paths is important for a PCM cooling system design to
take advantage of the primary heat flow path from the package. In a recent
study by Pal and Joshi (1996), a three-dimensional computational model
was used to predict the performance of PCM cooling of a plastic quad flat
package (PQFP). This study is next discussed in some detail.

Computational Model

A 208-pin plastic quad flat package was simulated. Details of the package
and materials were obtained from a study of the thermal model of a PQFP
by Rosten and Viswanath (1994). The dimensions of the package are shown
in Fig. 8.8. Several simplifications were made to model the package. The
lead frame was assumed to be in the same plane as the silicon chip. In order
to reduce the number of control volumes, every group of 5.2 leads of the
package were lumped as one equivalent lead, reducing the total lead count
from 208 to 40. The lead frame in the plastic was treated as a mixture with
uniform thermophysical properties. The solder joints for each lead were
neglected.

The computational domain considered is shown in Fig. 8.9. The
package is mounted on 1-mm-thick ceramic or FR-4 printed wiring board
(PWB) oriented in vertical direction. This assembly is placed in an enclosure of size 50 × 50 × 50 mm. A layer of PCM 10 mm thick is used under

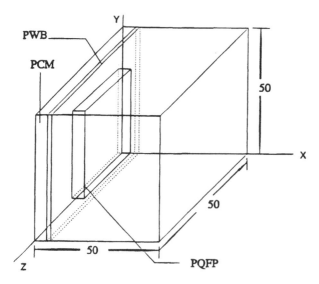

Figure 8.9 Computational domain (dimensions in mm).

the board. The reason for implementing the PCM under the board is that the primary heat flow path from the chip to the outside of the package is found to be through the leads by conduction. The right wall of the enclosure is assumed to be at a constant temperature of 25°C, and all other walls are adiabatic.

Governing Equations

The governing conservation equations for mass, momentum, and energy can be written as

continuity:

$$\frac{\partial}{\partial x}(\rho u) + \frac{\partial}{\partial y}(\rho v) + \frac{\partial}{\partial z}(\rho w) = 0 \tag{8.8}$$

x-momentum:

$$\frac{\partial}{\partial t}(\rho u) + \frac{\partial}{\partial x}(\rho u u) + \frac{\partial}{\partial y}(\rho v u) + \frac{\partial}{\partial z}(\rho w u)$$
$$= \frac{\partial}{\partial x}\left(\mu \frac{\partial u}{\partial x}\right) + \frac{\partial}{\partial y}\left(\mu \frac{\partial u}{\partial y}\right) + \frac{\partial}{\partial z}\left(\mu \frac{\partial u}{\partial z}\right) - \frac{\partial p}{\partial x} + S_x \tag{8.9}$$

y-momentum:

$$\frac{\partial}{\partial t}(\rho v) + \frac{\partial}{\partial x}(\rho u v) + \frac{\partial}{\partial y}(\rho v v) + \frac{\partial}{\partial z}(\rho w v)$$
$$= \frac{\partial}{\partial x}\left(\mu \frac{\partial v}{\partial x}\right) + \frac{\partial}{\partial y}\left(\mu \frac{\partial v}{\partial y}\right) + \frac{\partial}{\partial z}\left(\mu \frac{\partial v}{\partial z}\right) - \frac{\partial p}{\partial y} + S_y \tag{8.10}$$

z-momentum:

$$\frac{\partial}{\partial t}(\rho w) + \frac{\partial}{\partial x}(\rho u w) + \frac{\partial}{\partial y}(\rho v w) + \frac{\partial}{\partial z}(\rho w w)$$
$$= \frac{\partial}{\partial x}\left(\mu \frac{\partial w}{\partial x}\right) + \frac{\partial}{\partial y}\left(\mu \frac{\partial w}{\partial y}\right) + \frac{\partial}{\partial z}\left(\mu \frac{\partial w}{\partial z}\right) - \frac{\partial p}{\partial z} + S_z \tag{8.11}$$

energy:

$$\frac{\partial}{\partial t}(\rho c_p T) + \frac{\partial}{\partial x}(\rho u c_p T) + \frac{\partial}{\partial y}(\rho v c_p T) + \frac{\partial}{\partial z}(\rho w c_p T)$$
$$= \frac{\partial}{\partial x}\left(k \frac{\partial T}{\partial x}\right) + \frac{\partial}{\partial y}\left(k \frac{\partial T}{\partial y}\right) + \frac{\partial}{\partial z}\left(k \frac{\partial T}{\partial z}\right) + S_h \tag{8.12}$$

The same set of equations applies for various materials. The expressions for coefficients and source terms for various materials in these equations are

Table 8.2   Source Terms for Various Materials

	S_x	S_y	S_z	S_h
Air	0	$\rho g \beta (T - T_0)$	0	0
PCM	$\dfrac{C(1 - \varepsilon)^2}{\varepsilon^3 + b} u$	$\dfrac{C(1 - \varepsilon)^2}{\varepsilon^3 + b} v$ $+ \rho g \beta (T - T_m)$	$\dfrac{C(1 - \varepsilon)^2}{\varepsilon^3 + b} w$	$-\rho \dfrac{\partial (\Delta H)}{\partial t}$
Silicon	0	0	0	Q'''
Other solids	0	0	0	0

listed in Table 8.2. The thermophysical properties of various materials are provided in Table 8.3. The heat dissipation within the silicon chip was assumed to be volumetrically uniform. To handle conjugate conduction in the solids, the same single-domain approach described in the previous section was used.

Modeling of Phase Change

The phase change was handled using a single-domain, enthalpy porosity technique (Brent et al., 1988). In this method, the absorption of latent heat

Table 8.3   Thermophysical Properties Used for Computation

Materials	Thermal conductivity (W/m°C)	Density (kg/m³)	Specific heat (J/kg-K)	Dynamic viscosity (kg-m/s)
Air	0.0261	1.177	1005	1.85×10^{-5}
FR-4	0.35	1938	1600	10^{30}
Ceramic (C-786)	18.00	3875	840	10^{30}
Leads and paddle (copper)	385	8933	385	10^{30}
Encapsulant	0.31	1070	100	10^{30}
Lead frame encapsulant mixture	154.17	4215.2	214	10^{30}
Chip (silicon)	$154.86(300/T)^{4/3}$	691	2330	10^{30}
PCM (n-eicosene)	0.23	795	2050	3.57×10^{-3}

Melting point of PCM: 37°C   Latent heat per unit mass: 241 kJ/kg

during melting is included as a source term S_H in the energy equation
(Table 8.2). Latent heat content of each control volume in the PCM is
evaluated after each energy equation iteration cycle. Based on the latent
heat content, an effective porosity ($\varepsilon = \Delta H/L$) for each control volume is
determined. Control volumes that contain molten PCM have $\varepsilon = 1$ and
control volumes containing solid PCM have $\varepsilon = 0$. The control volumes
with values of ε between 0 and 1 are treated as mushy. Even though the
phase change is assumed to be isothermal, the idea of mushy zone is introduced to gradually "switch off" the velocities from liquid to solid at the
interface. The "switching off" is controlled by the source terms S_x, S_y, and
S_z in the momentum equations (Table 8.2). The solid-liquid interface is
assumed to correspond to the $\varepsilon = 0.5$ line.

Numerical Procedure and Validation

Grid sizes were selected carefully to resolve the internal details of the
package. Local grid refinement was required to model the chip, paddle, lead
frame, and leads of the package. The grid outside the package was relatively
coarse. Fine grid resolution in air was used near all solid walls to capture
the boundary layer effects. Based on grid size testing, a $34 \times 39 \times 39$ grid
was selected for all computations. Though the domain is symmetric in the
z-direction about the midplane, the analysis was performed for the entire
enclosure due to the fact that at higher Rayleigh numbers the flow structure
in the enclosure may be time dependent with no symmetry along the midplane.

A finite-volume-based numerical algorithm, SIMPLER (Patankar,
1980) is used to solve the governing conservation equations. This code is
fully implicit in time for transient computations. However, a control of time
step was necessary to achieve convergence due to the complex and transient
heat transfer stages involved. During the initial phase, when conduction is
the dominant mode of heat transfer, a time step of 10 s was used. However,
after the natural convection begins to dominate, the time step had to be
reduced to 5 s. A further reduction to 2 s was necessary when melting was
in progress. An optimum set of relaxation factors for best convergence was
obtained after trial and error. Convergence for a time step is assumed when
the sum of normalized residuals for temperature was less than 1×10^{-4}.
Simultaneously, a global energy balance residual of less than 1×10^{-3} is
prescribed for convergence.

Results for Computations with FR-4 Board

Computations were performed for six cases. These cases consider two different board materials and two different power levels to estimate their effects

Table 8.4 Summary of Parameters for Various Cases

Case	Power (watts)	PWB material	Computation
A	1	FR-4	With PCM
B	1	FR-4	Without PCM
C	1	Ceramic	With PCM
D	1	Ceramic	Without PCM
E	3	Ceramic	With PCM
F	3	Ceramic	Without PCM

on the performance of PCM cooling. Board materials were FR-4 and ceramic. A summary of various cases is presented in Table 8.4. Initially the computational domain was assumed to be at uniform temperature of 25°C. Figure 8.10 shows the timewise variation of maximum temperature of the chip and the temperature at the geometric center of the board for cases A and B. Three different stages of heat transfer can be observed for each curve. During the initial stage, a rapid temperature rise is seen. During this period, the heat transfer is largely by conduction from the chip to the lead frame, plastic encapsulating material, and surrounding air. The resulting rate of temperature rise for the board is slower than that of the chip. The second stage is characterized by a slowdown in the rate of temperature rise of the chip and the development of natural convection in the air. The air heated by the package and the PWB circulates and contacts the isothermal right wall. Heat is transferred from the hot air to the wall in the boundary layer adjacent to it. The third phase for case A begins when the heat conducted across the PWB raises the PWB/PCM interface temperature to the PCM melting point. During this phase, the PCM melts isothermally. At the initial times of the third stage, the melting is driven by conduction heat transfer from the board. This is characterized by a thin planar melt layer as shown in Fig. 8.10b, d. However, as melting progresses, natural convection begins to develop in the melt pool, due to which the solid liquid interface changes as seen in Figure 8.10f, h.

Due to the lower thermal conductivity of the FR-4 board, heat spreading along the y- and z-directions are relatively weaker than in x-direction, due to which the melting process is localized around the footprint area of the package and is not spread out uniformly along the board. After 1 h of operation, the melting process is still in progress and natural convection is dominant in the melt pool (Fig. 8.10g, h). During this phase, the maximum temperature of the package is stabilized at 61°C. For the case without PCM (case B), the third stage is a mere continuation of the second stage, with the package and board approaching a steady-state condition. However, it was

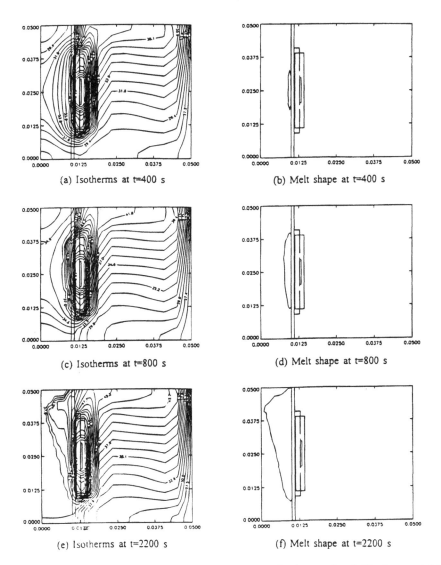

(a) Isotherms at t=400 s

(b) Melt shape at t=400 s

(c) Isotherms at t=800 s

(d) Melt shape at t=800 s

(e) Isotherms at t=2200 s

(f) Melt shape at t=2200 s

Figure 8.10   Isotherms, velocity vectors, and melt shapes for 1 W of power and FR-4 as the PWB material: (a) isotherms at $t = 400$ s, (b) melt shape at $t = 400$ s, (c) isotherms at $t = 800$ s, (d) melt shape at $t = 800$ s, (e) isotherms at $t = 2200$ s, and (f) melt shape at $t = 2200$ s.

Fig. continues

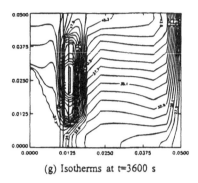

(g) Isotherms at t=3600 s

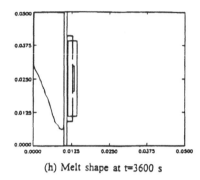

(h) Melt shape at t=3600 s

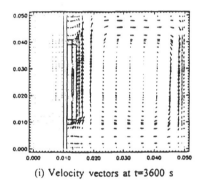

(i) Velocity vectors at t=3600 s

Figure 8.10 *Continued.* (g) Isotherms at $t = 3600$ s, (h) melt shape at $t = 3600$ s, and (i) velocity vectors at $t = 3600$ s.

found that after 1 h the package had not reached a steady-state condition. One run was performed with very large time step to assess the steady-state temperatures, and the maximum chip temperature was found to be 136°C. Figure 8.10i shows the velocity vector plot at the ($z = zl/2$) plane and indicates a natural convection flow in the enclosure. The velocity field shows a strong plume generated above the package which moves up and impinges on the upper surface and creates a circulation cell throughout the enclosure.

8.5 SUMMARY AND FUTURE DIRECTIONS

Traditional semiempirical techniques such as thermal network methodology or modified conduction analysis require input information in the form of convection coefficients. For most electronic cooling applications, applicable correlations do not exist. Conjugate simulations using the CFD/CHT approach do not require a priori specification of the convection coefficients at the component faces. This chapter has focused on the application of these techniques for the analysis of passively cooled electronic equipment. With the push toward reduced times to market of modern electronic products, the need for such analyses is likely to grow considerably.

While these techniques offer unprecedented potential capability for thermal characterization at the various levels of packaging, a number of issues need to be addressed to make them more useful for proactive design of electronic products. Currently, one of the most significant limitations of the CFD/CHT approach is the large computational effort involved in simulating a realistic electronic product. In order to accurately predict thermal performance of the complete electronic system, modeling at all three levels of packaging (i.e., component, board, and system) must be performed in an integrated and efficient manner. Innovative approaches for carrying out such global/local analyses are required. Also, very little experimentally validated information exists on the handling of transitional and turbulent transport in electronic cooling applications. The importance of thermal radiation in air-cooled systems and its incorporation in a general conjugate analysis needs more attention. Finally, one of the impediments in achieving better predictive capability with such techniques is the unavailability of thermophysical properties data for various electronic packaging materials. In many instances, for example for epoxy-fiberglass circuit boards, strong anisotropic behavior is found for the thermal conductivity. In-plane and out-of-plane values can be different by as much as a factor of 3. More research is needed for coming up with techniques to determine effective properties that can provide adequate accuracy in temperature predictions for such cases.

ACKNOWLEDGMENTS

The author acknowledges the assistance of Dr. D. Pal in the preparation of this chapter.

NOMENCLATURE

b constant in the porosity source term
c_p specific heat at constant pressure [J/kg-K]
C morphological constant

g gravitational acceleration [m/s²]

g gravitational acceleration $[m/s^2]$
h heat transfer coefficient $[W/m^2\text{-}K]$
h_c thickness of protrusion $[m]$
h_s thickness of substrate $[m]$
ΔH latent heat component of enthalpy $[kJ/kg]$
k thermal conductivity $[W/m\text{-}K]$
l protrusion length $[m]$
L latent heat of fusion $[J/kg\text{-}K]$
L_e enclosure length $[m]$
Nu $q_i L/k_f(T_i - T_c)$, Nusselt number
p pressure $[N/m^2]$
Pr v/α, Prandtl number of fluid
q_i heat flux crossing a solid surface $[W/m^2]$
Q heat generation rate $[W]$
Q''' volumetric heat generation rate $[W/m^3]$
Ra $g\beta Q l^2/v\alpha k_f$, Rayleigh number
R_c k_c/k_f, ratio of protrusion thermal conductivity to fluid thermal conductivity
R_s k_s/k_f, ratio of substrate thermal conductivity to fluid thermal conductivity
S source term $[W/m^3]$
T temperature $[K]$
T_c temperature at the cold enclosure wall parallel to the substrate $[K]$
u velocity component in x-direction $[m/s]$
v velocity component in y-direction $[m/s]$
w velocity component in z-direction $[m/s]$

Greek Symbols

α fluid thermal diffusivity $[m^2/s]$
β coefficient of volumetric expansion $[1/K]$
θ $(T - T_c)/(Q/k_f l)$ nondimensional temperature
μ dynamic viscosity $[kg\text{-}m/s]$
v kinematic viscosity $[m^2/s]$
ρ fluid density $[kg/m^3]$

Subscripts

0 ambient
a air
c protrusion (chip)

f fluid
H enthalpy
i condition at a solid surface
m melting point
p phase change material
s substrate

REFERENCES

Bar-Cohen A. Thermal management of electronic components with dielectric liquids. Proceedings of ASME/JSME Thermal Engineering Joint Conference. Vol 2. pp xv–xxxix.

Bergles AE, Bar-Cohen A. Direct liquid cooling of microelectronic components. In: Bar-Cohen A, Kraus AD, eds. Advances in Thermal Modeling of Electronic Components and Systems. Vol 2. ASME Press, 1990, pp 233–342.

Bentilla EW, Sterrett KF, Karre LE. Research and development study on thermal control by use of fusible materials. Northrop Space Laboratories, Contract No. NAS 8-11163, NASA Document No. N66-26691, 1966.

Brent AD, Voller VR, Reid KJ. Enthalpy-porosity technique for modeling convection-diffusion phase change: application to the melting of a pure metal. Numer Heat Transfer 13: 297, 1988.

Chan AMC, Banerjee S. Three-dimensional numerical analysis of transient natural convection in rectangular enclosures. ASME J Heat Transfer 101: 114, 1979.

Heindel TJ, Ramadhyani S, Incropera FP. Laminar natural convection in a discretely heated cavity I. Assessment of three-dimensional effects. ASME J Heat Transfer 117: 902, 1995a.

Heindel TJ, Ramadhyani S, Incropera FP. Laminar natural convection in a discretely heated cavity II. Comparisons of experimental and theoretical results. ASME J Heat Transfer 117 910, 1995b.

Ishizuka M, Fukuoka Y. Development of a new high density package cooling technique using low melting point alloys. Proceedings ASME/JSME Thermal Engineering Joint Conference. Vol 2, 1991, pp 375–380.

Joshi Y, Kelleher MD, Benedict TJ. Natural convection immersion cooling of an array of simulated electronic components in an enclosure filled with dielectric fluid. In Bergles AE, ed. Heat Transfer in Electronic and Microelectronic Equipment. Hemisphere, 1990, pp 445–468.

Joshi Y, Paje RA. Natural convection cooling of a ceramic substrate mounted leadless chip carrier in dielectric liquids. Int Comm Heat Mass Transfer 18: 39, 1991.

Joshi Y, Willson T, Hazard SJ III. An experimental study of natural convection from an array of heated protrusions on a vertical surface in water. ASME J Electron Pack 111: 121, 1989a.

Joshi Y, Willson T, Hazard SJ III. An experimental study of natural convection cooling of an array of heated protrusions in a vertical channel in water. ASME J Electron Pack 111: 33, 1989b.

Kelleher MD, Knock RH, Yang KT. Laminar natural convection in a rectangular enclosure due to a heated protrusion on one vertical. Part I. Experimental investigation. Proceedings of the 2nd ASME-JSME Joint Thermal Engineering Conference. Vol II. Honolulu Hawaii, 1987, pp 169–178.

Keyhani M, Chen L, Pitts DR. The aspect ratio effect on natural convection in an enclosure with protruding heat sources. Presented at the AIAA/ASME Thermophysics and Heat Transfer Conference, Seattle, WA, 1990.

Keyhani M, Prasad V, Cox R. An experimental study of natural convection in a vertical cavity with discrete heat sources. ASME J Heat Transfer 110: 616, 1988.

Kuhn D, Oosthuizen PH. Three-dimensional natural convective flow in a rectangular enclosure with localized heating. Proceedings of AIAA Thermophysics Conference, 1987, pp 55–62.

Lankhorst AM, Hoogendoorn CJ. Three-dimensional calculations of high Rayleigh natural convection flows in enclosed cavities. Proceedings of the National Heat Transfer Conference, Houston, TX. Vol 3. 1988, pp 463–470.

Lee JJ, Liu KV, Yang KT, Kelleher MD. Laminar natural convection in a rectangular enclosure due to a heated protrusion on one vertical wall Part II. Numerical simulations. Proceedings of the 2nd ASME-JSME Joint Thermal Engineering Conference, Vol II. Honolulu, Hawaii, 1987, pp 179–185.

Liu KV, Yang KT, Kelleher MD. Three-dimensional natural convection cooling of an array of heated protrusions in an enclosure filled with a dielectric fluid. Proceedings of the International Symposium on Cooling Technology for Electronic Equipment. Honolulu, Hawaii, 1987a, pp 486–497.

Liu KV, Yang KT, Wu YW, Kelleher MD. Local oscillatory surface temperature responses in immersion cooling of a chip array by natural convection in an enclosure. Proceedings of the Symposium on Heat and Mass Transfer in Honor of B. T. Chao. University of Illinois, Urbana-Champaign IL, 1987b.

Mallinson GD, de Vahl Davis G. Three-dimensional natural convection in a box: a numerical study. J Fluid Mech 83: 1, 1977.

Pal D, Joshi Y. Application of phase change materials for passive thermal control of plastic quad flat packages (PQFP): A computational study. Num Heat Transfer, Part A 30: 19, 1996.

Park KA, Bergles AE. Natural convection heat transfer characteristics of simulated microelectronic chips. ASME J Heat Transfer 109: 90, 1987.

Patankar SV. Numerical Heat Transfer and Fluid Flow. New York: Hemisphere/McGraw Hill, 1980.

Prasad V, Keyhani M, Shen R. Free convection in a discretely heated vertical enclosure: effects of Prandtl number and cavity size. ASME J Electron Pack 112: 63, 1990.

Product Manual Fluorinert Liquids 3M Corporation, Minneapolis, MN, 1985.

Rosten H, Viswanath R. Thermal modelling of the Pentium processor package. Proceedings of the Electronic Component and Technology Conference 1994.

Sathe S, Joshi Y. Natural convection arising from a heat generating substrate-mounted protrusion in a liquid-filled two-dimensional enclosure. Int J Heat Mass Transfer 34: 2149, 1991.

Semiconductor Industries Association Roadmap, 1994.

Snyder KW. An investigation of using a phase-change material to improve the heat transfer in a small electronic module for an airborne radar application. Proceedings of the International Electronics Packaging Conference. Vol 1. 1991, pp 276.

Viskanta R, Kim DM, Gau C. Three-dimensional numerical natural convection heat transfer of a liquid metal in a cavity. Int J Heat Mass Transfer 29: 475, 1986.

Witzman S, Shitzer A, Zvirin Y. Simplified calculation procedure of a latent heat reservoir for stabilizing the temperature of electronic devices. Proceedings of the Winter Annual Meeting of the ASME. HTD Vol 28. 1983, pp 29–34.

Wroblewski DE, Joshi Y. Computations of liquid immersion cooling for a protruding heat source in a cubical enclosure. Int J Heat Mass Transfer 36: 1201, 1993.

Yang HQ, Yang KT, Lloyd JR. Flow transition in laminar buoyant flow in a three-dimensional tilted rectangular enclosure. Proceedings of the 8th International Heat Transfer Conference, San Francisco, CA, 1986, pp 1495–1500.

9

Computational Fluid Dynamic Techniques in Air Quality Modeling

Mehmet T. Odman*

MCNC—Environmental Programs, Research Triangle Park, North Carolina

Armistead G. Russell

Georgia Institute of Technology, Atlanta, Georgia

Current affiliation: Georgia Institute of Technology, Atlanta, Georgia.

9.1 BACKGROUND AND INTRODUCTION

About one fourth of the population in the United States lives in areas which experience episodes of photochemical air pollution, and in 1990 almost 100 cities were out of compliance with the National Ambient Air Quality Standard (NAAQS) (NRC, 1992). Air pollution abatement would reduce the risks of adverse human health effects and vegetation damage and may slow the rate of material degradation as well. However, it is desirable to achieve these environmental standards in a cost-effective manner. Advanced computer models follow pollutant species emitted from anthropogenic and biogenic sources as they are transported downwind to receptor areas and simulate their complex chemical interactions with other species in the atmosphere. Currently these models represent the most scientifically sound foundation for testing alternative control strategies; thus, they are increasingly being used as the basis for regulations. Billions of dollars may be spent to comply with these regulations, so it is important to understand the uncertainty in the underlying models. While many (if not most) of the uncertainty in model predictions comes from uncertainty in the model inputs, there is some introduced by the model itself, e.g., by the numerical advection routines employed.

One of the major problems facing our society that has grown out of industrialization is the deterioration of air quality. In particular, problems such as acid deposition, smog, global climate warming, and stratospheric ozone depletion pose significant threats to both human health and welfare and related ecological damage. There are a number of facets to these problems, and the question is how do we address and mitigate them? Like most environmental problems, they are very complex. However, there is a tremendous payoff for identifying effective solutions. That is the role of air quality models. Here, we discuss the types of numerical techniques used by air quality models, particularly the more advanced models that follow atmospheric transport and chemistry, e.g., those describing urban and regional smog, acid deposition, global atmospheric chemistry, etc. While operating at different scales, and often being used for different problems, they have very similar characteristics. First, the problems are described by the same set of conservation equations. Second, the numerical techniques employed are similar to those used in more traditional computational fluid dynamics applications.

As an introduction, it is instructive to look at how we currently manage air pollution (Fig. 9.1). It is a feedback process, examining whether our current air quality meets the desired goals, identifying candidate control strategies (e.g., different ways to decrease emissions of air pollutants and their precursors), and testing how well those controls work. This last step is

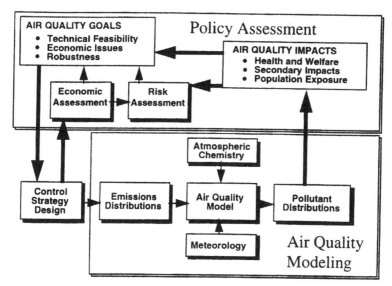

Figure 9.1 Schematic of the air quality control planning process. Air quality models are central to the process. Models make it possible to determine how emissions affect pollutant concentrations and the resulting environmental effects.

the determination of source-air quality relationships, and is generally done using an air quality model. To do an adequate job in this step, the model used must contain the applicable processes important to the problem at hand, and numerical routines that are commensurate in terms of accuracy and speed. For example, for urban smog, the model must treat both the chemistry and physics of the atmosphere. One of the more important processes is advection, and the models used to numerically follow the transport of species are very similar to those used in more traditional computational fluid dynamic (CFD) applications. However, the dominance of some of the processes can be quite different.

Before discussing the different numerical methods, it is instructive to develop the set of equations being solved. Atmospheric pollutant dynamics are mathematically described by the species conservation equation:

$$\frac{\partial c_i}{\partial t} + \nabla \cdot (\mathbf{U} c_i) = \nabla \cdot \left[\rho D_i \nabla \left(\frac{c_i}{\rho} \right) \right] + R_i(c_1, c_2, \ldots, c_n, T, t) + S_i(\mathbf{x}, t),$$

$$i = 1, 2, 3, \ldots, n \quad (9.1)$$

where c_i is the concentration of species i, $\mathbf{U}$ is the wind velocity vector, D_i is the molecular diffusivity of species i, R_i is the net production (depletion if negative) of species i by chemical reaction, $S_i (x, t)$ is the emission rate of i at location x, ρ is the air density; and n is the number of species. R can also be a function of the meteorological variables (e.g., temperature, T). In essence, this equation states that the time rate of change of a pollutant (term 1) depends on convective transport, diffusion, and chemical reaction of that pollutant. The result is a set of n coupled, nonlinear partial differential equations. The coupling and the nonlinearity are introduced by the reaction term. This equation is subject to the initial conditions on the species concentrations as discussed below. There are known analytical solutions for this equation for only rather simple wind fields, diffusivity profiles, and chemical interactions. Thus, solution is usually accomplished by computational methods. In actual practice, the situation is further complicated by the turbulent nature of the atmosphere, making the description (and our knowledge) of the wind velocities more problematic. For one, wind velocity fluctuations chaotically occur at time and spatial scales much smaller than we can computationally afford to follow over a typical domain of interest. Luckily, we usually are not interested in the concentration fluctuations at the very small scales.

As usual, the impact of turbulence is included by the use of Reynolds' decomposition, where the velocity components and concentrations are split into an average component ($\langle u \rangle$ and $\langle c \rangle$) and a fluctuating part (u', c'), e.g.,

$$u = \langle u \rangle + u' \tag{9.2}$$

Different averaging procedures are discussed in Seinfeld (1985). The result of this process is an equation for the dynamics of the averaged concentration component:

$$\frac{\partial \langle c_i \rangle}{\partial t} + \nabla \cdot (\langle U \rangle \langle c_i \rangle) = \nabla \cdot \left(\rho D_i \nabla \left(\frac{\langle c_i \rangle}{\rho} \right) - \langle U' c_i' \rangle \right)$$
$$+ R_i(\langle c_1 \rangle, \langle c_2 \rangle, \ldots, \langle c_n \rangle, T, t) + \langle S_i(x, t) \rangle, \qquad i = 1, 2, 3, \ldots, n \tag{9.3}$$

In practice, the turbulent diffusion term, $\langle u'c' \rangle$, is much larger than the molecular diffusion term, so the latter is usually neglected. However, calculating the turbulent diffusion is a challenge due to the closure problem. Thus, it is usually parameterized by the introduction of an effective turbulent diffusivity tensor, K_{ii}:

$$\rho K_{ii} \frac{\partial c/\rho}{\partial x_i} = -\langle u_i' c' \rangle \tag{9.4}$$

This is analogous to Fickian diffusion, and is called K-theory. The turbulent diffusivity is not a fluid property and is a function of the fluid dynamics. Use of this formulation introduces limitations into the applicability of the resulting model, primarily in a lower limit of the spatial scale. However, this is generally not a problem for regional models.

The resulting equation upon which most air quality models are based is

$$\frac{\partial \langle c_i \rangle}{\partial t} + \nabla \cdot (\langle \mathbf{U} \rangle \langle c_i \rangle) = \nabla \cdot \left[\rho K \nabla \left(\frac{\langle c_i \rangle}{\rho} \right) \right] + R_i(\langle c_1 \rangle, \langle c_2 \rangle, \ldots, \langle c_n \rangle, T, t)$$

$$+ \langle S_i(\mathbf{x}, t) \rangle, \quad i = 1, 2, 3, \ldots, n \quad (9.5)$$

and is often referred to as the atmospheric diffusion equation, or ADE (Seinfeld, 1985). Air quality models, based on the ADE, are applied to follow pollutant dynamics over a designated spatial domain for specific periods of time.

Examination of Eq. (9.5) shows that if there are no chemical reactions $(R = 0)$, or if R is linear in $\langle c_i \rangle$ and uncoupled, then a set of linear, uncoupled differential equations are formed for determining pollutant concentrations. This is the basis of transport only and transport with linear chemistry models (which, for brevity, will be called transport models). Transport models are used for studying the effects of sources of CO and primary particulate matter on air quality, and have been used for sulfate formation and deposition, but they are not suitable for studying reactive pollutants such as O_3, NO_2, HNO_3, and secondary organic species.

The solution of the ADE requires specification of initial and boundary conditions. The initial conditions generally used are those specifying the concentration at the beginning of a simulation:

$$c(\mathbf{x}, 0) = c^{\text{initial}}(\mathbf{x}) \tag{9.6}$$

Horizontal boundary conditions are usually taken as those concentrations at the modeling domain boundary:

$$c(\Omega, t) = c^{\text{boundary}}(t) \tag{9.7}$$

where Ω is the boundary of the modeling domain. The ground-level boundary condition is a statement of the balance between deposition, vertical diffusion, and ground-level emissions:

$$E - v_d \langle c \rangle = -K_{zz} \frac{\partial \langle c \rangle}{\partial z} \tag{9.8}$$

The upper-level boundary condition can take two forms. First is a specification of the species concentrations at the top of the domain, similar to the specification of the horizontal boundary conditions. However, this assumes knowledge of the atmospheric concentrations well above the ground which are often not known. An alternative boundary condition is to assume a negligible concentration gradient:

$$\frac{\partial c/\rho}{\partial z} = 0 \tag{9.9}$$

This, in essence, assumes that the best estimate of the species concentration (actually, the mixing ratio) above the modeling domain is the prediction just below the top. Use of the former upper-level condition (i.e. specifying the concentration) can lead to significant, artificial input (or loss) of pollutant mass by diffusion if the specified concentration is too high (or too low).

9.1.1 Eulerian Models

There are two distinct reference frames from which to view pollutant dynamics. The most natural is the Eulerian coordinate system which is fixed at the earth's surface. In that case, a succession of different air parcels are viewed as being carried by the wind past a stationary observer. The second is the Lagrangian reference frame which moves with the flow of air, in effect maintaining the observer in contact with the same air parcel over extended periods of time. Because pollutants are carried by the wind, it is often convenient to follow pollutant evolution in a Lagrangian reference frame, and this perspective forms the basis of Lagrangian trajectory. Both Eulerian and Lagrangian models have been used to understand pollutant dynamics, though Lagrangian models have very severe limitations in their formulation. Given those limitations (they do not capture wind shear, which is important in the atmosphere), and that those models do not use CFD-type techniques, we restrict the discussion below to Eulerian models.

Eulerian "grid" models are the most complex, but potentially the most powerful, air quality models, involving the least restrictive assumptions. They are also the most computationally intensive. Grid models solve a finite approximation to Eq. (9.5), including temporal and spatial variation of the meterological parameters, emission sources, and surface characteristics. Grid models divide the modeling region into a large number of cells, horizontally and vertically (Fig. 9.2), which interact with each other by simulating diffusion, advection, and sedimentation (for particles) of pollutant species. Most of the current regional photochemical models are Eulerian models. Input data requirements for grid models include spatially and temporally resolved emissions (by species), meteorology (e.g., wind velocities, temperatures, solar insolation, etc.), topographic features, initial and background pollutant concentrations (for initial and boundary conditions), and domain definition. Eulerian grid models predict pollutant

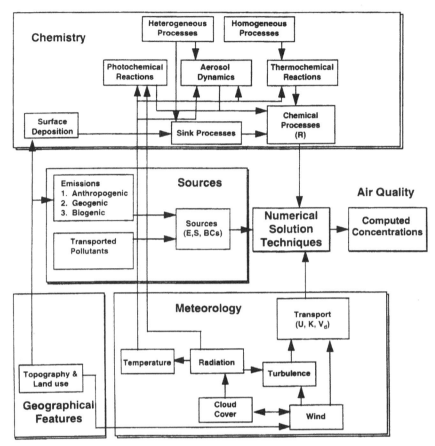

Figure 9.2 Schematic diagram of a typical air quality model, showing the model components and interactions.

concentrations throughout the entire modeling domain, usually including the airshed of interest and surrounding areas. Over successive time periods the evolution of pollutant concentrations and how they are affected by transport and chemical reaction can be tracked.

Most air quality models use the time-splitting approach (Yanenko, 1971; Marchuk, 1975) for the time integration of this equation. In this approach, the time-marching operator of ADE is decomposed into discrete operators for horizontal transport, vertical mixing, chemical transformations, and source and removal processes. Although the method of time splitting differs between models one such way is

$$c(t + \Delta t) = L_c(L_z(L_y(L_x(c(t)))))$$
$$c(t + 2\Delta t) = L_x(L_y(L_z(L_c(c(t + \Delta t))))) \tag{9.10}$$

where

$$L_x(c(t)) = \int_t^{t+\Delta t} \left(-\frac{\partial uc}{\partial x} + \frac{\partial}{\partial x} K_{xx} \frac{\partial c}{\partial x} \right) dt$$

$$L_y(c(t)) = \int_t^{t+\Delta t} \left(-\frac{\partial vc}{\partial y} + \frac{\partial}{\partial y} K_{yy} \frac{\partial c}{\partial y} \right) dt$$

$$L_z(c(t)) = \int_t^{t+\Delta t} \left(-\frac{\partial wc}{\partial z} + \frac{\partial}{\partial z} K_{zz} \frac{\partial c}{\partial z} \right) dt$$

$$L_c(c(t)) = \int_t^{t+\Delta t} (R_i(c) + S_i) \, dt \tag{9.11}$$

In essence, this means that at each time step the concentration field is first transported (diffused and advected) in the x-direction using the velocity field at time t. Next, the new concentration field (i.e., the one due to x-transport) is transported in the y-direction to get a new field. Next, the same is done in the z-direction. At this point the concentration field that existed at time t has been transported over the time step. Finally, the impact of chemistry over that time step is simulated. This gives the new field after one time step. Thus, the sequence of integrations is conducted as follows:

$$c^*(t + \Delta t) = L_x(c(t))$$
$$c^{**}(t + \Delta t) = L_y(c^*(t))$$
$$c^{***}(t + \Delta t) = L_z(c^{**}(t))$$
$$c(t + \Delta t) = L_c(c^{***}(t))$$

where c^*, c^{**}, and c^{***} are intermediate calculations. An appropriate numerical technique (e.g., a one-dimensional advection-diffusion technique or chemical kinetics solver) is used for integrating each operator over the chosen time step. Next, the process is repeated but in reverse order (to reduce biasing and also to reduce the amount of time spent in solving the chemical kinetics, which accounts for most of the computational time), as shown in Eq. (9.10). This gives the new concentration field after evolving for two time steps. Thus, splitting is between horizontal transport along the two principal axes, the vertical direction, and a chemistry/source operator. While horizontal transport is more advection dominated, diffusive transport usually dominates vertically. Solution of the chemistry and source operator uses much different techniques that are not discussed here. Of importance, though, is that the chemistry is the most intensive aspect of the solution, requiring about 85% of the computer time. Thus, it does not add significantly to the overall time to use a more accurate advection scheme. Further, the advection schemes, at present, appear to add the most error to the solution.

As stated, horizontal transport in the atmosphere is advection dominated and can be represented by the horizontal advection equation

$$\frac{\partial c}{\partial t} + \frac{\partial (uc)}{\partial x} + \frac{\partial (vc)}{\partial y} = 0 \tag{9.12}$$

where u and v are the advective velocities in the x- and y-directions, respectively. Equation (9.12) is known as the "flux form" of the advection equation. The same equation can be written in "advective form" as

$$\frac{\partial q}{\partial t} + u \frac{\partial q}{\partial x} + v \frac{\partial q}{\partial y} = 0 \tag{9.13}$$

where q is the mass mixing ratio related to the concentration c and density of air ρ as $q = c/\rho$. In the last two decades, the problem of advecting scalar fields (e.g., pollutant concentrations) has received significant attention, especially within the disciplines of computational fluid dynamics and meteorology. In a review article, Rood (1987) listed over 100 advection schemes for numerical solution of the advection equation. Several new schemes have since emerged; however, only a few are specifically designed for use in Eulerian air quality models.

Many models split the horizontal advection equation and solve two one-dimensional equations, one in each direction, using the solution of one as the initial condition of the other:

$$\frac{\partial c}{\partial t} + \frac{\partial (uc)}{\partial x} = 0$$
$$\frac{\partial c}{\partial t} + \frac{\partial (vc)}{\partial y} = 0 \tag{9.14}$$

These schemes will be referred to as 1-D. Others that solve the two-dimensional form directly will be referred to as 2-D. Although using 1-D schemes is very common, it has been found that problems can arise due to this additional splitting (Flatoy, 1993; Odman and Russell, 1993). For example, even if the flow field is divergence free (i.e., $\nabla \cdot \mathbf{u} = 0$), this does not mean that $\partial u/\partial x$ and $\partial v/\partial y$ are both equal to zero. This automatically introduces an error because the divergence (or convergence) in one direction affects the intermediate solution, which is used as the initial condition to the other direction, and the convergence (or divergence) in the other direction may not yield the desired two-dimensional solution. In this regard, 2-D schemes may be more desirable. All the schemes discussed here are 1-D schemes with the exception of the semi-Lagrangian scheme, which solves the advective form in Eq. (9.13). Two-dimensional schemes that solve the flux form are usually more difficult to implement and computationally less efficient; very few have been tested in air quality models. For an example of a 2-D scheme solving the flux form the reader is referred to Odman and Russell (1991b). The 1-D schemes discussed here can also be used for the solution of vertical advection; however, since vertical transport in the atmosphere is diffusion dominated, selection of solver for vertical advection is not as critical an issue as it is for horizontal advection.

There are several properties that an advection scheme must have in order to be useful in air quality models. As with all numerical methods, the numerical scheme for solving the advection equation must meet the convergence condition and correctly model the conservative, transportive, dissipative, and dispersive properties of the governing equation. A scheme is said to be convergent if the numerical solution approaches the true solution of the partial differential equation as the grid spacing and time-step size approach zero. Thus, if a scheme is convergent, one can obtain a numerical solution of any desired accuracy by reducing the grid spacing and the time-step size. For linear equations, consistency and stability are necessary and sufficient conditions for convergence (Lax's equivalence theorem). In practice, the computational resources available limit the grid spacing and the time-step size. Therefore, numerical errors associated with using limited grid spacing and time-step sizes must be of concern.

Due to the presence of pollutant sources and highly nonlinear chemical interactions between the concentration fields of different species, it is necessary to consider schemes with special properties: mass conservation, small numerical diffusion, and small phase errors. Advection schemes should be free of mass conservation errors to account accurately for pollutant sources and sinks and should have small numerical diffusion, since diffusion spreads a disturbance in every direction and smooths spatial gradients. They should also have small phase errors since disturbances that propagate at different speeds produce spurious oscillations such as "wiggles" or "ripples" in the numerical solution. These oscillations may lead to physically unrealistic negative concentrations, which are unacceptable in the presence of nonlinear chemistry in air quality models. Another source of spurious oscillations is the Gibbs phenomenon, which arises near steep gradients such as pollutant puffs or plumes due to the truncation of Fourier components of the solution. The nature of spurious oscillations leads to important concepts of "positive definiteness" and "monotonicity" in applying various schemes to air quality modeling. Schemes that do not allow negative concentrations are positive definite; however, these schemes may still allow oscillations such as overshoots in the solution. In general, monotonic schemes suppress all spurious oscillations. Since photochemical modeling starts with positive initial conditions, a monotonic scheme that does not create new extrema will always yield a positive solution. Thus, for this application, all monotonic schemes are naturally positive definite. While it is essential to have positive-definite schemes, that trait alone may be insufficient for air quality modeling and monotonic schemes may be more desirable.

Advection schemes fall under one of two categories: low-order schemes and high-order schemes. It is well known that the low-order schemes display considerable numerical diffusion which can easily outweigh physical horizontal diffusion. On the other hand, higher-order schemes generate spurious oscillations. Most schemes try to compromise between the dissipation error (or numerical diffusion) and oscillations related to dispersion (or phase shift) errors and the Gibbs phenomenon. The only way to minimize one of these errors without significant increase of the other is to introduce a nonlinear mechanism (Godunov, 1959). Typically, this mechanism is in the form of nonlinear flux corrections or nonlinear filtering. In advection schemes such adjustments are either applied implicitly or explicitly as a subsequent step to the linear solution.

Advection schemes with different properties introduce different errors, all of which are sources of uncertainty in air quality model predictions. It is critical to identify which of the above-mentioned properties a scheme possesses before recommending its use. Since an advection scheme with all the desired properties is currently not available, the problem becomes identifying the scheme with the most desirable properties and efficiency.

9.2 ADVECTION SCHEMES

9.2.1 Classification of Schemes

There are many different ways of classifying advection schemes. A common way is to classify the schemes based on the method used in their formulation (Rood, 1987). Since a wide variety of methods are used, any classification may fall short of being complete. The following is a fairly comprehensive list: (1) finite difference schemes, (2) finite volume schemes, (3) flux corrected schemes, (4) Lagrangian schemes, (5) finite element schemes, and (6) spectral schemes. Current trends in advection scheme development show a merging of the methods to take advantage of each approach's most desirable properties. For example, the characteristic-Galerkin method (Childs and Morton, 1990) combines the best of the finite element and Lagrangian methods. Flux corrections are being used in the framework of finite element and spectral schemes (Löhner et al., 1987). Also, the classical finite difference schemes are being abandoned in favor of modern finite volume schemes.

Finite Difference Schemes

Finite difference schemes are general approximation techniques to differential equations. The derivatives are approximated by differences such that differential equations are transformed into algebraic equations and continuous variables are represented at discrete points. The fifth-order compact upwind differencing scheme (Tolstykh, 1994) represents the state-of-the-art in finite difference schemes. It is mass conservative and it minimizes numerical diffusion and phase-shift errors. Also, the compact differencing operator makes the scheme more efficient compared to one that uses a full fifth-order differencing operator. The third-order Adams time integration described in Tolstykh (1994) requires information at several time steps. Multistep time integration methods are not feasible in air quality models because operators other than advection (chemistry, vertical diffusion) are applied between successive applications of the advection operator. A Runge-Kutta algorithm may also be used for time integration, but the resulting scheme does not provide a positive-definite solution and is computationally expensive. Therefore, although this scheme seems very accurate for moisture transport in general circulation models, it is not optimal for air quality modeling.

Finite Volume Schemes

Here, we focus on a particular family of schemes classified as "volume schemes" in Rood (1987). These schemes assume a piecewise continuous concentration distribution. While the concentration is continuous within each volume element or grid cell, there may be discontinuities at cell interfaces. The fluxes through the faces of the volume element are computed by integrating the subgrid distribution of concentration between the cell face and the departure point of the last particle that would leave the cell at the end of the time step.

Tremback et al. (1987) used higher-order area-preserving polynomials (e.g., a quartic with coefficients computed using area-preserving constraints) to represent the subgrid distribution of the advected scalar. Bott (1989a) normalized Tremback's advective fluxes in an attempt to reduce the phase-speed errors. Negative values of the transported quantity are suppressed by nonlinearly limiting the normalized fluxes. Recently, a monotonic version of the scheme was developed (Bott, 1992) and the time-splitting errors associated with the use of one-dimensional operators in multidimensional applications were reduced (Bott, 1993). A version of this algorithm with second-order polynomials for nonuniform grid spacing in the vertical also exists (Strand and Hov, 1993). These schemes are being increasingly used in atmospheric and air quality applications (Easter, 1993; Chlond, 1994).

In the piecewise parabolic method (PPM), the subgrid distribution of the advected quantity is represented by a parabola in each grid interval (Colella and Woodward, 1984). The approach is different from global spline methods (Emde, 1992) that try to fit a continuous curve to the advected quantity. A global fit is usually not well suited for representing localized, sharp transitions such as pollutant puffs. PPM not only provides a local fit of the data, but is monotonic and uses a special steepening procedure in the vicinity of sharp gradients. This ensures that positive quantities will remain positive and that sharp gradients will be handled correctly without the generation of spurious oscillations (Carpenter et al., 1990). This method has gained wide acceptance in the field of computational fluid dynamics (Bell et al., 1994) and has been also evaluated for atmospheric modeling (Carpenter et al., 1990).

In contrast to PPM, Emde (1992) developed a global method that uses a continuous curvature cubic spline method based on Purnell's scheme (Purnell, 1976). Purnell's scheme is a conservative, positive-definite algorithm that is second-order accurate in time and third-order accurate in space and has good amplitude and phase properties (Pielke, 1984). The positive-definite property is achieved by breaking the spline locally in grid cells where the concentrations become negative and fitting other curves (e.g., straight lines) in such a way that the mass is conserved.

Yamartino (1993) used piecewise cubic interpolands as a starting point for his higher-order scheme. In this scheme, the coefficients of a cell-centered cubic polynomial are constrained from the point of view of maintaining high-accuracy and low-diffusion characteristics while avoiding undesirable by-products (e.g., negative concentrations) associated with higher-order schemes but absent in low-order schemes. In addition, a filter is used for filling in undesired short-wavelength minima. One advantage of Yamartino's scheme is that it was designed to follow short-wavelength features (e.g., plumes from single grid point sources). This scheme has been specifically designed for air quality modeling.

The use of finite-volume schemes for atmospheric and air quality applications is very well developed, and these schemes can provide good accuracy fairly efficiently. Bott's scheme, Yamartino's scheme, and PPM are all highly recommended and will be further described shortly. While Bott's scheme is very simple and efficient, Yamartino's scheme is highly accurate in following short-wavelength features, which is a very desirable property in air quality modeling. PPM is the only truly monotonic scheme among the three, which is another desirable property. Further testing of these advection schemes is necessary to recommend any one of them over the others.

Flux-Corrected Transport Schemes

Flux-corrected transport (FCT) is a technique developed by Boris and Book (1973, 1976) and Book et al. (1975). The net transport flux is calculated as a weighted average of a flux computed by a low-order scheme and a flux computed by a higher-order scheme. The weighting is done in a manner that ensures that the higher-order flux is used to the greatest extent possible without introducing ripples (overshots and undershots). Zalesak (1979) generalized FCT to multidimensions. The performance of the more recent finite-volume schemes in the previous subsection are usually better than the FCT schemes.

Semi-Lagrangian Schemes

In a semi-Lagrangian method, one estimates the backward trajectory of a particle (or air parcel) that arrives at a certain grid point. Since the origin of a particle does not always coincide with a grid point, an interpolation scheme is necessary to estimate the original concentration. Once estimated, this concentration is assigned to the grid point of arrival. One advantage of the scheme is that it is not subject to the Courant stability condition, so large time steps can be used. However, the method is not mass conservative, and spurious oscillations may be generated (depending on the choice of interpolation scheme). Assuring mass conservation and obtaining positive definite or monotonic results requires special treatments. In the method described by Williamson and Rasch (1989) and Rasch and Williamson (1990), shape-preserving interpolands are used to maintain the monotonicity of the discrete data. Interpolation is unnecessary in the method described by Olim (1994), but it results in a scheme that is neither monotonic nor positive definite. Thus, of the semi-Lagrangian methods, Rasch and Williamson's method may be more appropriate for air quality modeling. In addition, this method has found acceptance in the community climate model (Williamson and Olson, 1994) and has been extensively tested and evaluated in that context.

Finite Element Schemes

The traditional Galerkin finite element approach uses a method of weighted residuals in which the solution is expanded in piecewise basis functions (e.g., chapeau functions for a one-dimensional problem). The residual is assumed to be orthogonal to the weighting functions. Pepper et al. (1979) indicated that this method combined with Crank-Nicolson time integration scheme preserves the peak value of a pulse quite well, but creates ripples behind the pulse. Raymond and Garder (1976) added a dissipative method that suppresses these ripples although the peak is also suppressed. This scheme was found to perform well by Chock and Dunker (1983): it has no restrictions on the Courant number, gives relatively accurate results, and is fairly efficient. Chock (1985) analyzed several different variations of this scheme and found that chapeau functions with the Forester filter (Forester, 1977) appears to be the best variation.

The use of special weighting functions that are different than the basis functions results in "upwinding" effects similar to those in finite difference methods. This technique is known as the Petrov–Galerkin method. Upwinding introduces some artificial diffusion that reduces spurious oscillations. Both Hughes and Brooks (1979) and Kelly et al. (1980) found that artificial diffusion, only in the direction of flow, was desirable and could be implemented, thereby eliminating the problem of crosswind diffusion. Hughes and Brooks (1979) and Brooks and Hughes (1982) proposed a formulation that modifies the usual weighting functions by a perturbation dependent upon the velocity field and the derivative of the basis functions. This scheme is called the streamline upwind/Petrov–Galerkin (SU/PG) method. Tezduyar et al. (1987) developed SU/PG for time-dependent advection-diffusion equations. Odman and Russell (1991a, 1991b) implemented SU/PG in an air quality model.

Donea (1984) describes an alternative algorithm for advective transport problems with improved stability properties. The time derivative is expanded in Taylor series; thus this method is called Taylor–Galerkin. The second- and third-order time derivatives are evaluated from the original equation in a manner similar to that employed in Hirt's heuristic stability analysis (Hirt, 1968). This process yields a generalized governing equation that is discretized in time only; the spatial variables are left continuous. Such an equation is successively discretized in space using the conventional Galerkin finite element method. One disadvantage of the Taylor–Galerkin method is that spurious oscillations are not completely suppressed. Parrot and Christie (1986) and then Löhner et al. (1987) linked this method with the concept of flux-corrected transport (Zalesak, 1979) and obtained a monotonic scheme. In this scheme, a low-order scheme (e.g., lumped mass Taylor–Galerkin scheme with added artificial diffusion) guaranteed to give monotonic results is combined with a higher-order scheme (e.g., consistent-mass Taylor–Galerkin scheme). Chock (1991) evaluated the one-dimensional Taylor–Galerkin method followed by a nonlinear filter. While Taylor–Galerkin is probably the best choice among finite element advection schemes, significant benefits are observed only when the scheme is used in two dimensions (possibly with flux correction). Two-dimensional codes have appeared recently (Löhner et al., 1987; Childs and Morton, 1990), but they are not readily accessible. Therefore, the Taylor–Galerkin method is not included in this study.

Pseudospectral Schemes

In a pseudospectral scheme the Taylor series expansion is used to replace the time derivative with space derivatives. Then fast Fourier transforms are used to approximate the space derivatives (Gazdag, 1973; Wengle and Seinfeld, 1978; Chock, 1991). This method is known as the accurate space derivative (ASD) scheme. This scheme is highly accurate; however, it needs to be coupled with a nonlinear filter (such as the Forester filter) to suppress the spurious oscillations that may exist in the solution. A disadvantage of the scheme is the requirement of periodic boundary conditions inherent to all spectral schemes. However, there are several solutions for this problem (Gazdag, 1973; Roache, 1978; Wengle and Seinfeld, 1978; Chock, 1991). Since the boundary conditions in air quality models are not periodic, special treatments are required at the boundary.

The pseudospectral scheme was found to be the most accurate of several methods tested by Chock (1985), but it required a long execution time. Chock (1991) compared several chapeau function methods, and ASD and found that the chapeau function with Taylor–Galerkin and a Forester filter was an excellent choice if one was concerned with execution time. However, the computationally intensive ASD method was significantly more accurate. Thus, the choice of advection scheme depends on the accuracy desired and the computational tools available. Recently, the parameters of the Forester filter were optimized for the ASD scheme by Dabdub and Seinfeld (1994).

9.2.2 Brief Description of Schemes

Generally, air quality models solve the advection-diffusion-reaction equation using the time-splitting approach (Yanenko, 1971). Each process is split into one-dimensional operators and the solution from one process is used as the initial condition for the next process. The advection of a tracer species is represented by the following conservation equation:

$$\frac{\partial c}{\partial t} + \nabla \cdot (\mathbf{u}c) = 0 \tag{9.15}$$

where c is the species concentration (in density units) and $\mathbf{u}$ is the velocity vector.

Time splitting leads to the following one-dimensional equation in Cartesian coordinates:

$$\frac{\partial c}{\partial t} + \frac{\partial (uc)}{\partial x} = 0 \tag{9.16}$$

To simplify the discussion, we consider a uniform (i.e., constant Δx) and staggered grid (c_j represents the grid cell average of the concentration, while $u_{j+1/2}$ is the advection velocity defined at grid cell interfaces).

In the finite-volume schemes discussed below (the piecewise parabolic method, the Bott scheme, and the Yamartino scheme), we use the explicit flux formula to advance the solution a single time step, Δt, from level n to level $n + 1$:

$$c_j^{n+1} = c_j^n - \frac{\Delta t}{\Delta x} (F_{j+1/2}^n - F_{j-1/2}^n) \tag{9.17}$$

where $F_{j+1/2}^n$ and $F_{j-1/2}^n$ are the advective fluxes of c through the right and left boundaries of grid cell j, respectively. Further, let us define a nondimensional coordinate x as $\xi = (x - x_j)/\Delta x$ so that, in grid cell j, $-1/2 \leq \xi \leq 1/2$. Now, suppose that the concentration has a certain distribution $c_j(\xi)$ in each grid cell. Depending on the direction of the velocity, the flux $F_{j+1/2}$ can be expressed as

$$F_{j+1/2} = \begin{cases} \dfrac{\Delta x}{\Delta t} \displaystyle\int_{1/2 - \varepsilon_{j+1/2}}^{1/2} c_j(\xi)\, d\xi, & u_{j+1/2} \geq 0 \\[3mm] \dfrac{\Delta x}{\Delta t} \displaystyle\int_{-1/2}^{-1/2 + \varepsilon_{j+1/2}} c_{j+1}(\xi)\, d\xi, & u_{j+1/2} < 0 \end{cases} \tag{9.18}$$

where $\varepsilon_{j+1/2} = |u_{j+1/2}| \Delta t/\Delta x$ is the Courant number at the right boundary of grid cell j. The departure point of the last particle that crosses the cell interface appears in the limits of the flux integrals as $\xi_{dp} = \pm 1/2 \mp \varepsilon_{j+1/2}$. In calculating this expression it was assumed that the velocity is uniform over the grid cell. Assuming linear variation of the winds between the two cell faces may lead to a more refined estimation of the mass fluxes through the cell faces. The computation of the departure point for linearly varying velocity can be found, for example, in Seibert and Morariu (1992). A practical correction for linearly varying fields is to build the divergence into the Courant number as

$$\tilde{\varepsilon} = \varepsilon \left(1 - \frac{\Delta u}{2} \frac{\Delta t}{\Delta x} \right) \tag{9.19}$$

Earlier, we mentioned that the conditions of higher-order accuracy and freedom from spurious oscillations cannot be achieved simultaneously. The only way to satisfy one of these conditions without significant violation of the other is to introduce a nonlinear mechanism. Typically, this mechanism is provided by nonlinear flux corrections or nonlinear filtering. In the advection schemes below, such adjustments are either applied implicitly through the solution or explicitly as a subsequent step to the linear solution. It is important to realize that the original problem described by Eq. (9.12) has a linear nature. Introduction of nonlinear mechanisms in order to improve certain aspects of the numerical solution modify this linear nature.

The schemes that will be described here were recently tried in atmospheric modeling. They are (1) the Smolarkiewicz scheme, (2) the piecewise parabolic method, (3) the Bott scheme, (4) the Yamartino scheme, (5) flux-corrected transport, (6) the semi-Lagrangian method, (7) the chapeau function scheme and Forester filter, and (8) the accurate space derivative (ASD) scheme.

Smolarkiewicz Scheme

The Smolarkiewicz scheme (Smolarkiewicz, 1983) is based on the first-order accurate upstream or "donor cell" method. In the donor cell method, the distribution of the advected quantity is assumed to be constant over the grid cell (i.e., $c_j(\xi) = c_j$). A second-order Taylor expansion of the upstream formula, assuming that u is constant over the whole domain, yields

$$\frac{\partial c}{\partial t}\bigg|_j^n + \frac{\partial(uc)}{\partial x}\bigg|_j^n = \frac{\partial}{\partial x}\left[\frac{1}{2}\left(|u|\,\Delta x - u^2\,\Delta t\right)\frac{\partial c}{\partial x}\right]_j^n + O(\Delta t^2, \Delta x^2) \qquad (9.20)$$

By comparison with Eq. (9.16), the right-hand side of Eq. (9.20) can be identified as the artificial diffusion introduced by the donor cell scheme. To increase accuracy, Smolarkiewicz (1983) reversed the effect of this artificial diffusion by defining an antidiffusion velocity:

$$\tilde{u}_{j+1/2} = \frac{1}{2}\left(|u_{j+1/2}|\,\Delta x - u_{j+1/2}^2\,\Delta t\right)\frac{1}{c}\frac{\partial c}{\partial x}\bigg|_{j+1/2} \qquad (9.21)$$

Though diffusion is physically irreversible, it is numerically possible to apply an antidiffusive step and partially recover what has been lost to diffusion; this resembles reversing a film that shows diffusion. It is customary to express the algorithm as a multistep scheme where Eq. (9.17) is applied first with the advective velocities and then successively with the antidiffusion velocities as

$$c_j^{**} = c_j^* - \frac{\Delta t}{\Delta x}\left[F_{j+1/2}^* - F_{j-1/2}^*\right] \qquad (9.22)$$

where c_j^* and c_j^{**} denote successive iterations. To optimize accuracy versus computer time, Smolarkiewicz (1983) recommends two iterative steps. Recently, the scheme has been extended to multiple dimensions and its oscillation damping properties have been improved (Smolarkiewicz and Grabowski, 1990).

Piecewise Parabolic Method

In PPM, the concentration distribution is assumed to be parabolic in any given grid cell. In terms of the grid cell average concentration c_j and the values of the parabola at the left and right boundaries of the cell, $c_{L,j}$ and $c_{R,j}$, this distribution can be written as

$$c_j(\xi) = c_{L,j} + \xi\left[(c_{R,j} - c_{L,j}) + 6\left(c_j^n - \frac{c_{L,j} + c_{R,j}}{2}\right)(1 - \xi)\right] \qquad (9.23)$$

Since the initial cell average is known, the construction of the parabola involves determination of the edge values. First, an approximation to c at $x_{j+1/2}$ is computed subject to the constraint that its value is within the range of values at the neighboring cells. For this purpose, a first guess for $c_{j+1/2}$ is

$$c_{j+1/2} = \frac{7}{12}(c_j^n + c_{j+1}^n) - \frac{1}{12}(c_{j+2}^n + c_{j-1}^n) \qquad (9.24)$$

In smooth parts of the solution away from extrema, $c_{L,j+1} = c_{R,j} = c_{j+1/2}$ so that the distribution is continuous at $x_{j+1/2}$. In other parts, the cell boundary values are further modified so that c is monotonic on each grid cell. Modification of boundary values introduces discontinuities at $x_{j+1/2}$. Thus, the global concentration distribution is piecewise continuous with continuous pieces in each cell and possible discontinuities at cell edges. There are two cases where the edge values are modified. First, if c_j is a local extremum, then the distribution is assumed to be constant instead of parabolic. The second case is when c_j is between $c_{L,j}$ and $c_{R,j}$, but sufficiently close to one of the values so that the parabola may take on values outside the range and lead to overshots or undershots. In this case, to make the parabola monotonic, one of the edge values is reset so that the derivative of $c(\xi)$ is zero at the opposite edge.

The most distinctive feature of this monotonic scheme is that the non-linear adjustments are purely geometric. The numerical diffusion introduced by this scheme may be slightly higher than in some other schemes discussed here, but its monotonic property may be very desirable for photochemical modeling purposes. The scheme has been used in meteorological modeling (Carpenter et al., 1990). It can be modified so that, in the neighborhood of a discontinuity, it produces a narrower profile. This feature, known as steepening, avoids the smearing of sharp gradients. Although Carpenter et al. (1990) did not recommend steepening for meterological modeling, this feature may be beneficial in air quality modeling practice, where sharp discontinuities such as point sources from single-grid cells are common.

Bott's Scheme

This is a positive definite scheme with small numerical diffusion (Bott, 1989a, b). The distribution of the concentration within the cell is represented by a polynomial of order l as

$$c_j(\xi) = \sum_{k=0}^{l} a_{j,k} \xi^k \tag{9.25}$$

The polynomial can be made area preserving by requiring

$$c_{j+1} = \int_i^{i+1} \sum_{k=0}^{l} a_{j,k} \xi^k \, d\xi, \qquad i = 0, \pm 1, \pm 2, \ldots, \pm \frac{l}{2} \tag{9.26}$$

over a stencil of $l + 1$ grid cells by varying the value of i. Note that in Eq. (9.26) the concentration distribution for cell j is extended to the $l/2$ neighboring cells to the left and to the right of cell j in order to create sufficient number of equations for the unknowns. The solution to this linear system of $l + 1$ equations yields the coefficients $a_{j,k}$. The coefficients obtained this way for a quadratic ($l = 2$) and quartic ($l = 4$) together with those of the donor cell (or upwind) scheme, and Tremback's scheme with second-order polynomials (Tremback et al., 1987), are listed in Table 9.1.

Using Eq. (9.18) and integrating the polynomial of Eq. (9.25) between appropriate limits, we arrive at a first estimate of the fluxes. Finally, to make the scheme positive definite, the total outflux from cell j is limited by requiring that it should be positive and less than what the available mass in the cell would allow:

$$0 \leq F_j^{out} \leq \frac{\Delta x}{\Delta t} c_j \tag{9.27}$$

The outflux F_j^{out} is a combination of the boundary fluxes and its expression depends on the sign of the velocities. Here, we used fourth-order polynomials, as recommended by Bott (1989a).

Recently, a monotonic version of the scheme was also developed (Bott, 1992) and the time-splitting errors associated with the use of one-dimensional operators in multidimensional applications were reduced (Bott, 1993). Monotonicity is obtained by directly replacing the positive-definite

Table 9.1 Coefficients of the Polynomials Used in Each Scheme

	Donor cell	Tremback-2	Bott-2	Bott-4
a_0	c_i	c_i	$-\frac{1}{24}(c_{i+1} - 26c_i + c_{i-1})$	$(9c_{i+2} - 116c_{i+1} + 2134c_i - 116c_{i-1} + 9c_{i-2})/1920$
a_1	—	$\frac{1}{2}(c_{i+1} - c_{i-1})$	$\frac{1}{2}(c_{i+1} - c_{i-1})$	$(-5c_{i+2} + 34c_{i+1} - 34c_{i-1} + 5c_{i-2})/48$
a_2	—	$\frac{1}{2}(c_{i+1} - 2c_i + c_{i-1})$	$\frac{1}{2}(c_{i+1} - 2c_i + c_{i-1})$	$(-c_{i+2} + 12c_{i+1} - 22c_i + 12c_{i-1} - c_{i-2})/16$
a_3	—	—	—	$(c_{i+2} - 2c_{i+1} + 2c_{i-1} - c_{i-2})/12$
a_4	—	—	—	$(c_{i+2} - 4c_{i+1} + 6c_i - 4c_{i-1} + c_{i-2})/24$

flux limiter of the original approach by new monotone flux limiters, as

$$\begin{aligned}
\min(c_{j-1}^n, c_j^n) \le c_j^{n+1} \le \max(c_{j-1}^n, c_j^n), & \quad \text{if } u \ge 0 \\
\min(c_{j+1}^n, c_j^n) \le c_j^{n+1} \le \max(c_{j+1}^n, c_j^n), & \quad \text{if } u < 0
\end{aligned} \tag{9.28}$$

Although the new flux-limited Bott scheme yields monotonic results, there is an inherent mass conservation problem. This problem is directly related to the flux step. Near the leading edge of a sharp wave, the use of second- or higher-order polynomials causes an underestimation of a certain downwind advective flux, say $F_{k-1/2}$. When this flux is not corrected, it is less than $F_{k+1/2}$ and an undershot occurs in cell k, as experienced with the original algorithm (Bott, 1989a, 1989b). The motivation for the monotone flux limitation is to avoid such undershots. However, there are cases when the monotone flux limiter leaves the underestimated flux intact [e.g., $c_{k-2} > c_{k-1}$, $F_{k-3/2} > F_{k-1/2}$, and $\alpha(c_{k-2} - c_{k-1}) > (F_{k-3/2} - F_{k-1/2})$]. Instead of increasing the underestimated flux, the limiter reduces the advective flux downwind, $F_{k+1/2}$, in order to avoid an undershot in cell k. This eventually reduces the net flux out of the domain, resulting in an accumulation of mass in the domain.

Flux-Corrected Transport Scheme

The FCT procedure of Zalesak (1979) can be summarized as follows:

1. Compute $F_{j+1/2}^L$, advective fluxes through cell faces given by a low-order scheme guaranteed to give monotonic solutions. We used the donor cell scheme as the low-order scheme.

2. Compute $F_{j+1/2}^H$, advective fluxes through cell faces given by a higher-order scheme. We used fourth-order area-preserving polynomials (i.e., Bott's scheme except the flux limiting) as the higher-order scheme.

3. The antidiffusive flux, $A_{j+1/2}$, is defined as the difference of higher- and low-order fluxes:

$$A_{j+1/2} = F_{j+1/2}^H - F_{j+1/2}^L \tag{9.29}$$

4. Compute the low-order solution as

$$c_j^L = c_j^n - \frac{\Delta t}{\Delta x} (F_{j+1/2}^L - F_{j-1/2}^L) \tag{9.30}$$

5. Limit the antidiffusive flux, $A_{j+1/2}$, with a limiting factor, $C_{j+1/2}$, in a manner such that the solution c_j^{n+1} as computed below is free of extrema not found in c_j^n or c_j^L:

$$A_{j+1/2}^C = C_{j+1/2} A_{j+1/2}, \qquad 0 \le C_{j+1/2} \le 1 \tag{9.31}$$

6. Compute the solution using limited antidiffusive fluxes:

$$c_j^{n+1} = c_j^L - \frac{\Delta t}{\Delta x} (A_{j+1/2}^C - A_{j-1/2}^C) \tag{9.32}$$

Although its details are omitted here, the limiting of the antidiffusive flux is the most important step. Detailed discussion of this can be found in Zalesak (1979). The accuracy of the flux-corrected schemes highly depends on the accuracy of the higher-order scheme employed.

The FCT concept has also been implemented in finite element methods. Löhner et al. (1987) linked this concept with the Taylor–Galerkin method of Donea (1984). In this scheme, element contributions from a low-order scheme (e.g., lumped-mass Taylor–Galerkin scheme with added artificial diffusion) guaranteed to give monotonic results are computed first. Then the element contributions from a higher-order scheme (e.g., consistent-mass Taylor–Galerkin scheme) are computed. Note that because the finite element method is being used, the term "flux" in Zalesak's original concept is replaced by "element contribution" here. In the next step, the antidiffusive element contributions are computed by taking the difference of higher-order and low-order element contributions. Finally, a low-order solution is computed, the antidiffusive element contribution is limited for monotonicity, and the low-order solution is updated by applying the limited antidiffusive element contribution. To our knowledge, this scheme has not been used in any atmospheric models.

Yamartino's Scheme

Yamartino's scheme (Yamartino, 1993) is another finite-volume scheme where the interpolating polynomial is a cubic spline:

$$c_j(\xi) = a_0 + a_1 \xi + a_2 \xi^2 + a_3 \xi^3 \tag{9.33}$$

where the coefficients are

$$a_0 = c_j$$
$$a_1 = d_j \Delta x$$
$$a_2 = -\frac{1}{4} (c_{j+1} - 2c_j + c_{j-1}) + \frac{3 \Delta x}{8} (d_{j+1} - d_{j-1})$$
$$a_3 = (c_{j+1} - c_{j-1}) - \frac{\Delta x}{6} (d_{j+1} + 10d_j + d_{j-1}) \tag{9.34}$$

The spline derivatives, d_j, are obtained from the tridiagonal system:

$$\alpha d_{j-1} + (1 - 2\alpha)d_j + \alpha d_{j+1} = \frac{c_{j+1} - c_{j-1}}{2 \Delta x} \tag{9.35}$$

with $\alpha = 0.22826$. Note that a value of $\alpha = 0$ would correspond to the explicit expressions of d_j.

The positivity of $c_j(\xi)$ is ensured by various mechanisms. First, when c_j is a local minimum, the donor cell scheme is used instead of the cubic spline. Second, the spline is spectrally limited by the relation

$$\left| \frac{a_k}{a_0} \right| \leq \frac{\pi^k}{k!} \tag{9.36}$$

where $k = 1, 2, 3$. Third, a mass conservative flux renormalization is applied where the fluxes are normalized with the ratio for the upwind cell of the cell concentration (i.e., concentration at $\xi = 0$) divided by the average concentration. Finally, a mildly diffusive filter is applied in an attempt to block the depletion of donor cells. Yamartino's scheme is not monotonic and can generate new maxima.

Chapeau Function Scheme

The chapeau function method is a classical weighted-residual finite element method. The solution is expanded in piecewise basis functions, $N_k(x)$, that look like hats ("chapeau" is French for hat) in one-dimensional space:

$$c(x, t) = \sum_{k=1}^{p} N_k(x) \cdot c_k(t)$$
$$u(x, t) = \sum_{k=1}^{p} N_k(x) \cdot u_k(t) \tag{9.37}$$

The residual is assumed to be orthogonal to the weighting functions, which may be the basis functions themselves, as is usually the case in the Bubnov–Galerkin methods:

$$\int N_i \left[\frac{\partial}{\partial t} (N_j c_j) + \frac{\partial}{\partial x} (N_k u_k N_j c_j) \right] dx = 0 \tag{9.38}$$

After a coordinate transformation $\xi = (x - x_j)/\Delta x$ where $-1/2 \leq \xi \leq 1/2$, the basis functions of a typical finite element with two nodes can be written as

$$N_1 = \frac{1}{2} - \xi, \qquad N_2 = \frac{1}{2} + \xi \tag{9.39}$$

and the weighted residual equation becomes

$$\int_{-1/2}^{+1/2} N_i \left[\Delta x \frac{\partial}{\partial t} (N_j c_j) + \frac{\partial}{\partial \xi} (N_k u_k N_j c_j) \right] d\xi = 0 \tag{9.40}$$

or

$$M_{ij} \frac{\partial c_j}{\partial t} + K_{ij} c_j = 0 \tag{9.41}$$

where

$$M_{ij} = \Delta x \int_{-1/2}^{+1/2} N_i N_j \, d\xi = \frac{\Delta x}{6} \begin{bmatrix} 2 & 1 \\ 1 & 2 \end{bmatrix}$$

$$K_{ij} = \int_{-1/2}^{+1/2} N_i \frac{\partial}{\partial \xi} (N_k u_k N_j) \, d\xi = \frac{1}{6} \begin{bmatrix} -4u_1 + u_2 & u_1 + 2u_2 \\ -2u_1 - u_2 & -u_1 + 4u_2 \end{bmatrix} \tag{9.42}$$

for each finite element.

If the Crank–Nicolson time-integration is used then Eq. (9.41) becomes

$$\left(\frac{1}{\Delta t} M_{ij} + \frac{1}{2} K_{ij} \right) c_j^{n+1} = \left(\frac{1}{\Delta t} M_{ij} + \frac{1}{2} K_{ij} \right) c_j^n \tag{9.43}$$

This equation results in the following tridiagonal system:

$$\left[1 - \frac{(2u_{i-1} + u_i) \Delta t}{2 \Delta x} \right] c_{i-1}^{n+1} + \left[4 + \frac{(u_{i+1} - u_{i-1}) \Delta t}{2 \Delta x} \right] c_i^{n+1}$$

$$+ \left[1 + \frac{(u_i + 2u_{i+1}) \Delta t}{2 \Delta x} \right] c_{i+1}^{n+1} = \left[1 + \frac{(2u_{i-1} + u_i) \Delta t}{2 \Delta x} \right] c_{i-1}^n$$

$$+ \left[4 - \frac{(u_{i+1} - u_{i-1}) \Delta t}{2 \Delta x} \right] c_i^n + \left[1 - \frac{(u_i + 2u_{i+1}) \Delta t}{2 \Delta x} \right] c_{i+1}^n \tag{9.44}$$

The numerical diffusion introduced by this scheme is quite small, but spurious oscillations are observed near steep gradients (Pepper et al., 1979). McRae et al. (1982) applied a smoothing filter to suppress these oscillations. This mass-conservative filter locally and selectively damps the short-wavelength noise by introducing artificial diffusion (Forester, 1977). Due to the local property of the filter, resolvable features of the linear solution such as peak pollutant concentrations are maintained. The filter can be applied successively until positive-definite results are obtained. Recently, Odman and Russell (1993) enhanced the performance of the Forester filter and generalized it for use with multidimensional finite element schemes.

Accurate Space Derivative Scheme

The accurate space derivative (ASD) scheme (Chock, 1991; Dabdub and Seinfeld, 1994) is classified as a spectral scheme and is highly accurate. However, it needs to be coupled with a nonlinear filter, such as the Forester filter, to suppress the spurious oscillations that may exist in the solution. Also, special treatments are necessary to satisfy the periodic boundary condition requirement since the boundary conditions in air quality models are not periodic. Computationally, the scheme is generally more expensive than others, but its high accuracy makes it particularly attractive.

First, the concentration is expanded as a truncated Taylor series in time. The order of accuracy of this explicit scheme can be increased by including more terms in the series. Here, we include terms up to the third-order time derivative:

$$c^{n+1} = c^n + \left(\frac{\partial c}{\partial t}\right)^n \Delta t + \left(\frac{\partial^2 c}{\partial t^2}\right)^n \frac{\Delta t^2}{2!} + \left(\frac{\partial^3 c}{\partial t^3}\right)^n \frac{\Delta t^3}{3!} + \cdots \qquad (9.45)$$

Then the time derivatives of the concentration are replaced by the space derivatives using the governing equation (i.e., advection equation). Note that the first time derivative already appears in this equation. Higher-order derivatives can be obtained by differentiating this equation. Assuming that the velocity is not a function of time, the spatial equivalents of time derivatives can be written in conservation law form, as

$$\frac{\partial c}{\partial t} = -\frac{\partial(uc)}{\partial x}$$

$$\frac{\partial^2 c}{\partial t^2} = \frac{\partial}{\partial x}\left[u\,\frac{\partial(uc)}{\partial x}\right]$$

$$\frac{\partial^3 c}{\partial t^3} = -\frac{\partial}{\partial x}\left\{u\,\frac{\partial}{\partial x}\left[u\,\frac{\partial(uc)}{\partial x}\right]\right\} \qquad (9.46)$$

However, to keep the derivatives as simplified as possible for the fast Fourier transform (FFT), the derivatives are computed in chain rule form:

$$\frac{\partial c}{\partial t} = -c\,\frac{\partial u}{\partial x} - \frac{\partial c}{\partial x}\,u$$

$$\frac{\partial^2 c}{\partial t^2} = c\left[u\,\frac{\partial^2 u}{\partial x^2} + \left(\frac{\partial u}{\partial x}\right)^2\right] + 3\,\frac{\partial c}{\partial x}\,u\,\frac{\partial u}{\partial x} + \frac{\partial^2 c}{\partial x^2}\,u^2$$

$$\frac{\partial^3 c}{\partial t^3} = -c\left[4u\,\frac{\partial u}{\partial x}\,\frac{\partial^2 u}{\partial x^2} + u^2\,\frac{\partial^3 u}{\partial x^3} + \left(\frac{\partial u}{\partial x}\right)^3\right]$$

$$\qquad - \frac{\partial c}{\partial x}\left[4u^2\,\frac{\partial^2 u}{\partial x^2} + 7u\left(\frac{\partial u}{\partial x}\right)^2\right] - 6\,\frac{\partial^2 c}{\partial x^2}\,u^2\,\frac{\partial u}{\partial x} - \frac{\partial^3 c}{\partial x^3}\,u^3 \qquad (9.47)$$

Next, the space derivatives of both c and u are obtained through the use of an FFT. To comply with the periodic boundary conditions requirement of the FFT, we add six more cells to the domain beyond the real boundaries. Then we create data for these extra cells utilizing a polynomial fit and forcing periodicity of the data and the derivatives. The polynomial is either parabolic or cubic, depending on the signs of the derivatives at the boundaries. If the derivatives at the boundaries are of opposite sign, then we use a parabolic fit; if the derivatives at the boundaries are of the same sign, we employ a cubic fit. Once the periodic boundary conditions are enforced, we transform the data to the Fourier space with forward FFT, take the derivatives in that space, and then transform the derivatives back to the original space using backward FFT. The computation of the derivative in the Fourier space assures high-order accuracy throughout the entire domain, including grid cells near the boundaries. We then substitute the derivatives into Eq. (9.45) and update the concentration field.

Due to its dispersive quality, the FFT approach tends to produce high-frequency noise. This noise is filtered and diffused using the nonlinear Forester filter:

$$c_j^{k+1} = c_j^k + \frac{1}{2} K_f[(c_{j+1} - c_j)(\chi_{j+1} + \chi_j) - (c_j - c_{j-1})(\chi_j + \chi_{j-1})]^k \quad (9.48)$$

where χ_j is a weight function determining the amount of filtering applied to cell j, and K_f is the coefficient of artificial diffusion. The filter is designed to be used in conjunction with second- and higher-order methods. Computational noise is minimized without incurring amplitude penalty. This is achieved by being able to select the desired frequencies of noise and adding diffusion locally at selected grid cells at which χ_j is set equal to one and held at zero elsewhere. This way, the high-frequency noise generated by the FFT can be diffused away.

Semi-Lagrangian Transport Scheme

In the semi-Lagrangian transport (SLT) scheme, one estimates the backward trajectory of a particle (or air parcel) that arrives at a certain grid point. Since the origin of a particle does not always coincide with a grid point, an interpolation scheme is necessary to estimate the original concentration. Once estimated, this concentration is assigned to the grid point of arrival. One advantage of the scheme is that it is not subject to the Courant stability condition, so large time steps can be used. However, the method is not mass conservative, and spurious oscillations may be generated (depending on the choice of interpolation scheme). Assuring mass conservation and obtaining positive definite or monotonic results requires special treatments. There are several semi-Lagrangian schemes listed in the literature (Rasch and Williamson, 1990; Smolarkiewicz and Grell, 1992). Semi-Lagrangian methods are being used particularly in global circulation and climate models (Williamson and Rasch, 1989).

The SLT scheme is based on the advective form in Eq. (9.13):

$$\frac{\partial q}{\partial t} + u \frac{\partial q}{\partial x} + v \frac{\partial q}{\partial y} = 0 \quad (9.49)$$

where q is the mass mixing ratio related to the concentration c and density of air ρ as $q = c/\rho$. The Lagrangian solution to this equation determines the departure point (x_D, y_D) of a particle at (x_A, y_A) as

$$x_D = x_A - \Delta t \, u$$
$$y_D = y_A - \Delta t \, v \tag{9.50}$$

The SLT scheme of Rasch and Williamson (1990) is described here. This scheme first determines the midpoint of the trajectory iteratively as

$$x_M^{k+1} = x_A - \frac{\Delta t}{2} \, u(x_M^k, y_M^k)$$
$$y_M^{k+1} = y_A - \frac{\Delta t}{2} \, v(x_M^k, y_M^k) \tag{9.51}$$

Four iterations are used for the very first time step, which starts with the arrival points as a first guess (for the midpoints) and one iteration thereafter where the midpoints from the previous time step are used as a first guess. The velocities at the midpoints are calculated using Lagrange cubic interpolation. The departure points are calculated as

$$x_D = x_A - \Delta t \, u(x_M, y_M)$$
$$y_D = y_A - \Delta t \, v(x_M, y_M) \tag{9.52}$$

Finally, the q field at the departure points are found by Hermite cubic interpolations and used to update the values at arrival points:

$$q(\xi) = (2\xi^3 - 3\xi^2 + 1)q_j + (-2\xi^3 + 3\xi^2)q_{j+1}$$
$$+ (\xi^3 - 2\xi^2 + \xi)d_j + (\xi^3 - \xi^2)d_{j+1} \tag{9.53}$$

where

$$d_j = \frac{1}{6\,\Delta x}(-2q_{j-1} - 3q_j + 6q_{j+1} - q_{j+2})$$
$$d_{j+1} = \frac{1}{6\,\Delta x}(q_{j-1} - 6q_j + 3q_{j+1} + 2q_{j+2}) \tag{9.54}$$

The arrival points are grid points. The reason for using the Hermite cubic interpolation is that the derivative terms are explicit and they may be limited to obtain monotonic results. A sufficient (but not necessary) condition for monotonicity is

$$0 \le \frac{d_j}{\Delta_j} \le 3$$
$$0 \le \frac{d_{j+1}}{\Delta_j} \le 3 \tag{9.55}$$

where Δ_j is the discrete slope defined as $\Delta_j = (q_{j+1} - q_j)/\Delta x$. Also, instead of using two-dimensional bicubic interpolants, the tensor products are employed where a series of interpolations in one dimension are followed by one interpolation in the other dimension. According to which direction is interpolated first, one may get a different answer, but the differences are minor according to Williamson and Rasch (1989).

9.3 EVALUATION OF ADVECTION SCHEMES

9.3.1 Description of Tests and Performance Measures Used

Typically, the performances of advection schemes are measured and compared with each other using test cases with idealized flow fields. These ideal flow tests have analytic solutions and are very useful for determining certain properties of the schemes. Here, we perform a preliminary evaluation to identify schemes with more desirable properties. As discussed earlier, the properties desirable in an advection scheme for use in photochemical modeling are mass conservation, small numerical diffusion and phase-speed errors, and freedom from Gibbs oscillations at least to the extent where they do not lead to negative concentrations and monotonicity.

In the figures and tables in this section, the schemes are abbreviated as follows:

ASD	accurate space derivative scheme
BOT	Bott's scheme
BOT-M	Bott's scheme with the monotonic limiter
BOT2D	two-dimensional version of Bott's scheme
FCT	flux-corrected transport scheme
HAT	chapeau function scheme
PPM	piecewise parabolic method
SLT	semi-Lagrangian transport scheme
SMO	Smolarkiewicz's scheme
YAM	Yamartino's scheme

The following tests are used for preliminary evaluation of the schemes:

1. *One-dimensional tests*: Tests where pulses of different shapes are advected with uniform velocity.
2. *Rotating cone test*: A cone-shaped puff is introduced in a rotational flow field and followed for a certain number of revolutions.
3. *Skew advection of a point-source plume*: The plume from a continuously emitting point source is advected diagonal to the grid.
4. *Shear flow tests*: The effects of wind shear are simulated with a stagnation or a vortex flow field.

5. *Rotating/reacting cones*: Cone-shaped puffs of NO_x and ROGs are rotated while they form ozone through a simplified nonlinear photochemical mechanism.

We based the evaluation and comparison of the schemes on the following performance measures:

1. *Peak ratio*: This ratio is a measure of the peak retention capability of an advection scheme. A value of 1.00 means that the scheme has perfect peak retention. The numerical diffusion introduced by some schemes may result in a peak clipping effect, resulting in a peak ratio that is smaller than unity. The ratio can also be larger than unity for some schemes that overshoot peak concentrations. The peak ratio is calculated here as

$$\frac{\max(c_i)}{\max(c_i^e)}, \qquad i = 1, 2, 3, \ldots, n \qquad (9.56)$$

where c_i^e is the exact solution for grid cell i, c_i is the predicted value, and n is the number of cells.

2. *Background-to-peak ratio*: In all tests, the background concentration is initialized to a nonzero value. If a scheme is not monotonic, it will create ripples near steep gradients. The ripples lead to predicted concentration values smaller than the background. To measure the magnitude of these ripples, the background-to-peak ratio is computed. The amplitude of the ripples is compared with the amplitude of the signal that should be resolved (e.g., peak concentration). If the ratio is negative, this means that the scheme is not positive definite. The perfect value for the background-to-peak ratio may vary from one test to another; we will state it for each case. The ratio is calculated as

$$\frac{\min(c_i)}{\max(c_i^e)} \qquad (9.57)$$

3. *Mass ratio*: This ratio is a measure of the mass conservation capability of the schemes. A value of 1.00 means that the scheme conserves mass. Nonconservative schemes may lose (mass ratio smaller than unity) or gain (mass ratio greater than unity) mass. Here, the mass ratio is calculated as

$$\frac{\sum c_i}{\sum c_i^e} \qquad (9.58)$$

4. *Distribution ratio*: This ratio is a measure of how well the mass is distributed in comparison to the exact solution. A value of 1.00

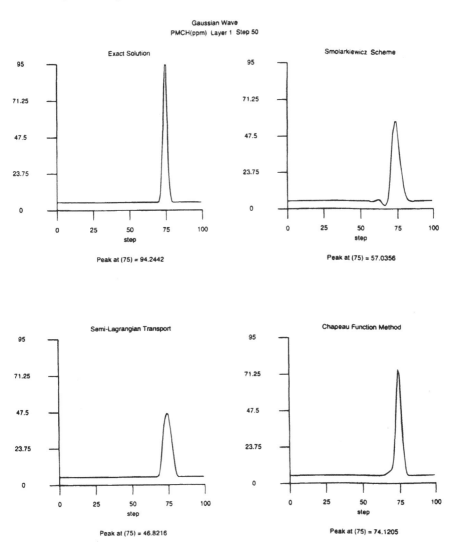

Figure 9.3 Solutions to the Gaussian wave test: exact solution (top left), SMO (top right), SLT (bottom left), and HAT (bottom right).

means that the scheme distributes the mass perfectly. The distribution ratio is calculated as

$$\frac{\sum c_i^2}{\sum c_i^{e2}} \tag{9.59}$$

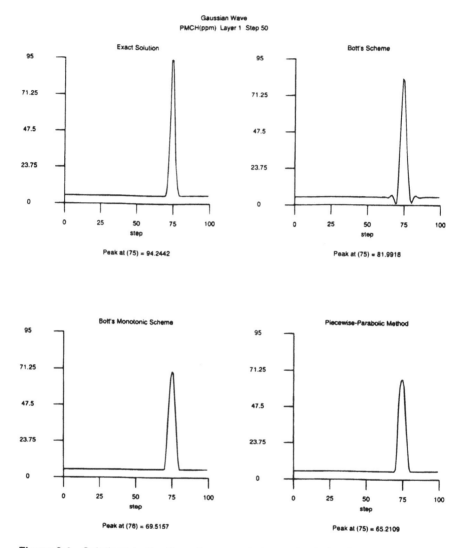

Figure 9.4 Solutions to the Gaussian wave test: exact solution (top left), BOT (top right), BOT-M (bottom left), and PPM (bottom right).

5. *Average absolute error*: This error is calculated as

$$\frac{1}{n} \sum |c_i - c_i^e| \tag{9.60}$$

and is an indicator of overall scheme performance.

6. *Root mean square (RMS) error*: This error is calculated as

$$\sqrt{\frac{1}{n} \sum \left(\frac{c_i - c_i^e}{c_i^e} \right)^2} \tag{9.61}$$

9.3.2 Evaluation Cases

Advection of One-Dimensional Pulses

In one-dimensional tests, pulses of various shapes are advected with constant velocity. Rectangular, triangular, and Gaussian-shaped pulses of inert species are very common (Muller, 1992). Usually, the one-dimensional tests cover a wide range of Courant numbers (a measure of how far the pulse will move in one time step compared to the grid size) for a full account of the stability and accuracy of the schemes. Traditional tests with long-wavelength pulses tend to show an advection scheme at its best. Adequate short-wavelength performance is also extremely important in photo-chemical models but is usually more difficult to obtain. Here, a Gaussian wave of wavelength equal to $8 \Delta x$ is advected from cell 25 to cell 75 of a 100-cell uniform grid.

Figures 9.3–9.5 show the shape of the Gaussian pulse after it has been advected a distance of $50\Delta x$ at a Courant number of 0.25 (i.e., after 200 time steps) as predicted by each scheme. Table 9.2 summarizes the values of various performance measures at the end of the test. The ASD and YAM schemes preserved the peak height very well. However, the distribution ratio has a lower value for YAM, indicating distortions of the pulse's shape. For the same reason, the average and RMS errors are larger than those of ASD. On all accounts, these two schemes perform much better than the

Table 9.2 Gaussian Wave Test

Performance measure	Scheme								
	ASD	BOT	BOT-M	FCT	HAT	PPM	SLT	SMO	YAM
Peak ratio	0.99	0.87	0.74	0.74	0.79	0.69	0.50	0.61	0.98
Background	0.05	0.01	0.05	0.05	0.05	0.05	0.05	0.02	0.05
Mass ratio	1.00	1.00	1.02	1.00	1.00	1.00	0.95	1.00	1.00
Distribution	0.99	0.93	0.83	0.81	0.77	0.79	0.52	0.66	0.92
Average error	0.08	0.87	1.38	0.98	1.01	1.16	2.13	2.21	0.51
RMS error	0.01	0.18	0.27	0.14	0.20	0.17	0.40	0.50	0.12

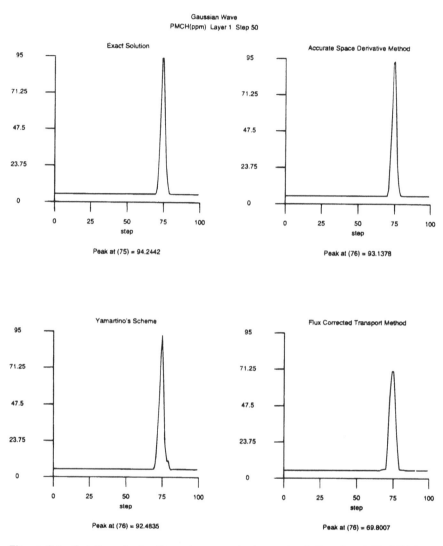

Figure 9.5 Solutions to the Gaussian wave test: exact solution (top left), ASD (top right), YAM (bottom left), and FCT (bottom right).

other schemes in this test. Bott's scheme (BOT) ranks third overall, but large ripples are observed at leading and trailing edges of the pulse as indicated by the values below the background (as much as 4% of the peak height). When the monotonic limiter is used (BOT-M), the ripples are

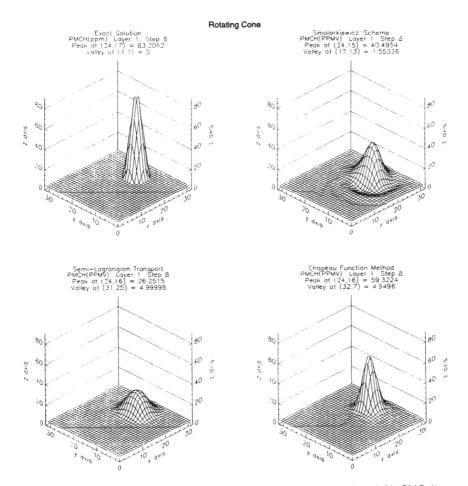

Figure 9.6 Solutions to the rotating cone test: exact solution (top left), SMO (top right), SLT (bottom left), and HAT (bottom right).

eliminated but the peak retention performance deteriorates. Also, a 2% increase in mass is observed. The FCT scheme has the same peak ratio as BOT-M while it conserves mass and leads to smaller errors than BOT-M. The chapeau function scheme (HAT) produces a higher peak than PPM and BOT-M, but the shape of the puff is not maintained as well; this is also indicated by a smaller distribution ratio. Smolarkiewicz's scheme (SMO) and SLT display poorer performance than other schemes. SMO produces ripples upwind from the pulse and leads to average and RMS errors larger than the SLT scheme's. However, 5% of the mass is lost with SLT and the peak retention is much worse than with SMO.

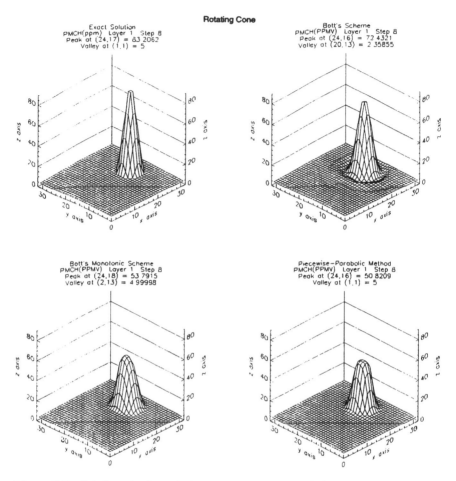

Figure 9.7 Solutions to the rotating cone test: exact solution (top left), BOT (top right), BOT-M (bottom left), and PPM (bottom right).

Rotating Cone Test

The rotating cone test is a routine way of measuring the long-wavelength advection fidelity of the schemes (Crowley, 1968; Chock, 1991; Odman and Russell, 1991b; Odman et al., 1995, 1996). In this test, a cone-shaped puff is introduced into a rotational flow field. The exact solution to this problem is a rigid-body rotation of the puff without any change to its original conical shape. Various errors are revealed in this test. For example, the numerical diffusion (or dissipation) error manifests itself in the drop of the peak height during rotation. Also, by observing the location of the peak, one can determine the leading or lagging phase-speed errors.

A 32×32 grid is used for this test (i.e., $-16\Delta x \leq x \leq +16\Delta x$; $-16\Delta y \leq y \leq +16\Delta y$; $\Delta x = \Delta y$). A cone-shaped puff with peak concentration equal to 100 ppm and a base radius of $4\Delta x$ is initialized such that its peak is located at $[+8\Delta x, 0]$. Note that the peak is not initially at a grid-cell center but at a cell corner (i.e., there are four cells around the peak with the same average concentration). The background concentration is set equal to 5 ppm. To obtain a counterclockwise rotation around an axis passing through the center of the domain, the wind field is defined as follows:

$$u = -\omega y, \qquad v = \omega x \qquad\qquad (9.62)$$

The angular velocity, ω, is adjusted so that the Courant number is approximately 0.28 at the peak of the puff.

Figures 9.6–9.8 show the solution after two full rotations as predicted by each scheme. The exact solution, which is nothing but the initial cone, is also shown. ASD and YAM maintain the peak height and the overall shape of the cone better than the others. BOT performs third best in this test, but it yields values below the background (as seen in the ring-shaped valley at the base of the cone). BOT-M, HAT, and PPM predict similar peak heights (65%, 70%, and 60%, respectively) but the shape distortions look very different in each case. PPM has the worst peak-clipping effect but the resulting shape has the smallest base span among the three schemes. SLT and SMO are clearly the most diffusive schemes; SMO also introduces a ripple upwind from the cone. Table 9.3 summarizes the performance measures at the end of two rotations. Since there is no shear in the flow field, BOT-M and the two-dimensional version of Bott's scheme (BOT2D) (Bott, 1993)

Table 9.3 Rotating Cone Test

Performance measure	Scheme								
	ASD	BOT	BOT-M	FCT	HAT	PPM	SLT	SMO	YAM
Peak ratio	0.99	0.87	0.65	0.69	0.71	0.61	0.32	0.49	0.99
Background	0.06	0.03	0.06	0.06	0.06	0.06	0.06	0.02	0.06
Mass ratio	1.00	1.00	1.02	1.00	1.00	1.00	0.94	1.00	1.00
Distribution	0.96	0.93	0.83	0.80	0.70	0.78	0.45	0.64	0.91
Average error	0.18	0.46	0.76	0.48	0.79	0.54	1.31	1.60	0.33
RMS error	0.05	0.16	0.30	0.14	0.22	0.18	0.31	0.51	0.13

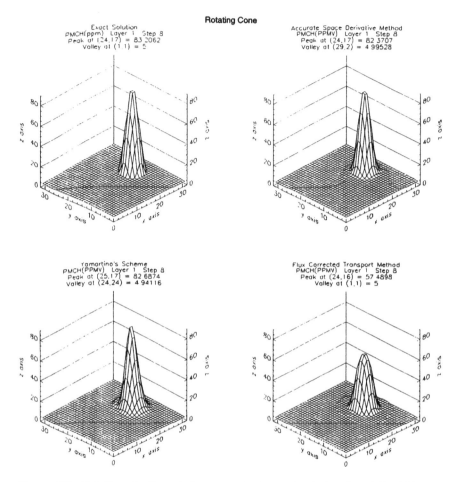

Figure 9.8 Solutions to the rotating cone test: exact solution (top left), ASD (top right), YAM (bottom left), and FCT (bottom right).

produce identical results. The mass conservation problem with these two schemes is revealed once again. SLT does not conserve mass either, with 6% of the mass lost. BOT preserves 87% of the peak height (third best after ASD and YAM), but it leads to ripples with an amplitude of 3% of the original peak height. FCT solution is free from ripples while the average and RMS errors are of the same order as BOT, but it has a peak ratio of only 69%. The performance of BOT-M, HAT, and PPM are comparable with HAT, predicting the highest peak but producing the lowest distribution ratio, and PPM gives the smallest average and RMS errors among the

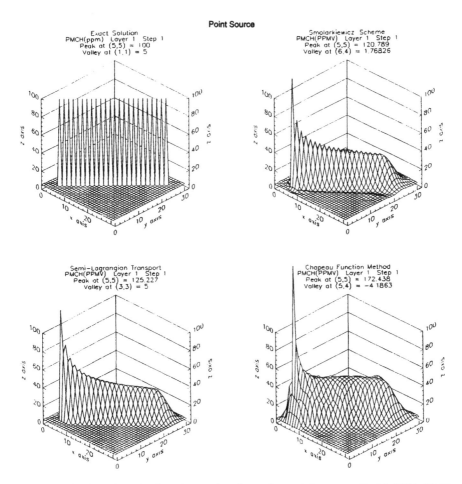

Figure 9.9 Solutions to the skew advection of a point-source plume test: exact solution (top left), SMO (top right), SLT (bottom left), and HAT (bottom right).

three. Notice that the comparison results obtained from this test are very similar to those of the Gaussian wave test above.

Skew Advection of a Point-Source Plume

Advection of a point-source plume is a test problem specifically designed to measure the short-wavelength performance of the schemes (Yamartino, 1993; Odman et al., 1995). In photochemical modeling, there is often a need to simulate the transport of plumes emitted into a single-grid cell. In this

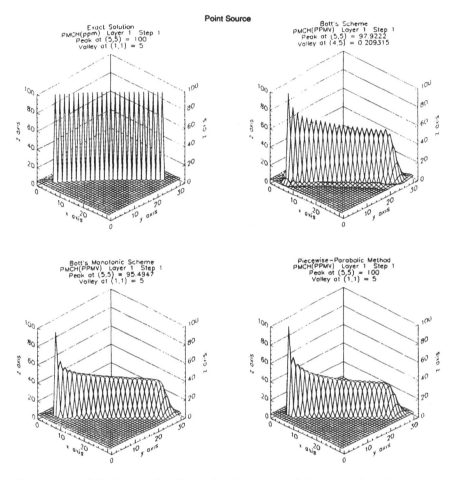

Figure 9.10 Solutions to the skew advection of a point-source plume test: exact solution (top left), BOT (top right), BOT-M (bottom left), and PPM (bottom right).

test, a point source located in a single-grid cell (cell [5,5] of a 32 × 32 grid) emits at a constant rate and the winds advect the plume diagonally (i.e., skew advection). The emission rate is adjusted to make up for the advected mass as predicted from the analytical solution, so that the concentration in the source cell remains constant at 100 ppm. All other cells are initialized with a background concentration of 5 ppm. The wind field is defined such that the plume is advected along the diagonal of the domain with a Courant number of 0.17 (0.12 in each coordinate direction). Ideally, the concentration along the diagonal should be constant and equal to the concentration in the source cell (i.e., 100 ppm). This stringent test also reveals how much crosswind diffusion is introduced by the schemes.

The plume as predicted by each scheme is shown in Figs. 9.9–9.11 at the instant when it reaches cell [28,28] according to the exact solution. A mass conservation problem is evident in the HAT solution. This was related to a smoothing treatment of the concentrations near the boundaries. After correcting this problem, however, we were still unable to get a stable solution from the HAT scheme even after trying various filtering parameters. Table 9.4 summarizes the results of this test for all the other schemes. It

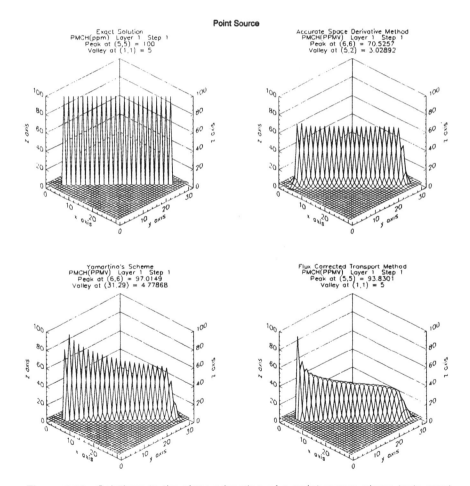

Figure 9.11 Solutions to the skew advection of a point-source plume test: exact solution (top left), ASD (top right), YAM (bottom left), and FCT (bottom right).

Table 9.4 Skew Advection of a Point-Source Plume

Performance measure	Scheme							
	ASD	BOT	BOT-M	FCT	PPM	SLT	SMO	YAM
Peak	0.71	0.98	0.95	0.94	1.00	1.25	1.21	0.97
Background	0.03	0.00	0.05	0.05	0.05	0.05	0.02	0.05
Mass	0.98	1.00	1.04	1.00	1.00	1.16	1.00	1.00
Distribution	0.62	0.72	0.62	0.58	0.57	0.79	0.55	0.72
Average error	1.81	2.51	2.98	2.91	3.10	4.30	3.64	1.39
RMS error	0.47	0.87	1.13	1.01	1.08	1.54	1.05	0.49
Cell [15,15]	0.68	0.62	0.46	0.45	0.42	0.46	0.44	0.72
Cell [25,25]	0.67	0.56	0.43	0.39	0.38	0.38	0.35	0.62

also gives the ratio of predicted solutions to exact solution at cells [15,15] and [25,25], which are approximately $14\Delta x$ and $28\Delta x$ downwind from the source. Note that, as in the rotating cone test, there is no shear in the flow field, so the results for BOT-M and BOT2D are the same and only BOT-M is shown.

In this test, YAM performs better than ASD due to its special short-wavelength capturing features. ASD underpredicts the peak height near the source, but further downwind the peak retention becomes better than all other schemes. At cell [15,15], ASD's peak ratio is second only to YAM's, but it is the highest (67%) at cell [25,25]. Mass conservation problems are seen with ASD (2% loss), BOT-M (4% increase), and SLT (16% increase). The average and RMS errors are smallest for YAM and ASD. SLT has the highest distribution ratio, but this is due to the added mass, so judging its performance based on this measure alone would be misleading. Also, note that the errors for SLT are larger than for any other scheme and that the peak is overpredicted by 25%. SMO also overpredicts the peak near the source by 21% and leads to the lowest concentration values further downwind. The distribution ratio is low and the average error is high due to this incorrect distribution of the mass in the plume. BOT predicts a peak concentration higher than both ASD's and YAM's near the source, but the values are much smaller downwind. Also, a ripple with an amplitude equal to 5% of the exact plume height is produced upwind from the source. However, the concentrations in this ripple do not go below zero due to the positive-definite nature of BOT. SMO produces a ripple of smaller magnitude (the amplitude is only 3% of the exact peak concentration). BOT-M does not lead to ripples, but achieves this at the expense of added artificial diffusion as seen in the reduced values of concentrations downwind. The

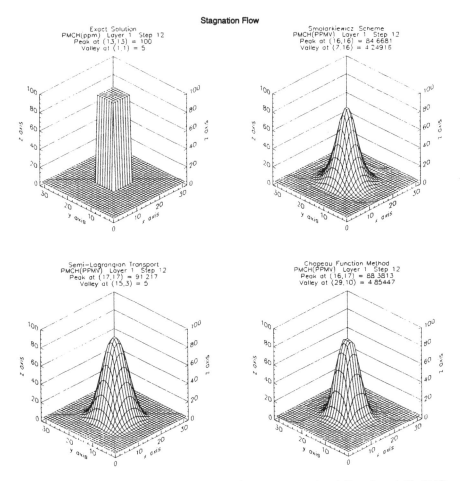

Figure 9.12 Solutions to the stagnation flow test: exact solution (top left), SMO (top right), SLT (bottom left), and HAT (bottom right).

performance of FCT is very close to BOT-M in this test. PPM performed sixth best, but it is the only scheme with perfect prediction of the peak concentration at the source.

Shear Flow Tests

The variability of wind speed in actual wind fields is another factor that may affect the performance of advection schemes. Staniforth et al. (1987) and Odman and Russell (1993) designed robust test cases and demonstrated that some advection schemes may perform poorly under more realistic

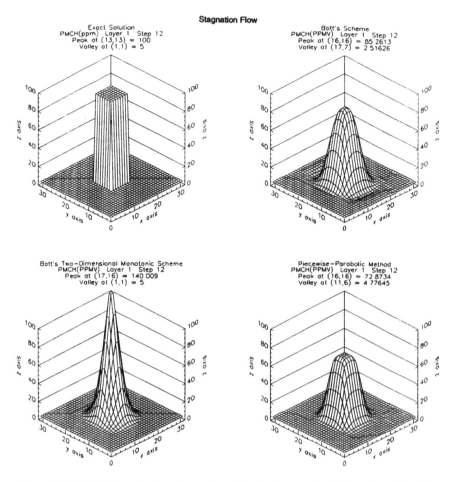

Figure 9.13 Solutions to the stagnation flow test: exact solution (top left), BOT (top right), BOT2D (bottom left), and PPM (bottom right).

meteorological conditions. Here, the effects of wind shear are simulated as follows: A square-shaped puff is released in a stagnation flow field, and the field is reversed (or rotated by 90°) each time the puff reaches an aspect ratio of 2:1. In other words, the velocity field is

$$u = \omega x, \qquad v = -\omega y \tag{9.63}$$

during the first and last quarters of a cycle, and

$$u = -\omega x, \qquad v = \omega y \tag{9.64}$$

during the second and third quarters such that the puff is first stretched in the x-direction, then in the y-direction and finally in the x-direction until the original square shape is exactly restored at the end of each cycle. The solution at intermediate stages can be found in Seibert and Morariu (1991) or Fogelson (1992).

The solutions to the stagnation flow problem as predicted by each scheme after three cycles are shown in Figs. 9.12–9.14 and the performance measures are listed in Table 9.5. In this test, the numerical diffusion of most schemes leads to mass loss through the boundaries when the puff is fully

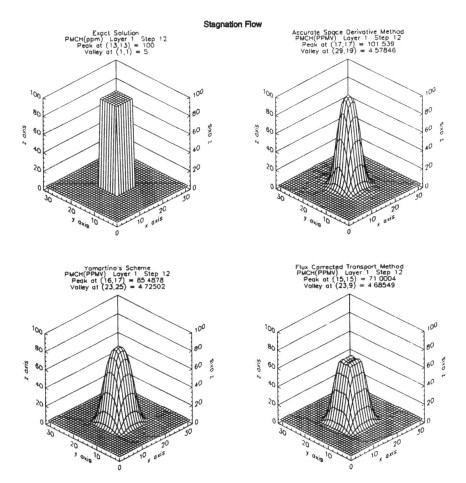

Figure 9.14 Solutions to the stagnation flow test: exact solution (top left), ASD (top right), YAM (bottom left), and FCT (bottom right).

Table 9.5 Stagnation Flow Test

Performance measure	Scheme								
	ASD	BOT	BOT2D	FCT	HAT	PPM	SLT	SMO	YAM
Peak	1.02	0.85	1.40	0.71	0.88	0.73	0.91	0.85	0.85
Background	0.05	0.03	0.05	0.05	0.05	0.05	0.05	0.04	0.05
Mass	0.93	1.00	1.00	0.97	0.96	0.97	1.12	0.94	0.96
Distribution	0.58	0.61	0.74	0.54	0.54	0.53	0.71	0.46	0.58
Average error	3.09	4.34	3.77	4.31	3.98	4.30	4.98	4.96	3.68
RMS error	0.77	1.46	1.11	1.41	1.09	1.34	1.80	1.23	1.24

stretched (aspect ratio of $2:1$). Due to this mass loss, some performance measures (e.g., mass ratio, distribution ratio, errors) may be misleading and should be used carefully in the evaluation. ASD slightly overpredicts the peak (2%), some ripples are observed in the background (Fig. 9.14), and 7% of the mass is lost. However, the average and RMS errors are smaller than with other schemes and the shape of the puff is in relatively good agreement with the exact solution. The performances of YAM and HAT are very close, but we rank HAT the second best performer in this test. Only 12% of the peak height and 4% of the mass are lost. YAM underpredicts the peak height by 15%, and loses 4% of the mass through the boundaries. Its average error is smaller than HAT's, but its RMS error is larger. BOT is the fourth best performer, followed by PPM. BOT introduces ripples with amplitudes as much as 2% of the peak height, and PPM loses 27% of the peak height. The performance of FCT is almost identical to PPM's. The solutions of BOT2D, SLT, and SMO do not compare well to the exact solution. BOT2D overpredicts the peak height by 40%, SLT increases the mass by 12%, and the SMO solution looks very diffusive.

Rotation of Reactive Cones

The advection tests described so far investigate certain measures such as the degree of preservation of the peaks and the generation of ripples. There are some properties of transport schemes, however, that an advection test alone may not reveal. The nonlinear chemistry alters the shape of a pollutant puff constantly; it can change its slope, making it less or more steep, or even invert a peak. The ability of the scheme to adapt to such changes can be seen and evaluated only in tests with chemistry. Hov et al. (1989) demonstrated that the aliasing errors inherent to the highly accurate pseudospectral scheme may be amplified in the presence of nonlinear chemistry. Odman and Russell (1991b), in a similar test, showed that sacrificing mass

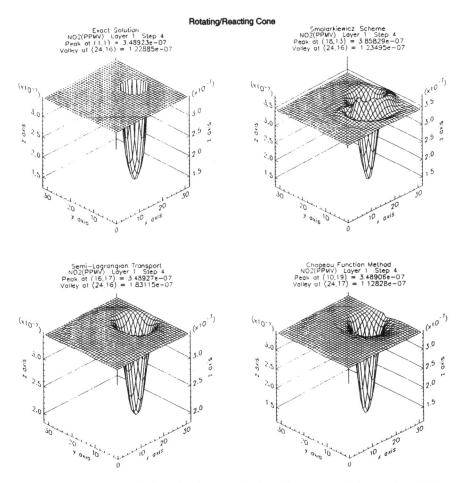

Figure 9.15 NO$_2$ predictions in the rotating/reacting cones test: exact solution (top left), SMO (top right), SLT (bottom left), and HAT (bottom right).

conservation may have important unfavorable consequences in modeling regional oxidant formation.

For testing purposes, the chemical mechanism is kept simple; this way, sound physical arguments can be made more readily. However, the mechanism still yields the kind of numerical difficulties encountered in photochemical models. One such description of the photochemistry was used by Hov et al. (1989) and later by Odman and Russell (1991b) and Chock and Winkler (1994a, b). The test consists of advecting puffs of different species while they react according to the simple chemical mechanism. The solution

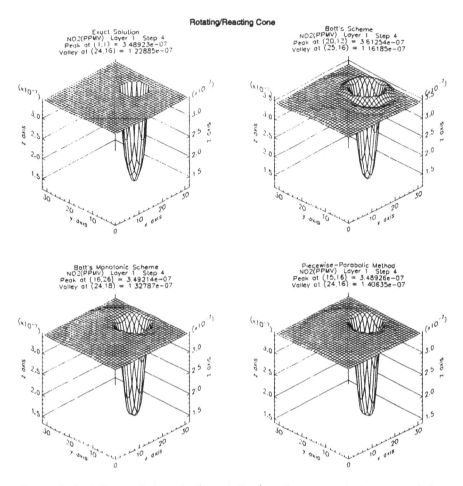

Figure 9.16 NO$_2$ predictions in the rotating/reacting cones test: exact solution (top left), BOT (top right), BOT-M (bottom left), and PPM (bottom right).

to the advection part is known: the concentration field is simply rotated. On the other hand, if no advection is taking place but the chemistry is activated, the concentrations will change and their values may be determined numerically by a very accurate method such as the Gear solver. If both advection and chemistry are activated, then the solution is simply a rotation of the fields one gets from performing the chemical integration. This solution (numerical chemistry and exact advection) is compared to the predictions from the advection schemes followed by chemistry. Four species (HC, HCHO, NO, and NO$_2$) are initialized as cones with the same geometry as in the rotating cone test above. The initial peak heights and background concentrations can be found in Odman and Russell (1991b).

Since the shapes of different pollutant puffs are altered differently by the chemistry, the errors are usually not the same for all species. In this simple mechanism, the errors for species such as NO, NO_2, HCHO, and O_3 are usually more pronounced. This test is very reliable in studying how the errors produced by the advection schemes would propagate in the model. The results for NO_2 and O_3 after 48 h of simulation are shown in Figs. 9.15–9.20, and some performance measured for HC, NO_2, and O_3 are summarized in Table 9.6. Note that all schemes with the exception of SMO and

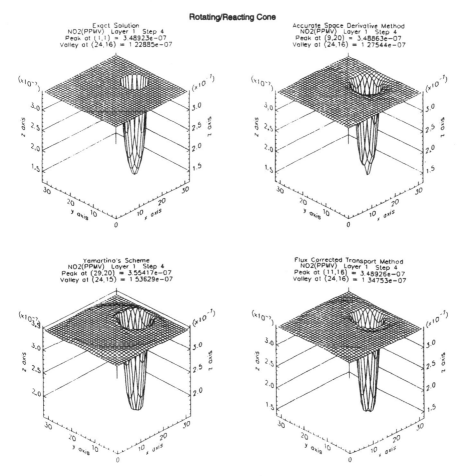

Figure 9.17 NO_2 predictions in the rotating/reacting cones test: exact solution (top left), ASD (top right), YAM (bottom left), and FCT (bottom right).

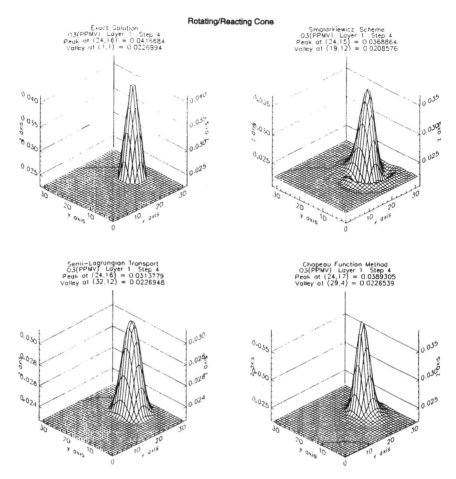

Figure 9.18 O₃ predictions in the rotating/reacting cones test: exact solution (top left), SMO (top right), SLT (bottom left), and HAT (bottom right).

SLT predict the HC peak within a few percent. BOT overpredicts this peak by 2.5% while SMO and SLT underpredict it by 7% and 9%, respectively. Also, the RMS error for SMO and SLT are the largest. ASD, YAM, and FCT perform better than other schemes in following the HC puff.

The NO₂ valley is best predicted by SMO (100%) but the shape of the solution in Fig. 9.15 and the large RMS error show that this may be fortuitous. The shape is severely distorted near the background due to the amplification of ripples in the presence of nonlinear chemistry. BOT produces similar ripples but they are not as amplified. In fact, the smallest

Rotating/Reacting Cone

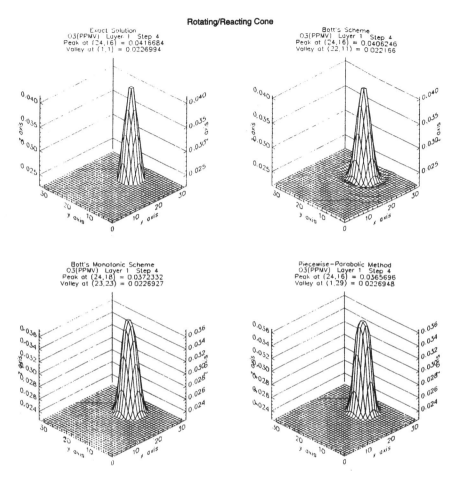

Figure 9.19 O_3 predictions in the rotating/reacting cones test: exact solution (top left), BOT (top right), BOT-M (bottom left), and PPM (bottom right).

RMS error and only a 5% underprediction of the valley may be signs that BOT's solution for NO_2 is the best among all the schemes. ASD over-predicts the NO_2 valley by approximately 4% but there is a slight shape distortion in Fig. 9.17. This distortion may be the reason for an RMS error slightly larger than BOT's. FCT overpredicts the valley by approximately 10% while the RMS error is smaller than ASD's. Other schemes perform reasonably, however YAM's overprediction of the valley by 25% is of concern. SLT overpredicts the valley by 49% and leads to the highest RMS error (7%). The poor performance of many schemes in this case may be due

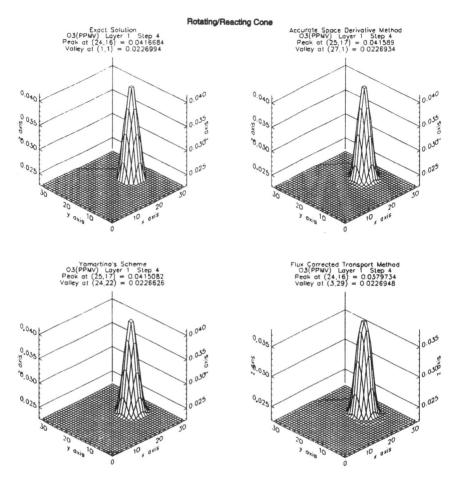

Figure 9.20 O_3 predictions in the rotating/reacting cones test: exact solution (top left), ASD (top right), YAM (bottom left), and FCT (bottom right).

to the extremely small magnitude of NO_2 concentrations after 48 h. It is important to note that one snapshot at the end of a 48-h simulation does not fully reflect how well the schemes followed NO_2 as ozone was being formed.

The best performance in predicting the O_3 peak is displayed by ASD and YAM. The RMS errors for these two schemes are also much smaller than the other schemes. The third best performer appears to be BOT, despite some ring-shaped ripples around the ozone puff. Recall that ASD and BOT also predicted both HC and NO_2 very well. The predictions for

Table 9.6   Rotating/Reacting Cones Test

	HC		NO$_2$		O$_3$	
Scheme	Peak (%)	RMS (%)	Valley (%)	RMS (%)	Peak (%)	RMS (%)
ASD	98.7	1.91	103.8	2.59	99.8	0.51
BOT	102.4	2.90	94.6	2.00	97.5	0.92
BOT-M	98.2	3.26	108.1	2.58	89.4	1.70
FCT	98.9	2.44	109.7	2.22	91.1	0.96
HAT	98.5	3.89	91.8	3.88	93.4	1.79
PPM	96.5	2.60	114.4	2.25	87.8	1.21
SLT	90.9	5.00	149.0	7.10	75.3	3.20
SMO	92.9	5.56	100.5	4.93	88.5	3.56
YAM	98.7	2.21	125.0	2.80	99.6	0.62

these ozone precursors are equally as important as the predictions of ozone concentrations. YAM's NO$_2$ predictions at the end of the 48-h simulation were not as good, but (as suggested earlier) this was probably due to the small magnitude of NO$_2$ at the end of the test; the predictions were much better when ozone was being formed. The next four best performers are FCT, BOT-M, HAT, and PPM, respectively. SMO predicts the O$_3$ peak very well but ripples in precursor fields also lead to large ripples in the O$_3$ field. For this reason, SMO's RMS error is large. SLT is the worst performer in predicting ozone, as expected based on its poor performance in predicting the precursors.

9.3.3   Results and Discussion

Eight advection schemes (not counting the variations of Bott scheme) were compared using test cases ranging from very simple one-dimensional advection to more robust two-dimensional shear and chemically reactive flows. Most schemes performed consistently in all the tests, but a few failed in some stressful tests. The differences between performance were more pronounced in some tests, and the ranking differed from test to test. However, important properties of the schemes were identified as a result of the diversity of the tests. This preliminary testing of the schemes resulted in the following findings:

- ASD has very high accuracy except for the smallest (i.e., $2\Delta x$) wavelengths, but even there the accuracy was still better than with most

other schemes. ASD is not strictly mass conservative and can cause
a slight mass imbalance. It may also lead to mild ripples and overshots (i.e., it is not monotonic), but usually yields a positive definite
result.

- BOT is highly accurate and mass conservative. It may create ripples
  and overshots, but it is positive definite. The phase-speed errors are
  usually small and the ripples do not seem to grow under the influence of nonlinear chemistry.
- BOT-M and BOT2D significantly reduce the accuracy of the original scheme. A mass conservation problem was discovered in the
  formulation of these schemes. The monotonic flux limitation artifi-
  cially adds mass downwind from steep negative gradients. This
  problem was also confirmed by Bott (personal communication,
  1995).
- FCT is fairly accurate and mass conservative. It does not create any
  ripples or overshots, but does so at the expense of diffusion to the
  background.
- HAT has fair accuracy and is mass conservative. However, it may
  lead to ripples that can grow and cause instabilities, especially near
  point sources close to the domain boundaries. Its performance is
  unpredictable under strong shear flow situations. The accuracy may
  be reduced under constant convergence or divergence, but a convergence followed by divergence, or vice versa, usually leads to high
  accuracy.
- PPM also has fair accuracy and is strictly mass conservative. It is
  monotonic, so it does not lead to ripples or overshots under any
  circumstances. While the diffusion to the background is very small,
  peak clipping can be significant. The steepening operator of PPM
  seems to transform even mild gradients to shock waves.
- SLT has poor accuracy and severe mass conservation problems.
  The only positive feature of SLT is that it does not lead to ripples.
  Though it was not obvious from the test here, it can conceptually
  withstand shear situations better than the schemes that split horizontal diffusion in two one-dimensional operators.
- SMO has relatively low accuracy. It is mass conservative but it may
  lead to ripples in concentration fields. Although the ripples do not
  cause negative concentrations (i.e., SMO is positive definite), they
  are associated with relatively large phase-speed errors. Also, in the
  presence of nonlinear chemistry, the ripples may be amplified.
- YAM has very high accuracy, even for the shortest wavelengths,
  and is mass conservative. It can lead to overshots under certain
  situations. YAM can also create mild ripples but it is positive defi-

Table 9.7 Computational Performance

	ASD	BOT	BOT-M	FCT	HAT	PPM	SLT	SMO	YAM
CPU time (Norm = SMO)	4.38	0.85	1.23	1.45	1.17	1.24	1.25	1.00	2.25

nite. Its performance in following point-source plumes is much better than that of the other schemes.

In addition to the properties discussed above, the computational performance of the schemes was also considered in the ranking. The CPU times for all the tests were averaged and normalized with respect to SMO (Table 9.7). The only scheme that is less CPU intensive than SMO is BOT; other schemes require more CPU time. ASD is the most CPU-intensive scheme as it requires more than four times more CPU than SMO does. However, the CPU time spent during horizontal advection is usually a small fraction of a photochemical model's total CPU time (most of the CPU time is spent in chemistry). Therefore, even with ASD, an air quality model's total CPU time would only increase by a fraction (most likely less than 50%).

9.4 EVALUATION OF ADVECTION SCHEMES IN AN AIR QUALITY MODEL AND THE IMPACT ON PREDICTIONS

The above tests show that in the cases studied, some, if not marked, differences are found in the results from the various methods. However, those cases have much more severe gradients than one might expect in an actual application. A logical question then is: will real and significant differences still be found in an actual case? This is tested by implementing some of the methods in an air quality model and applying the model to a typical situation. The application was to the Los Angeles, California, area. A three day smog event, August 27–29, 1987, was simulated. This period was from the Southern California Air Quality Study (SCAQS), a large field program designed to develop the data necessary to conduct more comprehensive air quality modeling. This is one of the SCAQS intensive monitoring periods and the episode has been extensively used for modeling purposes (SCAQMD, 1990; Harley et al., 1993). One of the most widely used air quality models is UAM-IV, and it was used here for the tests. Four schemes were tested: SMO, BOT, YAM, and ASD.

It should be noted that there is a major difference between air quality modeling and most other CFD applications as the solution of the advection portion of the problem takes a small portion of the total computational time. For example, in a current application, only about 10% of the total CPU time is spent solving the equations discussed above. Most of the time is spent solving the equations governing the chemical dynamics. Thus, using an advection solver that takes twice as long and is only slightly more accurate may be attractive since the total CPU time increases only about 10%. In light of the findings above, it is thus apparent that ASD is the only method that would appreciably change the required computational time in the applications below.

9.4.1 Tracer Experiments

Simulating a tracer experiment is another test that allows more realistic evaluations (Brost et al., 1988; Odman et al., 1995). All the test cases in section 9.3 are simple enough to have an analytical solution. They do not provide sufficient information about the behavior of the schemes under real atmospheric conditions. In actual simulations, there can be various factors not accounted for in simple test cases (e.g., variable winds, wide range of Courant numbers). Therefore, there is a need to evaluate the performance of the schemes, at least comparatively, in more realistic simulations. To simulate a tracer experiment, all the transport processes of the model must be activated. Evaluating the uncertainty due to the solution of horizontal advection is difficult because of all the other sources of uncertainties involved. In general, uncertainties are introduced by the prognostic or diagnostic wind fields, the computation of eddy diffusivities, and the tracer data itself. However, the test may be viable for a comparison of different schemes assuming that other uncertainties would affect the performance of different schemes in the same way.

Odman et al. (1996) simulated a tracer experiment from SCAQS using ASD, BOT, YAM, and SMO schemes. As expected, ASD, BOT, and YAM predicted a less diffused puff with a higher peak concentration than SMO. The trajectories of the plumes were very similar and the peaks were in the same grid cell. YAM predicted the highest peak concentration, as expected based on our experience with the skew advection of a point-source plume test. The second highest peak was predicted by BOT followed by ASD. Recall that the peak ratio for BOT was much higher that ASD, near the source, in the skew advection of a point source plume test. The puff did not travel a long enough distance in this experiment for ASD to show its performance in preserving the peaks farther downwind.

Table 9.8 Performance Evaluation of Ozone Predictions versus Observations for UAM Using the ASD, BOT, YAM, and SMO Advection Schemes

Metric	ASD	BOT	YAM	SMO
Peak (paired in time, space) [%]	−23.6	−24.0	−25.9	−31.4
Peak (paired in space) [%]	−23.6	−24.0	−25.9	−31.4
Peak (paired in time) [%]	−3.11	6.51	0.23	12.5
Peak (unpaired) [%]	22.0	21.8	18.0	27.4
Normalized bias [%]	14.4	6.60	8.93	8.43
Mean bias [pphm]	1.11	0.50	0.72	0.65
Normalized error [%]	47.3	49.8	49.3	50.76
Mean error [pphm]	3.56	3.73	3.70	3.80
Variance [%]	21.7	24.4	23.8	25.37

9.4.2 Evaluation of Advection Schemes with Full Chemistry

While the tracer experiment is one indication of how well a solver may work, the true test is one in which the conditions (such as meteorology and emissions) are similar to those in which it will be used to determine control strategies—that is, the set of emissions controls that an area should apply to reach attainment with the ambient air quality standards. Here, we investigate not only how closely the various methods compare in terms of the predictions to the base case (i.e., without emissions reductions) but also if the response to emissions controls vary with the choice of solver. If a bias is found in the response to controls, this would raise questions as to how confidently a method could be used.

Table 9.9 Performance Evaluation of Nitrogen Dioxide Predictions versus Observations for UAM Using the ASD, BOT, YAM, and SMO Advection Schemes

Metric	ASD	BOT	YAM	SMO
Peak (paired in time, (space) [%]	−68.1	−61.0	−65.0	−63.7
Peak (paired in space) [%]	−40.9	−31.8	−36.3	−22.5
Peak (paired in time) [%]	−19.1	−17.1	−20.4	−21.1
Peak (unpaired) [%]	4.11	10.1	8.02	21.1
Normalized bias [%]	−2.31	−6.2	−5.89	−5.19
Mean bias [pphm]	−0.71	−0.80	−0.79	−0.79
Normalized error [%]	40.4	38.9	39.0	39.7
Mean error [pphm]	1.94	1.92	1.93	1.94
Variance [%]	7.09	6.44	6.57	6.50

Table 9.10 Correlations of O_3 Predictions for Base Case and 50% ROG Reductions

Base Case			Independent (x)			
$y = mx + b$			ASD	BOT	SMO	YAM
d	ASD	m	1.000	0.973	0.954	0.982
e		b	0.000	0.427	0.447	0.306
p		R^2	1.000	0.951	0.951	0.967
e	BOT	m	0.978	1.000	0.974	0.996
n		b	−0.078	0.000	0.065	−0.030
d		R^2	0.951	1.000	0.987	0.989
e	SMO	m	0.997	1.014	1.000	1.011
n		b	−0.101	0.025	0.000	−0.014
t		R^2	0.951	0.987	1.000	0.979
(y)	YAM	m	0.984	0.993	0.968	1.000
		b	−0.066	0.107	0.164	0.000
		R^2	0.967	0.989	0.979	1.000

50% ROG			Independent (x)			
$y = mx + b$			ASD	BOT	SMO	YAM
d	ASD	m	1.000	1.001	0.943	1.015
e		b	0.000	0.268	0.463	0.128
p		R^2	1.000	0.890	0.898	0.922
e	BOT	m	0.889	1.000	0.929	0.986
n		b	0.378	0.000	0.273	0.020
d		R^2	0.890	1.000	0.980	0.979
e	SMO	m	0.952	1.055	1.000	1.043
n		b	0.145	−0.174	0.000	−0.172
t		R^2	0.898	0.980	1.000	0.966
(y)	YAM	m	0.908	0.993	0.926	1.000
		b	0.325	0.098	0.352	0.000
		R^2	0.922	0.979	0.966	1.000

Table 9.10 *Continued.*

50% NO$_x$			Independent (x)			
$y = mx + b$			ASD	BOT	SMO	YAM
d	ASD	m	1.000	0.962	0.960	0.970
e		b	0.000	0.356	0.324	0.259
p		R^2	1.000	0.966	0.962	0.975
e	BOT	m	1.003	1.000	0.994	0.999
n		b	−0.121	0.000	−0.012	−0.041
d		R^2	0.966	1.000	0.990	0.992
e	SMO	m	1.002	0.996	1.000	0.994
n		b	−0.061	0.081	0.000	0.041
t		R^2	0.962	0.990	1.000	0.982
(y)	YAM	m	1.005	0.993	0.987	1.000
		b	−0.086	0.094	0.083	0.000
		R^2	0.975	0.992	0.982	1.000

50% ROG/NO$_x$			Independent (x)			
$y = mx + b$			ASD	BOT	SMO	YAM
d	ASD	m	1.000	0.964	0.953	0.975
e		b	0.000	0.360	0.360	0.244
p		R^2	1.000	0.948	0.946	0.963
e	BOT	m	0.984	1.000	0.983	0.998
n		b	−0.029	0.000	0.033	−0.036
d		R^2	0.948	1.000	0.987	0.990
e	SMO	m	0.993	1.004	1.000	1.002
n		b	−0.016	0.047	0.000	0.006
t		R^2	0.946	0.987	1.000	0.978
(y)	YAM	m	0.988	0.991	0.975	1.000
		b	−0.007	0.106	0.134	0.000
		R^2	0.963	0.990	0.978	1.000

Table 9.11 Correlations of NO$_2$ Predictions for Base Case and 50% ROG Reductions

	Base Case		Independent (x)			
	$y=mx+b$		ASD	BOT	SMO	YAM
d	ASD	m	1.000	0.917	0.926	0.940
e		b	0.000	0.066	0.059	0.055
p		R^2	1.000	0.963	0.964	0.974
e	BOT	m	1.050	1.000	1.001	1.015
n		b	−0.042	0.000	−0.001	−0.003
d		R^2	0.963	1.000	0.985	0.991
e	SMO	m	1.041	0.984	1.000	0.999
n		b	−0.035	0.012	0.000	0.008
t		R^2	0.964	0.985	1.000	0.978
(y)	YAM	m	1.036	0.976	0.979	1.000
		b	−0.039	0.010	0.008	0.000
		R^2	0.974	0.991	0.978	1.000

	50% ROG		Independent (x)			
	$y=mx+b$		ASD	BOT	SMO	YAM
d	ASD	m	1.000	0.928	0.936	0.938
e		b	0.000	0.076	0.067	0.070
p		R^2	1.000	0.968	0.969	0.975
e	BOT	m	1.042	1.000	1.002	1.003
n		b	−0.051	0.000	−0.003	0.000
d		R^2	0.968	1.000	0.989	0.993
e	SMO	m	1.036	0.987	1.000	0.990
n		b	−0.044	0.013	0.000	0.013
t		R^2	0.969	0.989	1.000	0.981
(y)	YAM	m	1.039	0.989	0.991	1.000
		b	−0.051	0.006	0.003	0.000
		R^2	0.975	0.993	0.981	1.000

Table 9.11 *Continued.*

50% NO_x			Independent (x)			
$y = mx + b$			ASD	BOT	SMO	YAM
d	ASD	m	1.000	0.873	0.886	0.917
e		b	0.000	0.039	0.034	0.028
p		R^2	1.000	0.947	0.951	0.965
e	BOT	m	1.084	1.000	1.006	1.032
n		b	-0.026	0.000	-0.002	-0.007
d		R^2	0.947	1.000	0.986	0.985
e	SMO	m	1.072	0.980	1.000	1.012
n		b	-0.022	0.006	0.000	0.000
t		R^2	0.951	0.986	1.000	0.973
(y)	YAM	m	1.053	0.955	0.961	1.000
		b	-0.020	0.011	0.008	0.000
		R^2	0.965	0.985	0.973	1.000

50% ROG/NO_x			Independent (x)			
$y = mx + b$			ASD	BOT	SMO	YAM
d	ASD	m	1.000	0.894	0.905	0.925
e		b	0.000	0.040	0.035	0.031
p		R^2	1.000	0.960	0.960	0.973
e	BOT	m	1.074	1.000	1.005	1.022
n		b	-0.028	0.000	-0.002	-0.005
d		R^2	0.960	1.000	0.984	0.988
e	SMO	m	1.060	0.979	1.000	1.002
n		b	-0.023	0.008	0.000	0.002
t		R^2	0.960	0.984	1.000	0.974
(y)	YAM	m	1.051	0.967	0.972	1.000
		b	-0.023	0.009	0.007	0.000
		R^2	0.973	0.988	0.974	1.000

Table 9.12   Ozone Peak Predictions and Relative Reductions for Three Control
Strategies

	Base	50% ROG	Base–50% ROG	50% NO_x	Base–50% NO_x	50% ROG/NO_x	Base–50% ROG/NO_x
SMO	36.9	21.5	15.4	39.5	−2.6	29.1	7.8
ASD	35.4	20.3	15.1	36.5	−1.1	27.8	7.6
BOT	35.3	19.1	16.2	38.4	−3.1	27.5	7.8
YAM	34.2	18.1	16.1	38.4	−4.2	28.0	6.2

The results from applying the model for both the base case inventory and the control strategy inventories using the four advection schemes are given in Tables 9.8–9.12. As can be seen, the three more accurate methods (BOT, YAM, and ASD) compare well, while SMO shows some deviation from the rest. The response to controls was also very similar (Tables 9.10 and 9.11). The amount of reduction each advection solver predicts is approximately the same for the 50% ROG and the 50% ROG/NO_x cases. There is more variability between the solvers for the 50% NO_x case, but the numbers are all fairly small. Thus, the solvers may not predict the same absolute peak ozone, but the reductions for each control strategy are similar in most cases (Table 9.12). This test indicates that the choice of the solvers does not bias the choice of a control strategy.

In a more recent study (Krishnakumar et al., 1998), a case was found where the differences were more marked. In this case, UAM-IV was applied to the Pittsburgh, Pennsylvania region using the SMO and BOT routines. They found that the SMO scheme led to consistently higher ozone levels (a few percent) and that there were a few erroneously high predictions. Further, the peak ozone levels (e.g., those erroneously high predictions from using SMO) did not respond to controls in a similar fashion between the use of SMO and BOT. This is particularly of concern given how air quality modeling is used in a regulatory framework. Usually, greatest (if not complete) importance is placed on the peak ozone predicted anywhere in the domain. The results of that study raise questions as to the confidence one has in determining control strategies using the SMO.

9.5   SUMMARY

One of the most important components of an air quality model is the scheme used to integrate the advection portion of the governing equation. Horizontal transport in the atmosphere is dominated by advection, while vertical transport is dominated by diffusion. Thus, the horizontal transport solver must be able to handle the hyperbolic nature of the problem without introducing significant diffusion or dispersion. In the presence of the nonlinear chemistry, such errors can significantly degrade model results, particularly in test cases. A number of methods have evolved, and the current methods, including finite element, finite volume, pseudospectral, and advanced finite difference methods, give similar results in most test cases. One of the critical components of those methods is how to filter out the dispersive waves that lead to artificial minima and maxima (including negative concentrations) while not degrading the solution by adding excessive artificial diffusion.

There has been relatively little intercomparison of the methods in actual applications, i.e., comparison of how the methods work when applied to atmospheric modeling of specific events. Those few studies that have investigated the problem find that while the results are similar, there can be some problems. For example, the response to emissions controls, as well as peak predictions (which are of extreme importance in air quality modeling), can be in error if less accurate methods are used. This is particularly true with older methods that have been found to be diffusive. However, when a variety of methods that have been shown to perform well across a suite of test cases is used, the results in actual applications of those methods are similar.

REFERENCES

Anderson DA, Tannehill JC, Pletcher RH. Computational Fluid Mechanics and Heat Transfer. Washington, DC: Hemisphere, 1984.

ARB. Technical Guidance Document: Photochemical Air Quality Modeling. Modeling and Meteorology Branch, Technical Support Division, California Air Resources Board, Sacramento, CA, 1990.

Aron JD, Vicente F-G. The random choice method in the numerical solution of the atmospheric transport equation. Environ Software 9: 23–31, 1994.

Bell J, Berger M, Saltzman J, Welcome M. Three-dimensional adaptive mesh refinements for hyperbolic conservation laws. SIAM J Sci Comput 15(1): 127–138, 1994.

Book DL, Boris JP, Hain K. Flux-corrected transport. II. Generalizations of the method. J Comp Phys 18: 248–283, 1975.

Boris JP, Book DL. Flux-corrected transport. 1. SHASTA, A fluid transport algorithm that works. J Comp Phys 11: 38–69, 1973.

Boris JP, Book DL. Flux-corrected transport. III. Minimal error FCT algorithms. J Comp Phys 20: 397–431, 1976.

Bott A. A positive definite advection scheme obtained by nonlinear renormalization of the advective fluxes. Mon Wea Rev 117: 1006–1015, 1989a.

Bott A. Reply. Mon Wea Rev 117: 2633–2636, 1989b.

Bott A. The monotone area-preserving flux-form advection algorithm: reducing the time-splitting error in two-dimensional flow fields. Mon Wea Rev 121: 2637–2641, 1993.

Bott A. Monotone flux limitation in the area-preserving flux-form advection algorithm. Mon Wea Rev 120: 2592–2602, 1992.

Brooks AN, Hughes TJR. Streamline upwind/Petrov-Galerkin formulations for convection dominated flows with particular emphasis on the incompressible Navier-Stokes equations. Comp Meth Appl Mech Eng 32: 199–259, 1982.

Brost RA, Haagenson PL, Kuo Y-H. The effect of diffusion on tracer puffs simulated by a regional scale Eulerian model. J Geophys Res 93: 2389–2404, 1988.

Carpenter RL, Droegemeier KK, Woodward PR, Hane CE. Application of the piecewise parabolic method (PPM) to meteorological modeling. Mon Wea Rev 118: 586–612, 1990.

Childs PN, Morton KW. Characteristic Galerkin methods for scalar conservation laws in one dimension. SIAM J Numer Anal 27: 553–594, 1990.

Chlond A. Locally modified version of Bott's advection scheme. Mon Wea Rev 122: 111–125, 1994.

Chock DP. A comparison of numerical methods for solving the advection equation. II. Atmos Environ 19(4): 571–586, 1985.

Chock DP. A comparison of numerical methods for solving the advection equation. III. Atmos Environ 25A(5/6): 853–871, 1991.

Chock DP, Dunker AM. A comparison of numerical methods for solving the advection equation. Atmos Environ 17(1): 11–24, 1983.

Chock DP, Winkler SL. A particle grid air quality modeling approach. 2. Coupling with chemistry. J Geophys Res 99: 1033–1042, 1994a.

Chock DP, Winkler SL. A comparison of advection algorithms coupled with chemistry. Atmos Environ 28: 2659–2675, 1994b.

Chorin AJ. Random choice solution of hyperbolic systems. J Comp Phys 22: 517–533, 1976.

Colella P, Woodward PR. The piecewise parabolic method (PPM) for gas-dynamical simulations. J Comp Phys 54: 174–201, 1984.

Concus P, Proskuroski W. Numerical solution of a nonlinear hyperbolic equation by the random choice method. J Comp Phys 30: 153–166, 1979.

Crowley WP. Numerical advection experiments. Mon Wea Rev 96: 1–11, 1968.

Dabdub D, Seinfeld JH. Numerical advective schemes used in air quality models—sequential and parallel implementation. Atmos Environ 28(20): 3369–3385, 1994.

Donea J. A Taylor-Galerkin method for convective transport problems. Int J Num Meth Eng 20: 101–119, 1984.

Easter RC. Two modified versions of Bott's positive-definite numerical advection scheme. Mon Wea Rev 121: 297–304, 1993.

Emde KVD. Solving conservation laws with parabolic and cubic splines. Mon Wea Rev 120: 482–492, 1992.

Flatoy F. Balanced wind in advanced advection schemes when species with long lifetimes are transported. Atmos Environ 27A(12): 1809–1819, 1993.

Fogelson AL. Particle method solution of two-dimensional convection-diffusion equations. J Comput Phys 100: 1–16, 1992.

Forester CK. Higher order monotonic convective difference schemes. J Comp Phys 23: 1–22, 1977.

Gazdag J. Numerical convective schemes based on accurate computation of space derivatives. J Comp Phys 13: 100–113, 1973.

Glimm J. Solutions in the large for nonlinear hyperbolic systems of equations. Comm Pure Appl Math 18: 697–715, 1965.

Godunov SK. A difference scheme for numerical computation of discontinuous solution of hydrodynamic equations. Math Sb 47: 271–306, 1959 (in Russian).

Harley RA, Russell AG, McRae GJ, Cass GR, Seinfeld JH. Photochemical air quality modeling of the Southern California air quality study. Environ Sci Tech 27: 378–388, 1993.

Hirt CW. Heuristic stability theory for finite-difference equations. J Comp Phys 2: 339–355, 1968.

Hov O, Zlatev Z, Berkowicz R, Eliassen A, Prahm LP. Comparison of numerical techniques for use in air pollution models with nonlinear chemical reactions. Atmos Environ 23: 967–983, 1989.

Hughes TJR, Brooks AN. A multidimensional upwind scheme with no crosswind diffusion. In: Finite Element Methods for Convection Dominated Flows. ASME, 1979.

Hughes TJR. Recent progress in the development and understanding of SUPG methods with special reference to compressible Euler and Navier-Stokes equations. In: Finite Elements in Fluids. Vol 7. New York: Wiley, 1987, pp 273–287.

Kelly DW, Nakazawa S, Zienkiewicz OC, Heinrich JC. A note on upwinding and anisotropic balancing dissipation in finite element approximations to convective diffusion problems. Int J Num Meth Eng 15: 1705–1711, 1980.

Kreiss H-O. Some remarks about computational fluid dynamics. In: Hussaini MY, Kumar A, Salas MD, eds. Algorithmic Trends in Computational Fluid Dynamics. New York: Springer-Verlag, 1993.

Krishnakumar V, Wilkinson J, Russell A. Air quality modeling of Southwestern Pennsylvania: Errors in peak ozone predictions by the horizontal advection equation solver in the urban airshed model. J Air and Waste Manage Assoc (submitted).

Lohner R, Morgan K, Peraire J, Vahdati M. Finite element flux-corrected transport (FEM-FCT) for the Euler and Navier-Stokes equations. In: Finite Elements in Fluids. Vol 7. New York: Wiley, 1987, pp 105–121.

Marchuk GI. Methods of Numerical Mathematics. Springer-Verlag, 1975.

McRae GJ, Goodin WR, Seinfeld JH. Numerical solutions of the atmospheric diffusion equation for chemically reacting flows. J Comp Phys 45: 1–42, 1982.

Morris RE, Myers TC. User's guide for the urban airshed model. Vol I. User's manual for UAM (CB-IV). Research Triangle Park, NC: Environmental Protection Agency, 1990.

Morton KW, Stokes A. Generalized Galerkin methods for hyperbolic problems.

Proceedings of the Conference on Mathematics of Finite Elements and Applications IV. New York: Academic Press, 1982, pp 421–431.

Muller R. The performance of classical versus modern finite-volume advection schemes for atmospheric modeling in a one-dimensional test-bed. Mon Wea Rev 120: 1407–1415, 1992.

NRC. Rethinking the Ozone Problem in Urban and Regional Air Pollution. Washington, DC: National Academy Press, 1992.

Odman MT. Multiscale modeling of long range transport and fate of pollutants emitted from urban areas. PhD thesis, Carnegie Mellon University, Pittsburgh, PA, 1992.

Odman MT, Russell AG. A multiscale finite element pollutant transport scheme for urban and regional modeling. Atmos Environ 25A: 2385–2394, 1991b.

Odman MT, Russell AG. Multiscale modeling of pollutant transport and chemistry. J Geophys Res 96: 7363–7370, 1991a.

Odman MT, Russell AG. A nonlinear filtering algorithm for multi-dimensional finite element pollutant advection schemes. Atmos Environ 27A: 793–799, 1993.

Odman MT, Wilkinson JG, McNair LA, Russell AG, Ingram CL, Houyoux MR. Horizontal advection solver uncertainty in the urban airshed model. Final report, contract no. 93-722, California Air Resources Board, Sacramento, CA, 1996.

Odman MT, Xiu A, Byun DW. Evaluating advection schemes for use in the next generation of air quality modeling systems. In: Ranzieri AJ, Solomon PA, eds. Regional Photochemical Measurement and Modeling Studies. Pittsburgh, PA: Air & Waste Management Association, 1995, pp 1386–1401.

Olim M. A truly noninterpolating semi-Lagrangian Lax-Wendroff method. J Comp Phys 112: 253–266, 1994.

OTA (Office of Technology Assessment, U.S. Congress). Catching our breath—next steps for reducing urban ozone. OTA-O-412, Congressional Board of the 101st Congress. Washington, DC: U.S. Government Printing Office, 1989.

Pai P, Karamchandani P, Venkatram A. Performance of flux conserving and semi-Lagrangian advection schemes in simulating a photochemical episode. 21st NATO/CCMS International Technical Meeting on Air Pollution Modeling and Its Application, Baltimore, MD, November 6–10, 1995.

Parrott AK, Christie MA. FCT applied to the 2-D finite element solution of tracer transport by single phase flow in a porous medium. In: Numerical Methods for Fluid Dynamics. II. New York: Oxford University Press, 1986, pp 609–619.

Pepper DW, Kern CD, Long PE. Modeling the dispersion of atmospheric pollution using cubic splines and chapeau functions. Atmos Environ 13: 223–237, 1979.

Pielke RA. Mesoscale Meteorological Modeling. New York: Academic Press, 1984.

Purnell DK. Solution of the advective equation by upstream interpolation with a cubic spline. Mon Wea Rev 104: 42–48, 1976.

Rasch PJ, Wiliamson DL. On shape-preserving interpolation and semi-Lagrangian transport. SIAM J Sci Stat Comput 11(4): 656–687, 1990.

Raymond WH, Garder A. Selective damping in a Galerkin method for solving wave problems and variable grids. Mon Wea Rev 104: 1583–1590, 1976.

Roache PJ. A pseudo-spectral FFT technique for non-periodic problems. J Comp Phys 27: 204–220, 1978.

Roberts PT, Main HH. Ozone and particulate matter case study analysis for the Southern California air quality study. Prepared for the California Air Resources Board by Sonoma Technology, Inc., contract A932-050, 1993.

Rood RB. Numerical advection algorithms and their role in atmospheric transport and chemistry models. Rev Geophysics 25(1): 71–100, 1987.

Scheffe RD, Morris RE. A review of the development and application of the urban airshed model., Atmos Environ 27B: 23–39, 1993.

SCAQMD. Air quality management plan, technical report V-B, ozone modeling—performance evaluation. South Coast Air Quality Management District, 1990.

Seibert P, Morariu B. Improvements of upstream, semi-Lagrangian numerical advection schemes. J Appl Meteor 30: 117–125, 1991.

Seinfeld JH. Atmospheric Chemistry and Physics of Air Pollution. New York: John Wiley & Sons, 1985.

Seinfeld JH. Ozone air quality models: a critical review. JAPCA 38(5): 616–645, 1988.

Smolarkiewicz PK. A fully multidimensional positive definite advection transport algorithm with small implicit diffusion. J Comp Phys 54: 325–362, 1984.

Smolarkiewicz PK. A simple positive definite advection transport scheme with small implicit diffusion. Mon Wea Rev 111: 479–486, 1983.

Smolarkiewicz PK, Grabowski WW. The multidimensional positive definitive advection algorithm: nonoscillatory option. J Comp Phys 84: 355–375, 1990.

Smolarkiewicz PK, Grell GA. A class of monotone interpolation schemes. J. Comp Phys 101: 431–440, 1992.

Strand A, Hov O. A two-dimensional zonally averaged transport model including convective motions and a new strategy for the numerical solution. J Geophys Res 98: 9023–9037, 1993.

Staniforth A, Côté J, Pudykiewicz J. Comments on Smolarkiewicz's deformational flow. Mon Wea Rew 115: 894–900, 1987.

Tesche TW, et al. Improvement of procedures for evaluating photochemical models. Radian Corporation, Report to the California Air Resources Board, 1990.

Tezduyar TE, Park YJ, Deans HA, Finite element procedures for time-dependent convection-diffusion-reaction systems. In: Finite Elements in Fluids. Vol 7. New York: Wiley, 1987, pp 25–45.

Tolstykh MA. Application of fifth-order compact upwind differencing to moisture transport equation in atmosphere. J Comp Phys 112: 394–403, 1994.

Tremback GJ, Powell J, Cotton WR, Pielke RA. The forward-in-time upstream advection scheme: extension to higher orders. Mon Wea Rev 115: 540–555, 1987.

van Leer B. Towards the ultimate conservative difference scheme: IV. A new approach of numerical convection. J Comp Phys 23: 276–299, 1977.

Wengle H, Seinfeld JH. Pseudospectral solution of atmospheric diffusion problems.

J Comp Phys 26: 87–106, 1978.

Westerink JJ, Shea D. Consistent higher degree Petrov-Galerkin methods for the solution of the transient convective-diffusion equation. Int J Num Meth Eng 28: 1077–1101, 1989.

Williamson DL, Rasch PJ, Two-dimensional semi-Lagrangian transport with shape-preserving interpolation. Mon Wea Rev 117: 102–117, 1989.

Williamson DL, Olson JG. Climate simulations with a semi-Lagrangian version of the NCAR community climate model. Mon Wea Rev 122: 1594–1610, 1994.

Yamartino RJ, Scire JS, Carmichael GR, Chang YS. The CALGRID mesoscale photochemical grid model. I. Model formulation. Atmos Environ 26A: 1493–1512, 1992.

Yamartino RJ. Nonnegative, conserved scalar transport using grid-cell-centered, spectrally constrained Blackman cubics for applications on a variable-thickness mesh. Mon Wea Rev 121: 753–763, 1993.

Yanenko YN. The Method of Fractional Steps. New York: Springer, 1971.

Yu C, Heinrich JC. Petrov-Galerkin method for multidimensional, time-dependent, convective-diffusion equations. Int J Num Meth Eng 24: 2201–2215, 1987.

Yu C-C, Heinrich JC, Petrov-Galerkin methods for the time-dependent convective transport equation. Int J Num Meth Eng 23: 883–901, 1986.

Zalesak ST. Fully multidimensional flux-corrected transport algorithms for fluids. J Comp Phys 31: 335–362, 1979.

Appendix

Governing Equations in Various Systems

Vijay K. Garg

AYT Corporation/NASA Lewis Reseach Center, Cleveland, Ohio

A.1 GOVERNING EQUATIONS IN CARTESIAN COORDINATES

Coordinates x, y, z; velocity components in these directions u, v, w. The equation of continuity for a *compressible* flow,

$$\frac{\partial \rho}{\partial t} + \mathbf{V} \cdot (\rho \mathbf{V}) = 0 \tag{A.1}$$

can be written in Cartesian coordinates as

$$\frac{\partial \rho}{\partial t} + \frac{\partial (\rho u)}{\partial x} + \frac{\partial (\rho v)}{\partial y} + \frac{\partial (\rho w)}{\partial z} = 0 \tag{A.2}$$

For an *incompressible* flow, Eq. (A.2) reduces to

$$\frac{\partial u}{\partial x} + \frac{\partial v}{\partial y} + \frac{\partial w}{\partial z} = 0 \tag{A.3}$$

The momentum equations for a *viscous, compressible fluid* are

$$\rho \frac{Du}{Dt} = \rho g_x - \frac{\partial p}{\partial x} + \frac{\partial}{\partial x}\left[\mu\left(2\frac{\partial u}{\partial x} - \frac{2}{3}(\mathbf{V} \cdot \mathbf{V})\right)\right]$$
$$+ \frac{\partial}{\partial y}\left[\mu\left(\frac{\partial u}{\partial y} + \frac{\partial v}{\partial x}\right)\right] + \frac{\partial}{\partial z}\left[\mu\left(\frac{\partial w}{\partial x} + \frac{\partial u}{\partial z}\right)\right] \tag{A.4a}$$

$$\rho \frac{Dv}{Dt} = \rho g_y - \frac{\partial p}{\partial y} + \frac{\partial}{\partial y}\left[\mu\left(2\frac{\partial v}{\partial y} - \frac{2}{3}(\mathbf{V} \cdot \mathbf{V})\right)\right]$$
$$+ \frac{\partial}{\partial z}\left[\mu\left(\frac{\partial v}{\partial z} + \frac{\partial w}{\partial y}\right)\right] + \frac{\partial}{\partial x}\left[\mu\left(\frac{\partial u}{\partial y} + \frac{\partial v}{\partial x}\right)\right] \tag{A.4b}$$

$$\rho \frac{Dw}{Dt} = \rho g_z - \frac{\partial p}{\partial z} + \frac{\partial}{\partial z}\left[\mu\left(2\frac{\partial w}{\partial z} - \frac{2}{3}(\mathbf{V} \cdot \mathbf{V})\right)\right]$$
$$+ \frac{\partial}{\partial x}\left[\mu\left(\frac{\partial w}{\partial x} + \frac{\partial u}{\partial z}\right)\right] + \frac{\partial}{\partial y}\left[\mu\left(\frac{\partial v}{\partial z} + \frac{\partial w}{\partial y}\right)\right] \tag{A.4c}$$

where

$$\frac{D}{Dt} = \frac{\partial}{\partial t} + \mathbf{V} \cdot \mathbf{V} = \frac{\partial}{\partial t} + u\frac{\partial}{\partial x} + v\frac{\partial}{\partial y} + w\frac{\partial}{\partial z}$$

and $\mathbf{V} \cdot \mathbf{V}$ is defined by Eq. (A.3).

Equations (A.4) for a *viscous, incompressible fluid* reduce to

$$\frac{Du}{Dt} = g_x - \frac{1}{\rho}\frac{\partial p}{\partial x} + \nu\left(\frac{\partial^2 u}{\partial x^2} + \frac{\partial^2 u}{\partial y^2} + \frac{\partial^2 u}{\partial z^2}\right) \tag{A.5a}$$

$$\frac{Dv}{Dt} = g_y - \frac{1}{\rho}\frac{\partial p}{\partial y} + \nu\left(\frac{\partial^2 v}{\partial x^2} + \frac{\partial^2 v}{\partial y^2} + \frac{\partial^2 v}{\partial z^2}\right) \tag{A.5b}$$

$$\frac{Dw}{Dt} = g_z - \frac{1}{\rho}\frac{\partial p}{\partial z} + \nu\left(\frac{\partial^2 w}{\partial x^2} + \frac{\partial^2 w}{\partial y^2} + \frac{\partial^2 w}{\partial z^2}\right) \tag{A.5c}$$

If we let $\mu = 0$ in Eqs. (A.4) we obtain the momentum equations for an *inviscid fluid*.

The energy equation in Cartesian coordinates has the form

$$\rho \frac{D(c_p T)}{Dt} - \frac{Dp}{Dt} = \frac{\partial}{\partial x}\left(k\frac{\partial T}{\partial x}\right) + \frac{\partial}{\partial y}\left(k\frac{\partial T}{\partial y}\right) + \frac{\partial}{\partial z}\left(k\frac{\partial T}{\partial z}\right) + \Phi \qquad \text{(A.6)}$$

where

$$\Phi = 2\mu\left[\left(\frac{\partial u}{\partial x}\right)^2 + \left(\frac{\partial v}{\partial y}\right)^2 + \left(\frac{\partial w}{\partial z}\right)^2 + \frac{1}{2}\left(\frac{\partial u}{\partial y} + \frac{\partial v}{\partial x}\right)^2 \right.$$
$$\left. + \frac{1}{2}\left(\frac{\partial v}{\partial z} + \frac{\partial w}{\partial y}\right)^2 + \frac{1}{2}\left(\frac{\partial w}{\partial x} + \frac{\partial u}{\partial z}\right)^2 - \frac{1}{3}(\nabla \cdot V)^2\right]$$

For a *viscous, incompressible fluid* the stress functions are

$$\sigma_{xx} = 2\mu\frac{\partial u}{\partial x} - p, \quad \sigma_{yy} = 2\mu\frac{\partial v}{\partial y} - p, \quad \sigma_{zz} = 2\mu\frac{\partial w}{\partial z} - p \qquad \text{(A.7a)}$$

$$\sigma_{xy} = \mu\left(\frac{\partial u}{\partial y} + \frac{\partial v}{\partial x}\right) = \sigma_{yx}, \quad \sigma_{yz} = \mu\left(\frac{\partial w}{\partial y} + \frac{\partial v}{\partial z}\right) = \sigma_{zy}, \qquad \text{(A.7b)}$$

$$\sigma_{zx} = \mu\left(\frac{\partial u}{\partial z} + \frac{\partial w}{\partial x}\right) = \sigma_{xz} \qquad \text{(A.7c)}$$

A.2 GOVERNING EQUATIONS IN CYLINDRICAL COORDINATES

Coordinates r, θ, z; velocity components in these directions are V_r, V_θ, and V_z. The relationship between the cylindrical and Cartesian coordinates is illustrated in Fig. A.1.

For a *compressible fluid* the continuity equation becomes

$$\frac{\partial \rho}{\partial t} + \frac{\partial(\rho V_r)}{\partial r} + \frac{1}{r}\frac{\partial(\rho V_\theta)}{\partial \theta} + \frac{\partial(\rho V_z)}{\partial z} + \frac{(\rho V_r)}{r} = 0 \qquad \text{(A.8)}$$

For an *incompressible fluid* Eq. (A.8) reduces to

$$\frac{\partial V_r}{\partial r} + \frac{1}{r}\frac{\partial V_\theta}{\partial \theta} + \frac{\partial V_z}{\partial z} + \frac{V_r}{r} = 0 \qquad \text{(A.9)}$$

The momentum equations for a *compressible fluid* with *constant viscosity* are

$$\frac{DV_r}{Dt} - \frac{V_\theta^2}{r} = g_r - \frac{1}{\rho}\frac{\partial p}{\partial r} + \nu\left[\nabla^2 V_r - \frac{V_r}{r^2} - \frac{2}{r^2}\frac{\partial V_\theta}{\partial \theta} + \frac{1}{3}\frac{\partial}{\partial r}(\nabla \cdot V)\right] \qquad \text{(A.10a)}$$

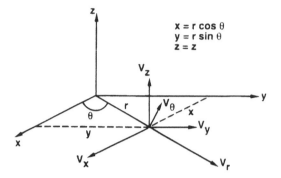

Figure A.1 Relationship between Cartesian and cylindrical coordinates.

$$\frac{DV_\theta}{Dt} + \frac{V_r V_\theta}{r} = g_\theta - \frac{1}{\rho r}\frac{\partial p}{\partial \theta} + v\left[\nabla^2 V_\theta - \frac{V_\theta}{r^2} + \frac{2}{r^2}\frac{\partial V_r}{\partial \theta} + \frac{1}{3}\frac{1}{r}\frac{\partial}{\partial \theta}(\nabla \cdot V)\right]$$

(A.10b)

$$\frac{DV_z}{Dt} = g_z - \frac{1}{\rho}\frac{\partial p}{\partial z} + v\left[\nabla^2 V_z + \frac{1}{3}\frac{\partial}{\partial z}(\nabla \cdot V)\right]$$

(A.10c)

where $\nabla \cdot V$ is given by Eq. (A.9), and

$$\nabla^2 = \frac{\partial^2}{\partial r^2} + \frac{1}{r}\frac{\partial}{\partial r} + \frac{1}{r^2}\frac{\partial^2}{\partial \theta^2} + \frac{\partial^2}{\partial z^2} = \frac{1}{r}\frac{\partial}{\partial r}\left(r\frac{\partial}{\partial r}\right) + \frac{1}{r^2}\frac{\partial^2}{\partial \theta^2} + \frac{\partial^2}{\partial z^2}$$

$$\frac{D}{Dt} = \frac{\partial}{\partial t} + V_r\frac{\partial}{\partial r} + \frac{V_\theta}{r}\frac{\partial}{\partial \theta} + V_z\frac{\partial}{\partial z}$$

For *incompressible flows* Eqs. (A.10) get simplified since $\nabla \cdot V = 0$. If we let $v = 0$ in Eqs. (A.10) we obtain the momentum equations for an *inviscid fluid*. The energy equation in cylindrical coordinates becomes

$$\rho\frac{D(c_p T)}{Dt} - \frac{Dp}{Dt} = \frac{1}{r}\frac{\partial}{\partial r}\left(kr\frac{\partial T}{\partial r}\right) + \frac{1}{r^2}\frac{\partial}{\partial \theta}\left(k\frac{\partial T}{\partial \theta}\right) + \frac{\partial}{\partial z}\left(k\frac{\partial T}{\partial z}\right) + \Phi$$

(A.11)

where

$$\Phi = 2\mu\left[\left(\frac{\partial V_r}{\partial r}\right)^2 + \left(\frac{1}{r}\frac{\partial V_\theta}{\partial \theta} + \frac{V_r}{r}\right)^2 + \left(\frac{\partial V_z}{\partial z}\right)^2 + \frac{1}{2}\left(\frac{\partial V_\theta}{\partial r} - \frac{V_\theta}{r} + \frac{1}{r}\frac{\partial V_r}{\partial \theta}\right)^2 \right.$$
$$\left. + \frac{1}{2}\left(\frac{1}{r}\frac{\partial V_z}{\partial \theta} + \frac{\partial V_\theta}{\partial z}\right)^2 + \frac{1}{2}\left(\frac{\partial V_r}{\partial z} + \frac{\partial V_z}{\partial r}\right)^2 - \frac{1}{3}(\nabla \cdot V)^2\right]$$

The stress functions for a *viscous, incompressible fluid* are

$$\sigma_{rr} = 2\mu\frac{\partial V_r}{\partial r} - p, \quad \sigma_{\theta\theta} = 2\mu\left(\frac{1}{r}\frac{\partial V_\theta}{\partial \theta} + \frac{V_r}{r}\right) - p, \quad \sigma_{zz} = 2\mu\frac{\partial V_z}{\partial z} - p$$

$$\sigma_{r\theta} = \mu\left(\frac{1}{r}\frac{\partial V_r}{\partial \theta} - \frac{V_\theta}{r} + \frac{\partial V_\theta}{\partial r}\right) = \sigma_{\theta r}, \quad \sigma_{\theta z} = \mu\left(\frac{1}{r}\frac{\partial V_z}{\partial \theta} + \frac{\partial V_\theta}{\partial z}\right) = \sigma_{z\theta}$$

$$\sigma_{zr} = \mu\left(\frac{\partial V_r}{\partial z} + \frac{\partial V_z}{\partial r}\right) = \sigma_{rz} \tag{A.12}$$

A.3 GOVERNING EQUATIONS IN SPHERICAL COORDINATES

Coordinates ω, ϕ, θ; velocity components in these directions are V_ω, V_ϕ, and V_θ. The relationship between the Cartesian, cylindrical, and spherical coordinates is illustrated in Figs. A.2 and A.3.

For a *compressible fluid* the continuity equation becomes

$$\frac{\partial \rho}{\partial t} + \frac{1}{\omega^2}\frac{\partial(\omega^2\rho V_\omega)}{\partial \omega} + \frac{1}{\omega \sin \phi}\frac{\partial(\sigma V_\phi \sin \phi)}{\partial \phi} + \frac{1}{\omega \sin \phi}\frac{\partial(\rho V_\theta)}{\partial \theta} = 0 \tag{A.13}$$

For an *incompressible fluid* Eq. (A.13) reduces to

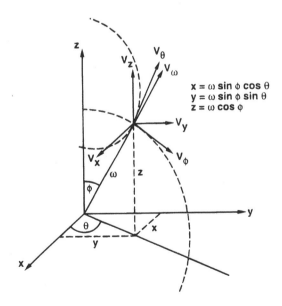

$$x = \omega \sin \phi \cos \theta$$
$$y = \omega \sin \phi \sin \theta$$
$$z = \omega \cos \phi$$

Figure A.2 Relationship between Cartesian and spherical coordinates.

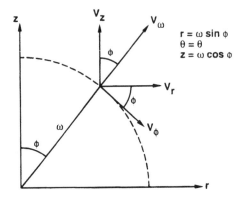

Figure A.3   Relationship between cylindrical and spherical coordinates.

$$\frac{1}{\omega}\frac{\partial(\omega^2 V_\omega)}{\partial\omega} + \frac{1}{\sin\phi}\frac{\partial(V_\phi \sin\phi)}{\partial\phi} + \frac{1}{\sin\phi}\frac{\partial V_\theta}{\partial\theta} = 0 \qquad (A.14)$$

The momentum equations for a *compressible fluid* with *constant viscosity* are

$$\frac{DV_\omega}{Dt} - \frac{V_\phi^2 + V_\theta^2}{\omega} = g_\omega - \frac{1}{\rho}\frac{\partial\rho}{\partial\omega}$$

$$+ \nu\left[\nabla^2 V_\omega - 2\frac{V_\omega}{\omega^2} - \frac{2}{\omega^2}\frac{\partial V_\phi}{\partial\phi} - \frac{2V_\phi \cot\phi}{\omega^2} - \frac{2}{\omega^2 \sin\phi}\frac{\partial V_\theta}{\partial\theta}\right]$$

$$+ \frac{\nu}{3}\left[\frac{\partial^2 V_\omega}{\partial\omega^2} + \frac{2}{\omega}\left(\frac{\partial V_\omega}{\partial\omega} - \frac{V_\omega}{\omega}\right) + \frac{1}{\omega}\frac{\partial^2 V_\phi}{\partial\omega\,\partial\phi} - \frac{1}{\omega^2}\left(\frac{\partial V_\phi}{\partial\phi} + V_\phi \cot\phi\right)\right.$$

$$\left. + \frac{\cot\phi}{\omega}\frac{\partial V_\phi}{\partial\omega} + \frac{1}{\omega\sin\phi}\left(\frac{\partial^2 V_\theta}{\partial\omega\,\partial\theta} - \frac{1}{\omega}\frac{\partial V_\theta}{\partial\theta}\right)\right] \qquad (A.15a)$$

$$\frac{DV_\phi}{Dt} + \frac{V_\omega V_\phi}{\omega} - \frac{V_\theta^2 \cot\phi}{\omega} = g_\phi - \frac{1}{\rho\omega}\frac{\partial p}{\partial\phi}$$

$$+ \nu\left[\nabla^2 V_\phi + \frac{2}{\omega^2}\frac{\partial V_\omega}{\partial\phi} - \frac{V_\phi}{\omega^2 \sin^2\phi} - \frac{2\cos\phi}{\omega^2 \sin^2\phi}\frac{\partial V_\theta}{\partial\theta}\right]$$

$$+ \frac{\nu}{3}\left[\frac{1}{\omega^2}\frac{\partial^2 V_\phi}{\partial\phi^2} + \frac{1}{\omega}\left(\frac{\partial^2 V_\omega}{\partial\phi\,\partial\omega} + \frac{2}{\omega}\frac{\partial V_\omega}{\partial\phi} + \frac{\cot\phi}{\omega}\frac{\partial V_\phi}{\partial\phi}\right)\right.$$

$$\left. + \frac{1}{\omega^2 \sin\phi}\left(\frac{\partial^2 V_\theta}{\partial\phi\,\partial\theta} - \cot\phi\frac{\partial V_\theta}{\partial\theta} - V_\phi\right)\right] \qquad (A.15b)$$

$$\frac{DV_\theta}{Dt} + \frac{V_\theta V_\omega}{\omega} + \frac{V_\phi V_\theta \cot\phi}{\omega} = g_\theta - \frac{1}{\rho\omega\sin\phi}\frac{\partial p}{\partial\theta}$$

$$+ \nu\left[\nabla^2 V_\theta - \frac{V_\theta}{\omega^2\sin^2\phi} + \frac{2}{\omega^2\sin\phi}\frac{\partial V_\omega}{\partial\theta} + \frac{2\cos\phi}{\omega^2\sin^2\phi}\frac{\partial V_\phi}{\partial\theta}\right]$$

$$+ \frac{\nu}{3}\frac{1}{\omega\sin\phi}$$

$$\left[\frac{1}{\omega\sin\phi}\frac{\partial^2 V_\theta}{\partial\theta^2} + \frac{\partial^2 V_\omega}{\partial\theta\,\partial\omega} + \frac{2}{\omega}\frac{\partial V_\omega}{\partial\theta} + \frac{1}{\omega}\frac{\partial^2 V_\phi}{\partial\theta\,\partial\phi} + \frac{\cot\phi}{\omega}\frac{\partial V_\phi}{\partial\theta}\right]$$

$$\text{(A.15c)}$$

where

$$\nabla^2 = \frac{1}{\omega^2}\frac{\partial}{\partial\omega}\left(\omega^2\frac{\partial}{\partial\omega}\right) + \frac{1}{\omega^2\sin\phi}\frac{\partial}{\partial\phi}\left(\sin\phi\frac{\partial}{\partial\phi}\right) + \frac{1}{\omega^2\sin^2\phi}\frac{\partial^2}{\partial\theta^2}$$

$$\frac{D}{Dt} = \frac{\partial}{\partial t} + V_\omega\frac{\partial}{\partial\omega} + \frac{V_\phi}{\omega}\frac{\partial}{\partial\phi} + \frac{V_\theta}{\omega\sin\phi}\frac{\partial}{\partial\theta}$$

For the *incompressible fluid* Eqs. (A.15) reduce to

$$\frac{DV_\omega}{Dt} - \frac{V_\phi^2 + V_\theta^2}{\omega} = g_\omega - \frac{1}{\rho}\frac{\partial p}{\partial\omega}$$

$$+ \nu\left[\nabla^2 V_\omega - 2\frac{V_\omega}{\omega^2} - \frac{2}{\omega^2}\frac{\partial V_\phi}{\partial\phi} - \frac{2V_\phi\cot\phi}{\omega^2} - \frac{2}{\omega^2\sin\phi}\frac{\partial V_\theta}{\partial\theta}\right] \quad \text{(A.16a)}$$

$$\frac{DV_\phi}{Dt} + \frac{V_\omega V_\phi}{\omega} - \frac{V_\theta^2\cot\phi}{\omega} = g_\phi - \frac{1}{\rho\omega}\frac{\partial p}{\partial\phi}$$

$$+ \nu\left[\nabla^2 V_\phi + \frac{2}{\omega^2}\frac{\partial V_\omega}{\partial\phi} - \frac{V_\phi}{\omega^2\sin^2\phi} - \frac{2\cos\phi}{\omega^2\sin^2\phi}\frac{\partial V_\theta}{\partial\theta}\right] \quad \text{(A.16b)}$$

$$\frac{DV_\theta}{Dt} + \frac{V_\theta V_\omega}{\omega} + \frac{V_\phi V_\theta\cot\phi}{\omega} = g_\theta - \frac{1}{\rho\omega\sin\phi}\frac{\partial p}{\partial\theta}$$

$$+ \nu\left[\nabla^2 V_\theta - \frac{V_\theta}{\omega^2\sin^2\phi} + \frac{2}{\omega^2\sin\phi}\frac{\partial V_\omega}{\partial\theta} + \frac{2\cos\phi}{\omega^2\sin^2\phi}\frac{\partial V_\phi}{\partial\theta}\right] \quad \text{(A.16c)}$$

If we let $\nu = 0$ in Eqs. (A.15) we obtain the momentum equations for an *inviscid fluid*.

The energy equation in spherical coordinates becomes

$$\rho\frac{D(c_p T)}{Dt} - \frac{Dp}{Dt} = \frac{1}{\omega^2}\frac{\partial}{\partial\omega}\left(\omega^2 k\frac{\partial T}{\partial\omega}\right) + \frac{1}{\omega^2\sin\phi}\frac{\partial}{\partial\phi}$$

$$\times\left(k\frac{\partial T}{\partial\phi}\sin\phi\right) + \frac{1}{\omega^2\sin^2\phi}\frac{\partial}{\partial\theta}\left(k\frac{\partial T}{\partial\theta}\right) + \Phi \quad \text{(A.17)}$$

where

$$\Phi = 2\mu\left[\left(\frac{\partial V_\omega}{\partial \omega}\right)^2 + \left(\frac{1}{\omega}\frac{\partial V_\phi}{\partial \phi} + \frac{V_\omega}{\omega}\right)^2 \right.$$
$$+ \left(\frac{1}{\omega \sin \phi}\frac{\partial V_\theta}{\partial \theta} + \frac{V_\omega}{\omega} + \frac{V_\phi \cot \phi}{\omega}\right)^2$$
$$+ \frac{1}{2}\left\{\omega\frac{\partial}{\partial \omega}\left(\frac{V_\phi}{\omega}\right) + \frac{1}{\omega}\frac{\partial V_\omega}{\partial \phi}\right\}^2$$
$$+ \frac{1}{2}\left\{\frac{\sin \phi}{\omega}\frac{\partial}{\partial \phi}\left(\frac{V_\theta}{\sin \phi}\right) + \frac{1}{\omega \sin \phi}\frac{\partial V_\phi}{\partial \theta}\right\}^2$$
$$\left.+ \frac{1}{2}\left\{\frac{1}{\omega \sin \phi}\frac{\partial V_\omega}{\partial \theta} + \omega\frac{\partial}{\partial \omega}\left(\frac{V_\theta}{\omega}\right)\right\}^2\right] - \frac{2}{3}\mu(\nabla \cdot V)^2$$

The stress functions for a *viscous, incompressible fluid* are

$$\sigma_{\omega\omega} = 2\mu\frac{\partial V_\omega}{\partial \omega} - p, \qquad \sigma_{\phi\phi} = 2\mu\left(\frac{1}{\omega}\frac{\partial V_\phi}{\partial \phi} + \frac{V_\omega}{\omega}\right) - p$$

$$\sigma_{\theta\theta} = 2\mu\left(\frac{1}{\omega \sin \phi}\frac{\partial V_\theta}{\partial \theta} + \frac{V_\omega}{\omega} + \frac{V_\phi \cot \phi}{\omega}\right) - p$$

$$\sigma_{\omega\phi} = \mu\left[\omega\frac{\partial}{\partial \omega}\left(\frac{V_\phi}{\omega}\right) + \frac{1}{\omega}\frac{\partial V_\omega}{\partial \phi}\right] = \sigma_{\phi\omega} \qquad\qquad \text{(A.18)}$$

$$\sigma_{\phi\theta} = \mu\left[\frac{\sin \phi}{\omega}\frac{\partial}{\partial \phi}\left(\frac{V_\theta}{\sin \phi}\right) + \frac{1}{\omega \sin \phi}\frac{\partial V_\phi}{\partial \theta}\right] = \sigma_{\theta\phi}$$

$$\sigma_{\theta\omega} = \mu\left[\frac{1}{\omega \sin \phi}\frac{\partial V_\omega}{\partial \theta} + \omega\frac{\partial}{\partial \omega}\left(\frac{V_\theta}{\omega}\right)\right] = \sigma_{\omega\theta}$$

A.4   ORTHOGONAL CURVILINEAR COORDINATES
AND VECTOR FIELD EQUATIONS

Let x_1, x_2, and x_3 be a set of *orthogonal* curvilinear coordinates; i.e., they are mutually at right angles to each other at their point of intersection. The relation between Cartesian coordinates (x, y, z) and curvilinear coordinates may be written as

$$x = x(x_1, x_2, x_3)$$
$$y = y(y_1, y_2, y_3) \qquad\qquad\qquad\qquad\qquad\qquad \text{(A.19)}$$
$$z = z(z_1, z_2, z_3)$$

In regions where the Jacobian

$$J = \begin{vmatrix} \dfrac{\partial x}{\partial x_1} & \dfrac{\partial x}{\partial x_2} & \dfrac{\partial x}{\partial x_3} \\ \dfrac{\partial y}{\partial y_1} & \dfrac{\partial y}{\partial y_2} & \dfrac{\partial y}{\partial y_3} \\ \dfrac{\partial z}{\partial z_1} & \dfrac{\partial z}{\partial z_2} & \dfrac{\partial z}{\partial z_3} \end{vmatrix} \neq 0 \qquad (A.20)$$

we can invert the transformation (A.19) to get

$$\begin{aligned} x_1 &= x_1(x, y, z) \\ x_2 &= x_2(x, y, z) \\ x_3 &= x_3(x, y, z) \end{aligned} \qquad (A.21)$$

If $J = 0$ this implies that x_1, x_2, and x_3 are not independent, but are connected by some functional relationship of the form

$$f(x_1, x_2, x_3) = 0$$

Curvilinear coordinates may be interpreted geometrically as in Fig. A.4.

The element ds of the curve passing through the points P and Q in Cartesian coordinates is

$$ds = i\,dx + j\,dy + k\,dz, \qquad i, j, k \text{ are unit vectors} \qquad (A.22)$$

Equation (A.22) may be written

$$(ds)^2 = (dx)^2 + (dy)^2 + (dz)^2 \qquad (A.23)$$

Additionally

$$(ds)^2 = (h_1\,dx_1)^2 + (h_2\,dx_2)^2 + (h_3\,dx_3)^2 \qquad (A.24)$$

where

$$\begin{aligned} h_1^2 &= \left(\frac{\partial x}{\partial x_1}\right)^2 + \left(\frac{\partial y}{\partial x_1}\right)^2 + \left(\frac{\partial z}{\partial x_1}\right)^2 \\ h_2^2 &= \left(\frac{\partial x}{\partial x_2}\right)^2 + \left(\frac{\partial y}{\partial x_2}\right)^2 + \left(\frac{\partial z}{\partial x_2}\right)^2 \\ h_3^2 &= \left(\frac{\partial x}{\partial x_3}\right)^2 + \left(\frac{\partial y}{\partial x_3}\right)^2 + \left(\frac{\partial z}{\partial x_3}\right)^2 \end{aligned} \qquad (A.25)$$

h_1, h_2, and h_3 are termed metric coefficients or scale factors. It should be noted that unit vectors in the curvilinear system are also transformed and

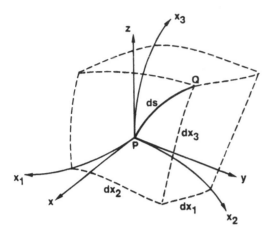

Figure A.4 Curvilinear coordinates.

change with the system. In order to express the governing equations in any orthogonal coordinate system, vector field functions involving the metric coefficients in curvilinear coordinates are needed.

A.4.1 Vector Field Functions

i_1, i_2, i_3 are unit vectors in the directions of x_1, x_2, and x_3 positive. In the following, ϕ is a scalar function, p, q are arbitrary vectors with components p_1, p_2, p_3, q_1, q_2, q_3 in the x_1, x_2, x_3 directions. The "directional sense" is the conventional right-handed system. The following relations hold, the functions being invariant; i.e., they do not depend on any particular coordinate system.

$$\text{Grad } \phi \equiv \nabla\phi = \frac{1}{h_1}\frac{\partial\phi}{\partial x_1}i_1 + \frac{1}{h_2}\frac{\partial\phi}{\partial x_2}i_2 + \frac{1}{h_3}\frac{\partial\phi}{\partial x_3}i_3 \tag{A.26}$$

$$\text{Div } p \equiv \nabla \cdot p = \frac{1}{h_1 h_2 h_3}\left[\frac{\partial}{\partial x_1}(h_2 h_3 p_1) + \frac{\partial}{\partial x_2}(h_3 h_1 p_2) + \frac{\partial}{\partial x_3}(h_1 h_2 p_3)\right] \tag{A.27}$$

$$\text{Curl } p \equiv \nabla \times p = \frac{1}{h_1 h_2 h_3}\left\{h_1\left[\frac{\partial(h_3 p_3)}{\partial x_2} - \frac{\partial(h_2 p_2)}{\partial x_3}\right]i_1\right.$$
$$+ h_2\left[\frac{\partial(h_1 p_1)}{\partial x_3} - \frac{\partial(h_3 p_3)}{\partial x_1}\right]i_2$$
$$\left.+ h_3\left[\frac{\partial(h_2 p_2)}{\partial x_1} - \frac{\partial(h_1 p_1)}{\partial x_2}\right]i_3\right\}$$

$$= \frac{1}{h_1 h_2 h_3} \begin{vmatrix} h_1 \mathbf{i}_1 & h_2 \mathbf{i}_2 & h_3 \mathbf{i}_3 \\ \dfrac{\partial}{\partial x_1} & \dfrac{\partial}{\partial x_2} & \dfrac{\partial}{\partial x_3} \\ h_1 p_1 & h_2 p_2 & h_3 p_3 \end{vmatrix} \tag{A.28}$$

$$\mathbf{\nabla} \cdot (\mathbf{\nabla}\phi) \equiv \nabla^2 \phi = \frac{1}{h_1 h_2 h_3} \left[\frac{\partial}{\partial x_1} \left(\frac{h_2 h_3}{h_1} \frac{\partial \phi}{\partial x_1} \right) \right.$$
$$\left. + \frac{\partial}{\partial x_2} \left(\frac{h_3 h_1}{h_2} \frac{\partial \phi}{\partial x_2} \right) + \frac{\partial}{\partial x_3} \left(\frac{h_1 h_2}{h_3} \frac{\partial \phi}{\partial x_3} \right) \right] \tag{A.29}$$

$$\nabla^2 \mathbf{p} \equiv \mathbf{\nabla}(\mathbf{\nabla} \cdot \mathbf{p}) - \mathbf{\nabla} \times (\mathbf{\nabla} \times \mathbf{p})$$
$$= \left\{ \frac{1}{h_1} \frac{\partial}{\partial x_1} (\mathbf{\nabla} \cdot \mathbf{p}) + \frac{1}{h_2 h_3} \left[\frac{\partial}{\partial x_3} \left(\frac{h_2}{h_3 h_1} \left(\frac{\partial(h_1 p_1)}{\partial x_3} - \frac{\partial(h_3 p_3)}{\partial x_1} \right) \right) \right. \right.$$
$$\left. \left. - \frac{\partial}{\partial x_2} \left(\frac{h_3}{h_1 h_2} \left(\frac{\partial(h_2 p_2)}{\partial x_1} - \frac{\partial(h_1 p_1)}{\partial x_2} \right) \right) \right] \right\} \mathbf{i}_1$$
$$+ \left\{ \frac{1}{h_2} \frac{\partial}{\partial x_2} (\mathbf{\nabla} \cdot \mathbf{p}) + \frac{1}{h_3 h_1} \left[\frac{\partial}{\partial x_1} \left(\frac{h_3}{h_1 h_2} \left(\frac{\partial(h_2 p_2)}{\partial x_1} - \frac{\partial(h_1 p_1)}{\partial x_2} \right) \right) \right. \right.$$
$$\left. \left. - \frac{\partial}{\partial x_3} \left(\frac{h_1}{h_2 h_3} \left(\frac{\partial(h_3 p_3)}{\partial x_2} - \frac{\partial(h_2 p_2)}{\partial x_3} \right) \right) \right] \right\} \mathbf{i}_2$$
$$+ \left\{ \frac{1}{h_3} \frac{\partial}{\partial x_3} (\mathbf{\nabla} \cdot \mathbf{p}) + \frac{1}{h_1 h_2} \left[\frac{\partial}{\partial x_2} \left(\frac{h_1}{h_2 h_3} \left(\frac{\partial(h_3 p_3)}{\partial x_2} - \frac{\partial(h_2 p_2)}{\partial x_3} \right) \right) \right. \right.$$
$$\left. \left. - \frac{\partial}{\partial x_1} \left(\frac{h_2}{h_3 h_1} \left(\frac{\partial(h_1 p_1)}{\partial x_3} - \frac{\partial(h_3 p_3)}{\partial x_1} \right) \right) \right] \right\} \mathbf{i}_3 \tag{A.30}$$

Note the distinction between the vector operator ∇^2 and the scalar (div grad).

$$(\mathbf{p} \cdot \mathbf{\nabla})\mathbf{q} = \frac{1}{h_1} \left[\left(p_1 \frac{\partial q_1}{\partial x_1} + p_2 \frac{\partial q_2}{\partial x_1} + p_3 \frac{\partial q_3}{\partial x_1} \right) - \frac{p_2}{h_2} \left(\frac{\partial(h_2 q_2)}{\partial x_1} - \frac{\partial(h_1 q_1)}{\partial x_2} \right) \right.$$
$$\left. + \frac{p_3}{h_3} \left(\frac{\partial(h_1 q_1)}{\partial x_3} - \frac{\partial(h_3 q_3)}{\partial x_1} \right) \right] \mathbf{i}_1$$
$$+ \frac{1}{h_2} \left[\left(p_1 \frac{\partial q_1}{\partial x_2} + p_2 \frac{\partial q_2}{\partial x_2} + p_3 \frac{\partial q_3}{\partial x_2} \right) - \frac{p_3}{h_3} \left(\frac{\partial(h_3 q_3)}{\partial x_2} - \frac{\partial(h_2 q_2)}{\partial x_3} \right) \right.$$
$$\left. + \frac{p_1}{h_1} \left(\frac{\partial(h_2 q_2)}{\partial x_1} - \frac{\partial(h_1 q_1)}{\partial x_2} \right) \right] \mathbf{i}_2$$
$$+ \frac{1}{h_3} \left[\left(p_1 \frac{\partial q_1}{\partial x_3} + p_2 \frac{\partial q_2}{\partial x_3} + p_3 \frac{\partial q_3}{\partial x_3} \right) - \frac{p_1}{h_1} \left(\frac{\partial(h_1 q_1)}{\partial x_3} - \frac{\partial(h_3 q_3)}{\partial x_1} \right) \right.$$
$$\left. + \frac{p_2}{h_2} \left(\frac{\partial(h_3 q_3)}{\partial x_2} - \frac{\partial(h_2 q_2)}{\partial x_3} \right) \right] \mathbf{i}_3$$
$$\tag{A.31}$$

The components of the stress tensor can be expressed as

$$\sigma_{x_1x_1} = -p + \frac{2}{3}\,\mu(2e_{x_1x_1} - e_{x_2x_2} - e_{x_3x_3})$$

$$\sigma_{x_2x_2} = -p + \frac{2}{3}\,\mu(2e_{x_2x_2} - e_{x_3x_3} - e_{x_1x_1})$$

$$\sigma_{x_3x_3} = -p + \frac{2}{3}\,\mu(2e_{x_3x_3} - e_{x_1x_1} - e_{x_2x_2}) \tag{A.32}$$

$$\sigma_{x_1x_2} = \sigma_{x_2x_1} = \mu e_{x_1x_2}$$

$$\sigma_{x_2x_3} = \sigma_{x_3x_2} = \mu e_{x_2x_3}$$

$$\sigma_{x_3x_1} = \sigma_{x_1x_3} = \mu e_{x_3x_1}$$

where the expressions for the rates of strain are

$$e_{x_1x_1} = \frac{1}{h_1}\frac{\partial u_1}{\partial x_1} + \frac{u_2}{h_1 h_2}\frac{\partial h_1}{\partial x_2} + \frac{u_3}{h_1 h_3}\frac{\partial h_1}{\partial x_3}$$

$$e_{x_2x_2} = \frac{1}{h_2}\frac{\partial u_2}{\partial x_2} + \frac{u_3}{h_2 h_3}\frac{\partial h_2}{\partial x_3} + \frac{u_1}{h_2 h_1}\frac{\partial h_2}{\partial x_1}$$

$$e_{x_3x_3} = \frac{1}{h_3}\frac{\partial u_3}{\partial x_3} + \frac{u_1}{h_3 h_1}\frac{\partial h_3}{\partial x_1} + \frac{u_2}{h_3 h_2}\frac{\partial h_3}{\partial x_2}$$

$$e_{x_1x_2} = \frac{h_2}{h_1}\frac{\partial}{\partial x_1}\left(\frac{u_2}{h_2}\right) + \frac{h_1}{h_2}\frac{\partial}{\partial x_2}\left(\frac{u_1}{h_1}\right) = e_{x_2x_1} \tag{A.33}$$

$$e_{x_2x_3} = \frac{h_3}{h_2}\frac{\partial}{\partial x_2}\left(\frac{u_3}{h_3}\right) + \frac{h_2}{h_3}\frac{\partial}{\partial x_3}\left(\frac{u_2}{h_2}\right) = e_{x_3x_2}$$

$$e_{x_3x_1} = \frac{h_1}{h_3}\frac{\partial}{\partial x_3}\left(\frac{u_1}{h_1}\right) + \frac{h_3}{h_1}\frac{\partial}{\partial x_1}\left(\frac{u_3}{h_3}\right) = e_{x_1x_3}$$

u_1, u_2, and u_3 are the velocity components in the x_1, x_2, and x_3 directions, respectively.

The components of $\nabla \cdot \sigma_{ij}$ are

$$x_1: \frac{1}{h_1 h_2 h_3}\left[\frac{\partial}{\partial x_1}(h_2 h_3 \sigma_{x_1x_1}) + \frac{\partial}{\partial x_2}(h_3 h_1 \sigma_{x_1x_2}) + \frac{\partial}{\partial x_3}(h_1 h_2 \sigma_{x_1x_3})\right]$$

$$+ \sigma_{x_1x_2}\frac{1}{h_1 h_2}\frac{\partial h_1}{\partial x_2} + \sigma_{x_1x_3}\frac{1}{h_1 h_3}\frac{\partial h_1}{\partial x_3} - \sigma_{x_2x_2}\frac{1}{h_1 h_2}\frac{\partial h_2}{\partial x_1} - \sigma_{x_3x_3}\frac{1}{h_1 h_3}\frac{\partial h_3}{\partial x_1}$$

$$x_2: \frac{1}{h_1 h_2 h_3}\left[\frac{\partial}{\partial x_1}(h_2 h_3 \sigma_{x_1x_2}) + \frac{\partial}{\partial x_2}(h_3 h_1 \sigma_{x_2x_2}) + \frac{\partial}{\partial x_3}(h_1 h_2 \sigma_{x_2x_3})\right]$$

$$+ \sigma_{x_2x_3}\frac{1}{h_2 h_3}\frac{\partial h_2}{\partial x_3} + \sigma_{x_1x_2}\frac{1}{h_1 h_2}\frac{\partial h_2}{\partial x_1} - \sigma_{x_3x_3}\frac{1}{h_2 h_3}\frac{\partial h_3}{\partial x_2} - \sigma_{x_1x_1}\frac{1}{h_1 h_2}\frac{\partial h_1}{\partial x_2}$$

$$x_3: \frac{1}{h_1 h_2 h_3} \left[\frac{\partial}{\partial x_1} (h_2 h_3 \sigma_{x_1 x_3}) + \frac{\partial}{\partial x_2} (h_3 h_1 \sigma_{x_2 x_3}) + \frac{\partial}{\partial x_3} (h_1 h_2 \sigma_{x_3 x_3}) \right]$$

$$+ \sigma_{x_1 x_3} \frac{1}{h_1 h_3} \frac{\partial h_3}{\partial x_1} + \sigma_{x_2 x_3} \frac{1}{h_2 h_3} \frac{\partial h_3}{\partial x_2} - \sigma_{x_1 x_1} \frac{1}{h_1 h_3} \frac{\partial h_1}{\partial x_3} - \sigma_{x_2 x_2} \frac{1}{h_2 h_3} \frac{\partial h_2}{\partial x_3}$$

$$(A.34)$$

The dissipation function becomes

$$\Phi = \mu \left[2(e_{x_1 x_1}^2 + e_{x_2 x_2}^2 + e_{x_3 x_3}^2) + e_{x_2 x_3}^2 + e_{x_1 x_3}^2 + e_{x_1 x_2}^2 \right.$$

$$\left. - \frac{2}{3} (e_{x_1 x_1} + e_{x_2 x_2} + e_{x_3 x_3})^2 \right]$$

$$(A.35)$$

The above relations can be used to derive the governing equations in any orthogonal curvilinear coordinate system.

A.5   METRIC COEFFICIENTS IN VARIOUS ORTHOGONAL COORDINATE SYSTEMS

A.5.1   Circular Cylinder Coordinates

$$dx_1 = dr \qquad dx_2 = d\theta \qquad dx_3 = dz$$

$$h_1 = 1 \qquad h_2 = r \qquad h_3 = 1$$

$$x = r \cos \theta \qquad y = r \sin \theta \qquad z = z$$

A.5.2   Elliptic Cylinder Coordinates

$$dx_1 = d\eta \qquad dx_2 = d\xi \qquad dx_3 = dz$$

$$h_1 = h_2 = c(\sinh^2 \xi + \sin^2 \eta)^{1/2} \qquad h_3 = 1$$

$$x = c \cosh \xi \cos \eta \qquad y = c \sinh \xi \sin \eta \qquad z = z$$

$$c = \text{positive constant}$$

A.5.3   Bipolar Cylinder Coordinates

$$dx_1 = d\eta \qquad dx_2 = d\xi \qquad dx_3 = dz$$

$$h_1 = h_2 = \frac{c}{\cosh \eta - \cos \xi} \qquad h_3 = 1$$

$$x = \frac{c \sinh \eta}{\cosh \eta - \cos \xi} \qquad y = \frac{c \sin \xi}{\cosh \eta - \cos \xi} \qquad z = z$$

$c = $ positive constant

A.5.4 Parabolic Cylinder Coordinates

$dx_1 = d\eta \qquad dx_2 = d\xi \qquad dx_3 = dz$

$h_1 = h_2 = 2c(\xi^2 + \eta^2)^{1/2} \qquad h_3 = 1$

$x = c(\xi^2 - \eta^2) \qquad y = 2c\xi\eta \qquad z = z$

$c = $ positive constant

A.5.5 Spherical Coordinates

$dx_1 = d\omega \qquad\qquad dx_2 = d\phi \qquad\qquad dx_3 = d\theta$

$\quad h_1 = 1 \qquad\qquad\qquad h_2 = \omega \qquad\qquad\qquad h_3 = \omega \sin \phi$

$\quad x = \omega \sin \phi \cos \theta \qquad y = \omega \sin \phi \sin \theta \qquad z = \omega \cos \phi$

A.5.6 Prolate Spheroidal Coordinates

$dx_1 = d\xi \qquad dx_2 = d\eta \qquad dx_3 = d\phi$

$h_1 = h_2 = c(\sinh^2 \xi + \sin^2 \eta)^{1/2} \qquad h_3 = c \sinh \xi \sin \eta$

$x = c \sinh \xi \sin \eta \cos \phi \qquad y = c \sinh \xi \sin \eta \sin \phi \qquad z = c \cosh \xi \cos \eta$

A.5.7 Oblate Spheroidal Coordinates

$dx_1 = d\xi \qquad dx_2 = d\eta \qquad dx_3 = d\phi$

$h_1 = h_2 = c(\sinh^2 \xi + \cos^2 \eta)^{1/2} \qquad h_3 = c \cosh \xi \sin \eta$

$x = c \cosh \xi \sin \eta \cos \phi \qquad y = c \cosh \xi \sin \eta \sin \phi$

$z = c \sinh \xi \cos \eta$

Index

T - #0026 - 111024 - C0 - 229/152/25 - PB - 9780367400453 - Gloss Lamination